Survival & Rescue Equipment of World War II:

Army Air Forces and U.S. Navy

VOLUME 4

Small Arms
Knives/Machetes
Compasses
Fishing Kits
Survival Literature

Dustin Clingenpeel

EGP

ISBN: 978-1-958407-24-0

Elm Grove Publishing
San Antonio, Texas, USA
www.elmgrovepublishing.com

Contents

Introduction

You can't know too much! With knowledge, aircraft pilots perform better and are better prepared to meet unforeseen conditions if they thoroughly understand an operation. Also, a pilot's future value is enhanced if he fully understands the operation in which he has participated. Thus, both training and debriefing are critical to pilot success.

Highlighted in Volumes 1 and 2 are the history and role of the equipment officer Personal Equipment Officer for the USAAF, and Aviation Equipment Survival Officer for the BuAer. Each position was borne from the necessities of war. Additionally, inauguration was that of survival training from cadet leading to deployment to the forward areas. The Air Combat Intelligence Officer (ACIO) was integral to combat operations.

The primary responsibilities of the ACIO were: briefing pilots prior to missions; interrogating pilots after combat; preparing all action reports; preparing reports covering aircraft performance, objectives, targets and enemy ships; drafting daily intelligence summaries; and exchanging information between intelligence centers; studying the organization and disposition of enemy air forces (especially their air bases, with a focus on runways, taxiways, revetments, anti-aircraft batteries, search lighting and other features) and other data.

The duties of the ACIO extended to cooperation with the Flight Surgeon in identifying pilots and aircrew who should be grounded, and working with the Equipment Officer in training pilots in air-sea rescue, evasion and survival. ACIOs also censored mail, maintained squadron diaries, wrote up citations for awards and decorations, conducted Prisoner of War (P/W) orientation, and visual aircraft recognition.

Despite all these responsibilities, the ACIO was a pawn in the massive and complex intelligence apparatus of the US Government. The role is highlighted here due to its direct relation to survival, evasion, escape, P/W, and air-sea rescue work. The ACIO of both the BuAer and USAAF provided similar functions to their respective services but have distinctive origins.

In 1882, the Office of Naval Intelligence (ONI) was established, designated Op-16. It was combined with the Navy Department Library under the direction of the Office of the Chief of Naval Operations (CNO).

This office was comprised of Naval Attaches assigned to foreign stations to buy and gather information on foreign navies. These early intelligence officers were primarily translators of foreign books and periodicals, resulting in manuals and 3D models.

The first three intelligence officers representing the BuAer were sent to Europe as Assistant Naval Attaches just prior to US involvement in WWI. Their objective was to support aerial photographic reconnaissance efforts. From then and leading up to 1939, information reached the BuAer from naval and military air attaches via the ONI. Data was also collected by foreign entities and passed through such sources as the National Advisory Committee for Aeronautics and US-based aircraft manufacturers. All intelligence during this period was processed though the Technical Information Section, BuAer.

As of 26 November 1940, a permanent BuAer liaison officer was assigned to ONI, providing a more fluent movement of intelligence. This was by direction of the CNO for all bureaus and divisions of the Navy Department. It decentralized the intelligence gathering and distribution network for the bureaus and divisions, better assisting them to cater to their own needs.

As a result, and because air intelligence consumed much of the workload of the Technical Information Section, the Aviation Intelligence Branch of BuAer was constituted on 29 September 1941. The AIB consisted of two departments: the Air Intelligence Section and the Technical Applications Section. Combined, they would prepare all tactical information for the fleet air arm.

The AIB collected and processed intelligence on both Allied and enemy aviation activities from liaison officers within the ONI. Information was indexed and distributed within the BuAer and to Naval air stations. At the time, air stations had intelligence officers on staff.

By January 1942, such a significant amount of information was flowing from ONI and British sources that the BuAer restructured the AIB, creating four sections to properly disseminate the data. They quickly realized that air intelligence officers in the forward area lacked proper training and a pipeline for transmittal of information, a deficiency left over from a peacetime structure. It became clear that intelligence needed to be collected, analyzed and distributed by agents in the field. In addition, clear and concise action reports from the fleet needed to be drafted and disseminated by the AIB, to provide guidance for improving tactics for future air combat operations. In short, a full reciprocal intelligence exchange was necessary, and a new requirement for a specially trained officer for this duty was essential.

In February 1942, the Navy began to hand-pick Reserve Officers to attend a newly instituted training course for air combat intelligence. These officers were selected from the graduates of the Indoctrination School at NAS Rhode Island, where they had received two months of training on general naval matters. They were required to have up to two years of active-duty service, either with the fleet or at onshore bases.

In April, 1942, the BuAer formally established the Naval Air Combat Intelligence School at NAS Rhode Island where it welcomed its first 53 officers, who graduated on 16 June 1942. The course was originally two months, later extended to ten weeks. On completion, the officers were versed in aviation material, recognition of ships and aircraft, proper methods of briefing and interrogating pilots and a variety of other intelligence subjects. ACIOs were then rushed to the forward area to staff carrier air groups, squadrons, and shore installations.

Within two years, air combat intelligence evolved to such an extent that the Air Combat Intelligence Group was incorporated as one of the four major subdivisions of the ONI, designated as Op-16-V, on 24 January 1944. As of October 1944, up to 1,100 ACIOs were deployed across the world with the fleet, marines, and with Allied units in Europe, Africa, China, India, and the Pacific.

The USAAC, later the USAAF, had within its structure a basic staff under the Chief and Deputy of the Air Staff. These divisions included A-1, A-2, A-3 and A-4; A-2 was Intelligence. As part of the US Army, its role was reconnaissance and support of ground forces. From WWI to 1942, its intelligence derived from the General Staff, US Army designated as G-2, its Military Intelligence Division.

Like the BuAer, the USAAC A-2 division collected and disseminated information on foreign armies. The few air intelligence officers trained before the war had attended G-2 schools. Some attended the Command and General Staff School at Fort Leavenworth, Kansas and the Engineer School at Fort Belvoir, Virginia. The school at Fort Leavenworth trained staff officers for infantry divisions, and Fort Belvoir taught aspects of the Corps of Engineers. Neither of the schools' curricula focused on education of air intelligence practices, but rather taught students to become intelligence officers for ground forces.

Beginning in 1939, the US overhauled and expanded its entire military organization, which included redefining the role of the USAAC. Groups and squadrons began to be equipped with new aircraft and to build up defenses, first in the Atlantic, then the Pacific. A new intelligence organization was required for these new operational demands. In January 1942, the USAAF established an air intelligence school at facilities provided by the University of Maryland. By April, it had moved to a new government facility in Harrisburg, Pennsylvania. The curriculum was not as extensive as the BuAer, as it first emphasized photographic interpretation, but it expanded its courses as needs arose. The course was eight weeks, and eventually evolved to cover the same topics as the BuAer. Once graduated, the officers became group or squadron S-2 Air Intelligence Officers.

In April 1944, the intelligence school moved from Harrisburg to the AAF school of Applied Tactics in Orlando, Florida, where S-2 officers worked directly with bomber and fighter cadres in their advanced operational training. S-2s also attended classes on emergency equipment as honorary PEOs, medical training fundamentals as support to the Flight Surgeon, and air-sea rescue indoctrination tactics.

Initially, the ACIO concerned itself with operational information: objectives, targets, enemy ships, aircraft, and results of air action. Duties were expanded as the war progressed. In 1942, the first ACIOs helped promote the combat area air-sea rescue structure. Coordination did not yet exist, but as air-sea rescue began to evolve, ACIOs applied that information to their briefings. Similarly, evasion and escape awareness were added.

Liaison work with British intelligence (MI9)introduced a new curriculum: bringing home downed airmen. In October 1942, the Military Intelligence Service, Section-X, was established, an agency solely dedicated to facilitating all matters pertaining to evasion and escape. As this organization established stations in forward areas, it briefed ACIOs entering that theater of operation. In turn, the new information was relayed to aircrew by their squadron ACIO.

Eventually, this became a typical scenario: After the respective squadrons had been informed of the number of planes they would be expected to provide for a strike, schedules were drawn up and the appropriate pilots and aircrew reported for a briefing. An hour before the briefing, pilots dropped in at the briefing hut or ready room and carefully studied photos of the target assigned and its surrounding area. At the briefing, the ACIO, who had earlier made a detailed study of the target, lectured on the approaches to the target, the appearance of the target, defenses around the target, interception to be expected, weather information, and survival and evasion information. Depending on the target location, the ACIO might also review locations of Coastwatcher operations, situations with local native populations, winds and currents, and location of crash boats, lifeguard submarines and other surface craft. ACIOs also distributed appropriate evasion aids and discuss personal emergency equipment. The Operations Officer then detailed the tactics to be used. He discussed takeoff, rendezvous, bomb loads, manner of approach to target, specific target for each division or section, route of retirement and place of rally, route home, and other support. These briefings could extend from an hour and a half to two hours.

(Above) Founding fundamentals for ACIO was that of aerial photographic interpretation. Relevant intelligence data would be extrapolated and applied to daily updates for areas. This information collected would transfer to printing of target, pilot and strip maps used on operational flights and briefings. (Left) Naval ACIO studies recently developed photographs recording pertinent data. (Right) P-47 pilots of the 7th AF study details provided in aerial photographs for planned bombing and strafing attack over Pagan Island, Marianas, c.1944. (National Archives)

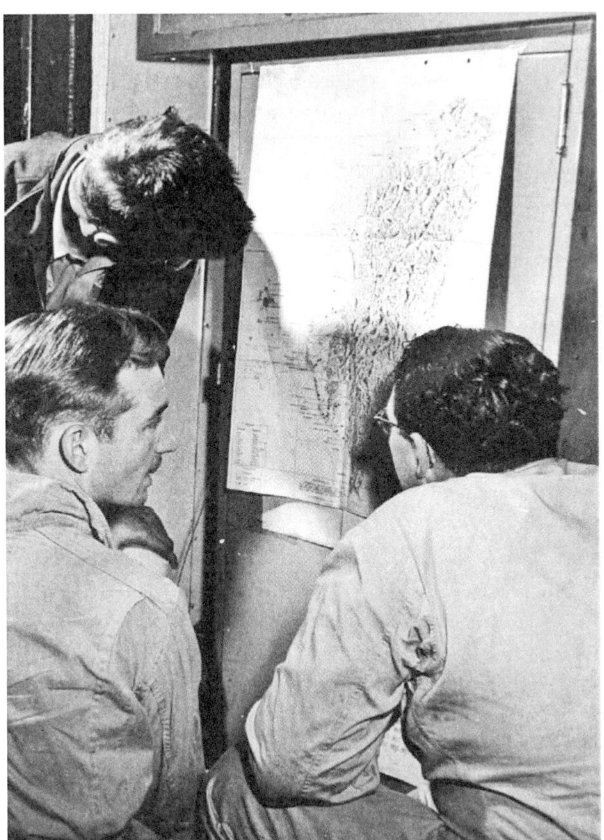

(Left) Pilots of VF-11 study map of Formosa, February 1945. Maps were typical feature in ready rooms for evaluation and orientation at all times. Such charts often illustrate highlighted areas of enemy concentrations and friendly areas to be used if in evasion and survival scenario. ACIO's would use such maps with topography discussing routes of travel. (National Archives)

(Below) Maps of Europe in the briefing room of the 410th BG, 9th AF, England, May 1944. Provided on orientation board among other reference material for memorization. (National Archives)

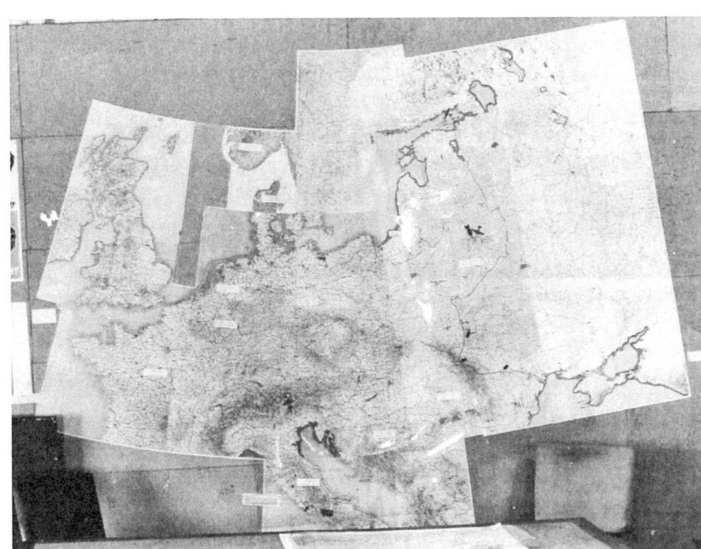

(Above) Lt. Col. Jointilla briefs pilots of the 353rd FG, England, January 1944. Extracted from aerial photographs and other intelligence, exacting details of each mission are passed onto combat crews. Left is of time table of mission to include air-sea rescue, emergency homing channels and call signs. Air-sea rescue procedure on blackboard from left shoulder of Jointilla. On orientation map, emergency routes established. S-2 officer or assigned Escape Officer would discuss optimal areas of bail-out and evasion procedure. (National Archives)

(Left) A look at B-25 combat aircrews of the 820th BS in the briefing hut on Tawara, January 1944. (National Archives)

(Left) Lt. Col. Bailey briefs pilots of the 352nd FS, England, January 1944. Same type of oversized orientation map. Significance to note are the training aids posters on wall. Here, PEO would use Type-K Dinghy poster and other visual aids for orientation on emergency equipment. S-2 Officer would reference posters for aircraft recognition to additionally use maps of Europe to discuss evasion routes and highlight Safe Areas, to include discussion on evasion aids. (National Archives)

(Above) ACIO of VF-17, Lt. B.D. Henning catches up on paper work, Bougainville, February 1944. (National Archives)

(Right) LCDR William Miles briefs pilots of VT-87 aboard the USS Ticonderoga, July 1945. Observed in ready room behind, on the wall and in the laps are variable types of maps for orientation. (National Archives)

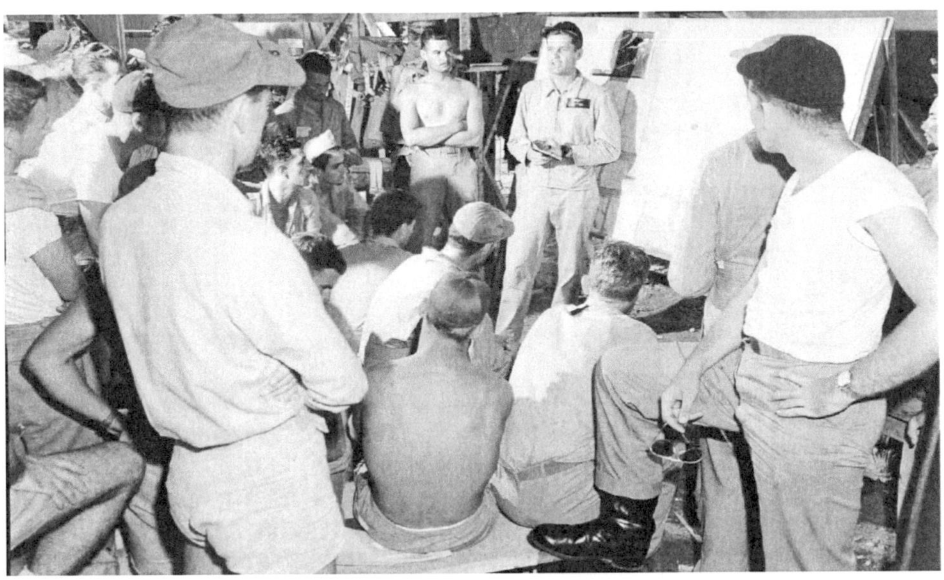

(Left) Intelligence briefing for VMF-221, Barakoma Field, Vella Lavella, October 1943. As part of strike briefing, the ACIO will cover jungle and ocean survival. He will note, on the map board, where Coastwatcher networks are and location of friendly natives. He will conduct cursory overview of personal emergency equipment along with distribution of evasion type aids. Often as part of these briefings, the ACIO will read from survivor reports as helpful hints on what to do and not. (National Archives)

(Left) Martin PBM Mariner pilots of VBP-205 receive briefing on Okinawa from their base of operations, Tanapag, Saipan, April 1945. All details of Okinawa and surrounding area would be discussed: topography, harbors, enemy airfields, shipping lanes, attitudes of indigenous peoples, air-sea rescue procedure and positions of the Allied fleet. The map being referenced has highlighted areas of enemy concentration. An additional training aids poster for aerial recognition is on the left. (National Archives)

"Our aviators and seaman, with confidence that they will fly and sail again tomorrow, dare to face greater odds in the war today. Vision our Services, ingenuity of industry has provided survival and rescue equipment, which has lessened the hazards, improved the safety, and given our men greater courage. May we never be content with present equipment, but constantly improve it with experience, continued study and cooperative effort."

Admiral Russell R. Waesche,
Commandant of the United States Coast Guard

SMALL ARMS

Prior to the US entry into WWII, issuing of sidearms for the US Navy was relegated to amphibious forces, shore parties, or patrols. In 1940, the US Navy Department underwent a sizable procurement and supply reorganization to meet projected wartime requirements. For example, two divisions of the NAF were created to handle field requisitions and one to maintain field inventory. These two sections were then combined into the Aviation Supply Office in April of 1941. The ASO was located at the newly established Naval Aviation Supply Depot in Philadelphia; it had two primary distribution points, known as Aviation Supply Annexes, for aviation materiel located at Norfolk, Virginia and Oakland, California, which served a multitude of Naval Air Stations.

The .45 1911 automatic pistol had been the standard sidearm for the US Navy, issued by ship and station armories with the M-1916 hip holster, a pistol belt, and double magazine pouch. Sidearms were not part of allowance for Naval flight personnel but they were viewed as essential wartime equipment for some fleet activities. In 1942, the commanding officer of the USS Enterprise, Captain Osborne Hardison, reported that automatic pistols were carried by all pilots and gunners, issued through the ship. He stated this created accountability for ship supply issues as units and personnel got transferred. To remedy the situation, he recommended sidearms be issued through the same process as flight clothing.

The US Navy soon foresaw a potential for shortage of automatic pistols and, in preparation, two categories were created to regulate the issuance of sidearms. The decision was made to reserve the .45 for fighting forces overseas with an alternate weapon issued to security personnel in the continental United States.

(Right) Ens. De Clinton Nichols of VB-4, December 1942 is eating a cup of ice cream on the wing of his SBD Dauntless after returning from a mission. On his right hip is a .45 automatic with M-1916 holster. He supported landings for Operation Torch.

(Below) VB-6 pilot Walter Finger and rear-seat gunner Jack Glass ARM2/c wear issued .45 automatics from ships armory, USS Enterprise, February 1943. (National Archives)

(Below right) Silver Star recipient Lt. Wallace Moore of VGF-29 wears standard M-1916 hip holster with pistol belt for his .45 1911, USS Suwannee, January 1943; another example of issuing personal sidearm for Naval aviators in combat theater. (National Archives)

(Above) Milson Ball AMM./1c, Ens. Ned Barrett and Carl Kovak AMM/1c of VS-66 crew the J2F Duck rescue plane on Betio Island, Tarawa Atoll, October 1944. Barrett and Kovak were equipped with the .45 auto and M-1916 holster. (National Archives)

(Below) TBF/TBM pilot discusses his plane's operation with ground crew member after action over Saipan, June 1944. He wears M-1916 hip holster with .45 automatic as personal sidearm. (National Archives)

(Above) Triple Ace Lt. jg Ike Kipford of VF-17 prefers his .45 with a M-1916 hip holster, Bougainville, March 1944. (National Archives)

(Below) VT-14 flying officers aboard the USS Wasp, August 1944. Ens D.K. Drake and Ens. Jack Coffey both utilize .45 autos with M-1916 holsters. (National Archives)

In early 1941, the US Navy made a limited procurement of 3,000 Smith & Wesson .38 cal. Special with a four-inch barrel to supply the Naval Civilian Police Corps. Subsequent procurements armed security personnel of Naval facilities, civilian production plants, USCG shore patrols and the Merchant Marine.

Upon entering the war, the demand for sidearms increased exponentially, quickly consuming existing stocks. Additional quantities were procured by three bureaus: the US Navy Bureau of Ordnance, US Army Ordnance Department and the Defense Supply Corporation. The BuOrd made two procurements: one on 2 March 1942 for 20,000 revolvers and an extension on 30 June 1942 for an additional 45,000 S&W .38 special revolvers with four-inch barrels. Most of these revolvers included the V-prefix on the serial number. Serial number V1 shipped from the factory on 24 April 1942. These direct Navy procurements were completed in January of 1943 with the final delivery of 9,500 revolvers. All orders after June of 1942 were requisitioned through the Army Supply Program. The US Army Ordnance Department then took over all procurement of the S&W .38 cal. Victory model revolvers.

The Victory model was officially known as the Cal. 38 Special Revolver, S&W Military and Police maintained through two Navy distribution points: The Naval Supply Depot's NSD Norfolk and NSD Oakland. The bulk of the direct US Navy contract revolvers were marked US NAVY on the top strap above the cylinder. US Navy requisitions from the Army Supply Program were marked US PROPERTY and the inspector's initials GHD (Guy H. Drewry, Army Inspector of Ordnance), showing no distinction between service use.

On 12 December 1942, US Army Ordnance awarded a contract on behalf of BuOrd for a quantity of 20,000 .38 Specials. As prescribed for the order they were to be property marked US Navy boosting the potential total to 85,000 Victory models with the US Navy property mark. This contract was executed to meet BuAer needs. Deliveries commenced in February 1943.

These revolvers would include a four-inch barrel and what is referred to as a sandblast "Black Magic," or parkerized, finish. There were many subtle variations of the S&W .38 Special procured by the US government. Just over 350,000 S&W .38 Specials were procured by the US Army and US Navy in a multitude of barrel lengths, finishes and markings from 1941 through 1945.

In October 1942, a request to issue sidearms to flight personnel was submitted by the Commander, Air Forces, Pacific Fleet. Stated, the weapons were not for combat purposes, but to sustain life after forced landings in isolated regions or at sea. Initial recommendations suggested the pistol not be greater than .38 cal. and issued with a shoulder holster, to be distributed as part of the squadron allowance list. The BuAer further recommended that the sidearm be provided for self-defense purposes only. Based on these requests, the BuOrd found the .38 Special revolver suitable for issuance to fleet activities and it was officially placed under squadron allowance lists on 21 January 1943, per BuOrd Circular Letter G1-43. That letter stated that one .38 cal. revolver with shoulder holster, 100 rounds of ball ammunition and a five-inch sheath knife would be issued to all fleet aircrew members with orders to operate outside the continental limits of the United States, when preparing to depart within 30 days. Stock would be requisitioned from Naval Supply Depots Norfolk and Oakland and maintained for issuance to aircrews by BuAer central supply points at Naval Air Stations Pearl Harbor, San Diego, Alameda, Seattle, Norfolk and Quonset Point. The revolvers would then be accountable to the squadron commander or senior Naval aviator for flight personnel. It was their responsibility to requisition the adequate allowance plus ten percent for loss.

This process relieved the burden on ship armories and better managed procurement schedules for the .45 1911 automatic. Upon distribution of the circular letter, the BuOrd took immediate steps to secure stock for issuance to fleet activities, excluding the shoulder holster, which was not yet in production.

Use of .38 revolvers by squadrons was reported by VF-9 as early as September 1942 and by VGS-11 (Escort Scouting Squadron 11) in October 1942. On 27 April 1944 the BuOrd distributed a servicewide letter stating that the .38 Special revolver was reserved for aviation flight crews only and that any such sidearms in service outside the continental United States were to be turned in and exchanged for the .45 1911.

Cal. .38 Special Revolver, S&W Military and Police. Property marks stamped on top strap, another variant includes red inlay Property Of US Navy milled into left side of frame, this property mark was not factory applied.

Holsters used by security forces, for whom the .38 Special was originally intended, were from commercial sources. Following the increased procurements in 1942, a suitable holster to be supplied with the .38 revolver was developed and placed into production. A brown russet leather holster cut for a S&W four-inch barrel was placed in procurement in 1942 under BuOrd contract NXSO-14430, awarded to the Brauer Bros. MFG. Co. for 55,000 holsters at a cost of $.95 each. This contract spanned from September 1942 to March 1943. Its official BuOrd nomenclature was Holster, Belt Type, Dwg. No. 20-17-103, available under stock number 2-H-1468. Due to the unavailability of a shoulder holster, a hip holster, web pistol belt, and web ammunition pouch were distributed to aircrew. The BuAer extended contract NXSO-14430 for an additional 17,270 holsters to be delivered with the initial revolver shipments for fleet activities. The US Army Ordnance Department referenced it as the Holster, Hip, Revolver, Cal. .38, T1; procurement quantities unknown. Just over 70,000 hip holsters were manufactured for the US Navy.

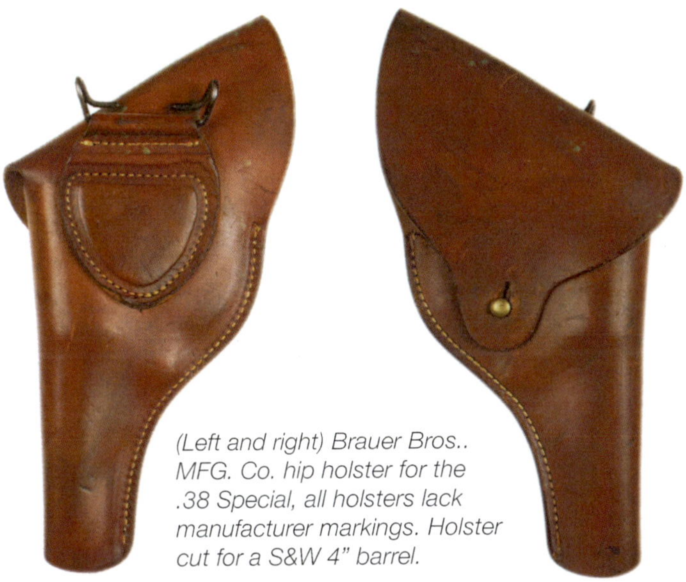

(Left and right) Brauer Bros.. MFG. Co. hip holster for the .38 Special, all holsters lack manufacturer markings. Holster cut for a S&W 4" barrel.

(Above) Pilot Ens. Clyde Mitchell of VS-57 is wearing .38 revolver with hip holster, Florida Island, March 1943. VS-57 operated OS2U Kingfisher floatplanes typically assigned to reconnaissance and anti-submarine patrols. (National Archives)

(Left) VB-9 Lt. Cmdr. Arthur Decker utilizes hip holster. USS Essex, March 1944. CAG-9 deployed aboard the Essex in March 1943. (National Archives)

(Below) Gunner stands on wing of Dauntless dive bomber talking with his pilot, wearing the hip holster aboard the USS Independence, May 1943. Note jackknife adjacent to holster. (National Archives)

(Above) TBF crew of VC-86, USS Bismarck Sea, January 1945. Edward Lodzinsk ARM2/c wears hip holster for his .38 Special. (National Archives)

(Above right) Brauer Bros. hip holster worn by pilot Ens. Clyde Gillham of VB-5, USS Yorktown, August 1944. (National Archives)

(Right) PV-1 aircrew W.F Barre AMM2/c of FAW-4 with .38 hip holster with modification of leg tie. Aleutian Is., September 1943. (National Archives)

(Below) Even after the shoulder holsters became available, some pilots preferred the hip type as evidenced by VF-40 pilot Lt. jg Levi Monteau in the ready room of the USS Suwanee, April 1945. (National Archives)

(Below right) VPB-118 PB4Y-2 Privateer pilot Lt. jg Charles Reeves. Loose, to his left, pistol belt has Bruaer Bros.. hip holster attached with .38 Special. (National Archives)

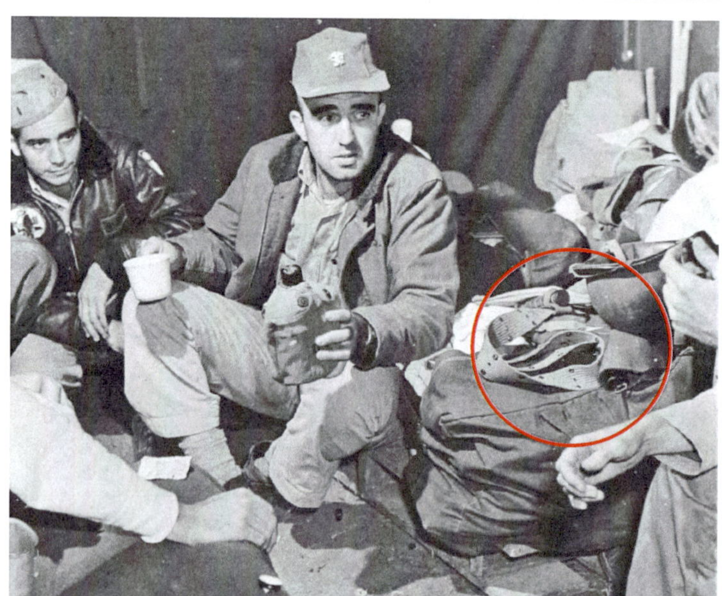

The khaki web ammunition pouch supplied as an accessory for the .38 Special S&W held six rounds and attached to the pistol belt. The BuOrd noted these were intended for security forces but were the only standard method for securing extra ammunition. They were referenced as Pouches, Web, .38 cal. referenced under stock number 2-B-1500. Contracts were awarded to the A.L. Siegel Co. Inc. at the time of procurement for the hip type holster for an equal amount of 55,000 pouches. Though there was a BuAer extension on the hip holsters, it was not stated if they produced the web ammunition pouch in an extended quantity to match that order. In BuOrd Catalog of Navy Materials, July 1945, the standard QMC first aid pouch was listed as an alternate ammunition carrier, available in US Navy supply under stock number 2-P-265.

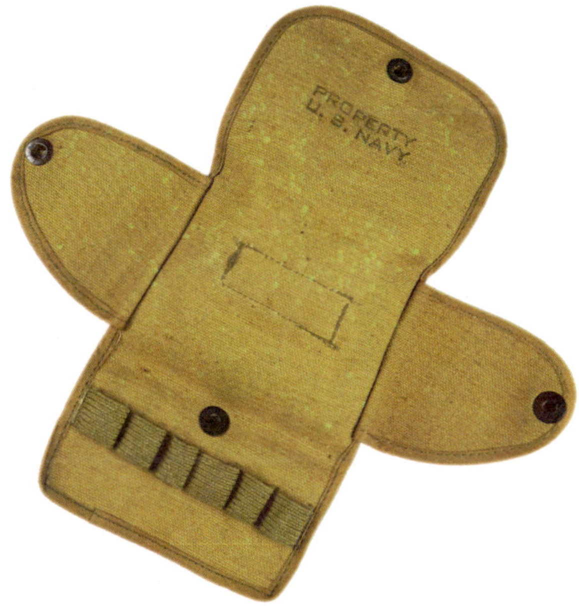

(Above and right) A.L. Siegel six round web ammunition pouch, US Navy.

(Left) Standard M-1924 first aid pouch. Other types include M-1910 and M-1942.

(Left and below left) Locally modified M-1924 first aid pouch to accommodate 28 rounds. Elastic strap for the repair of flying goggles used for loops. (Paul Faulkner collection)

(Below) Two unidentified VS-57 aircrew wear .38 Special hip holsters and web ammunition pouches 2-B-1500. Scouting Squadron 57 was en route to South Pacific aboard SS President Monroe, February 1943.

(Above) Lt. Bob Koberg of VPB-202 wears the basic complement issued to pilots and aircrew, Saipan, c.1944. Included is a web pistol belt, 5" hunting knife, web ammunition pouch and hip holster obscured by his right arm. The revolver and knife were considered emergency tools.

(Above) USMC Capt. J.P. Newlands of VMF-222 uses hip holster with the six round ammunition carrier, partially obscured by his left hand, c.1943. (Courtesy Gregory Pons)

(Below) Pilot of VF-11, USS Hornet, January 1945. Accompanying hip holster is web ammunition pouch. (National Archives)

(Below) VB-83 aircrew member shows hip holster for his .38 revolver; attached to his pistol belt are two web ammunition pouches 2-B-1500, USS Essex, May 1945. (National Archives)

The use of a sidearm had been service-tested on a limited basis in the combat zone through 1942 and was found to be better suited to use in a shoulder holster. Hip holsters interfered with movement and snagged in cramped crew positions, problematic in emergencies. Coinciding with the potential decision to adopt the .38 S&W, the BuAer conducted field tests from October through November 1942 at NAS Norfolk of shoulder holsters and the revolver for aircraft personnel.

(Below) Tested, commercially available shoulder harness with spring clip holster. Exhibited, shoulder holster tested at NAS Norfolk. The tests recorded users' likes and dislikes over two weeks, after which the BuAer approved the issue of the .38 S&W revolver with a shoulder holster. BuAer delegated the design and procurement of the holster to BuOrd, taking an advisory position. (National Archives)

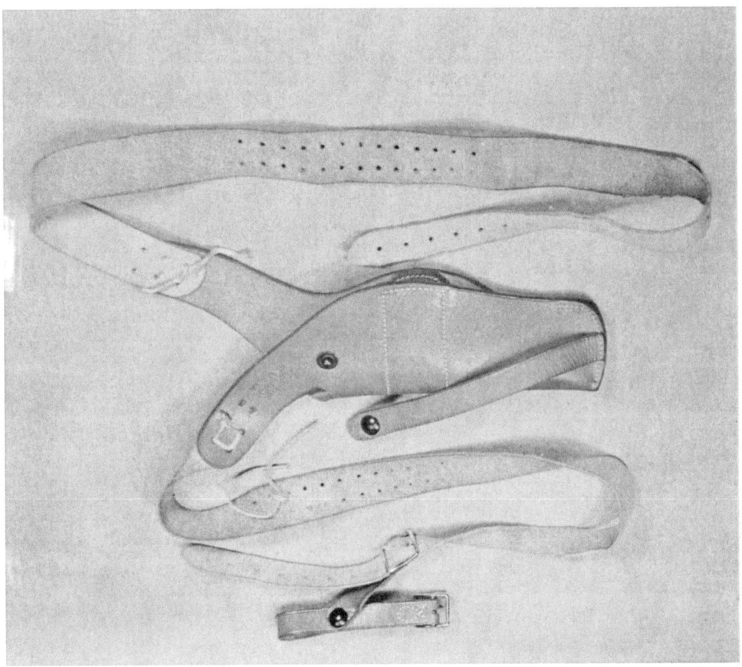

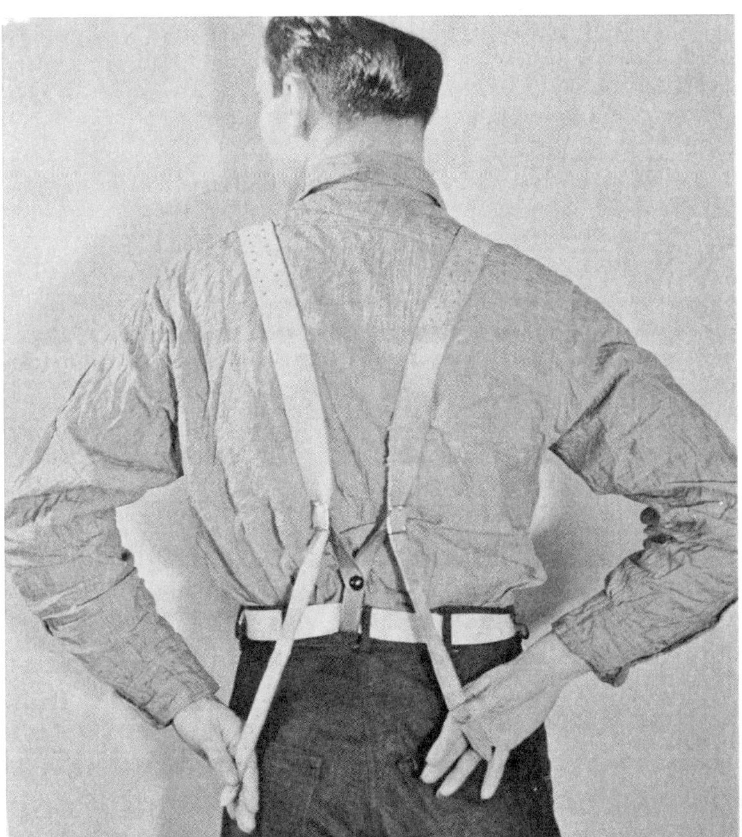

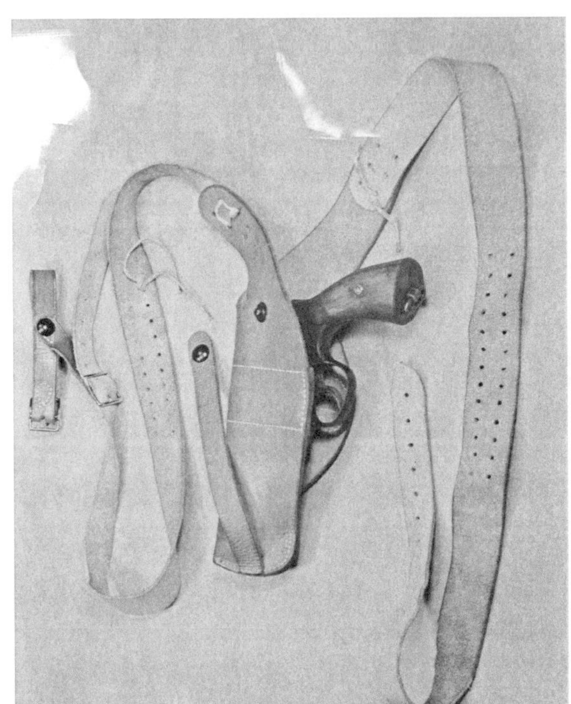

In November 1942, the BuOrd extended requests to various manufacturers to submit shoulder holster designs for testing and evaluation for possible standardization. Eight companies sent samples, which underwent a demonstration at NAS Anacostia on 4 January 1943. One of the eight submissions was from the Boyt Harness Co., which simply provided the US Army Ordnance shoulder holster M3 already in production. Upon conclusion of the demonstration, the most satisfactory holster in both strap arrangement and holster design was this Army Ord. type. The BuAer negotiated with Boyt to modify the cut to accept the .38 Special and widen the shoulder strap to 1-¼ inches, the M3 holster included a 3/4 inches wide shoulder strap at that time.

The new holster was approved on 20 January 1943. Contracts to manufacture 35,000 shoulder holsters of the modified M3 Army type were awarded to the Boyt Harness Co. for $1.35 each. Production commenced 6 February 1943 with the first deliveries 20 days later, contract projected to be completed 60 days after start of production. An extended contract was awarded in December 1943 for 55,000 more shoulder holsters under contract NXSO-45649, contract was completed at the end of March 1944. The shoulder holsters were made of a brown russet leather with US embossed on the front and USN stamped on the back. The official nomenclature was Holster, Shoulder Type, Dwg. No. 20-17-111, available via stock number 2-H-1470. No "Mark" designations were assigned to either the hip or shoulder type holsters.

(Right) The .38 shoulder holster is seen in service by early summer of 1943, slowly replacing the hip model. Here, VC-13 pilot Lt. jg. Brown wears the shoulder holster as prescribed, USS Core, August 1943. Chino khaki trousers were part of the standard uniform at the time, and retaining strap at the bottom of the holster was intended to secure to trouser type belt. (National Archives)

(Left and below) Shoulder Holster as procured by BuOrd, Dwg. No. 20-17-111 supplied with USN/Boyt/43 or 44, reflecting year of manufacture.

(Left) VB squadron aircrew members hold after-action discussions following attacks in the Gilbert Islands, September 1943. Radioman on left tells of his narrow escape while returning to the carrier. An oil line had been shattered by anti-aircraft fire, which blinded the pilot who temporarily lost control of the airplane. As the plane dove toward the water, the radioman shouted to the pilot who pulled up at the last second a few feet before impact. Each is wearing newly issued shoulder holsters for the .38 revolver.

(Below left) VOF-1 Hellcat pilot Lt. jg Clifford Case sports a brand new shoulder holster aboard the USS Tulagi, July 1944. VOF-1 along with VF-74 supported invasion of Southern France, Operation Anvil-Dragoon in August 1944. (National Archives)

(Below) PV-1 Ventura pilot of VPB-135, Attu Is., Alaska, May 1944. Lt. J.W. Brownlie wears his shoulder holster under his arctic flying clothing. Attached by ring to belt retaining strap is a flashlight. He and crew had just recently returned from a night bombing mission over Paramushiro in the Kurils. (National Archives)

Prior to the adoption of the leather shoulder holster for the .38 Special, a shoulder holster assembly was proposed by the NAF in the summer of 1942. At the dawn of 1942, there were only a handful of Carrier Air Groups that operated in the Pacific up through the Battle of Midway. Those Carrier Air Groups include: CAG-2, CAG-3, CAG-5, CAG-6 and CAG-8, to include other air detachments. To some extent, the .45 1911 was distributed amongst these air groups soon after the declaration of war as they deployed to the combat area. Evidence can be found in the narrative from the Introduction in VOL-3 as well from images of the time period. Photographs illustrate both the M-1916 holster and what could be best described as locally manufactured shoulder holsters. Other commands reflected the same sentiment as Captain Hardison of the USS Enterprise strongly suggesting issuing pistols as a standard complement for flight personnel. The unanimous comment was that the sidearms were better suited worn with a shoulder holster. In response to fleet requirements from an accumulation of after-action reports and interviews, the NAF attempted to design and incorporate a shoulder holster harness for the standard automatic pistol as part of an assembly with the emergency back-pad kit NAF-69233 for pararaft borne individuals.

The harness designed by the NAF was made from 1½ inch webbing, adjusted and secured by tie tapes across chest. A holster for the .45 1911 was made from waterproof canvas and a standard magazine pouch was secured opposite the holster. A hunting knife was intended to be secured to right shoulder strap but was dropped as it interfered with parachute harness, prescribed to worn elsewhere.

(Below) Front view of harness designed by NAF to secure .45 1911 and magazine pouch. Strap to secure hunting knife can be seen on right shoulder. (National Archives)

(Right) Back view of harness. (National Archives)

(Below right) Front view with NAF-69233 emergency back pad kit and shoulder holster harness as complete assembly. (National Archives)

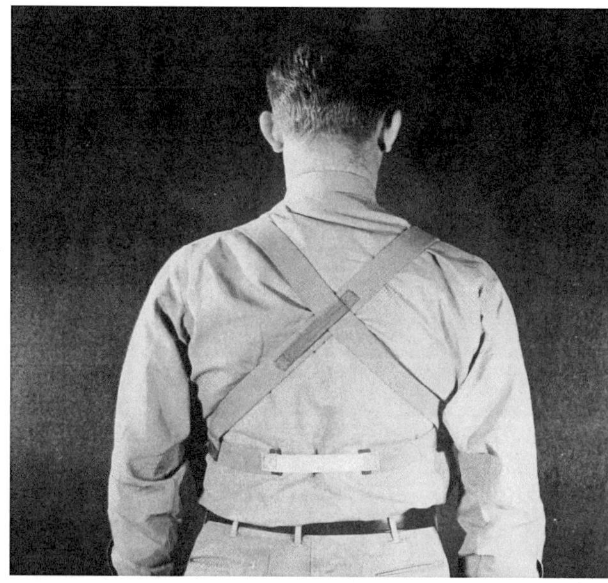

(Above) USMC fighter ace and posthumous Medal of Honor recipient Lt. Robert Hanson of VMF-215 wearing an NAF shoulder holster harness. These were distributed with back-pad kit NAF-69233. (National Archives)

(Above) Marine squadron member of VMF-222 wearing same harness. (Courtesy Gregory Pons)

(Below) VMF-213 pilot and ace Lt. Milton Vedder wearing NAF shoulder holster harness and back-pad kit during interview by Air Combat Intelligence Officer aboard a PBY after rescue in April 1943. (National Archives)

Differences between the laboratory photos of the NAF shoulder harness and photos of it in service can be noted, documentation was limited from concept to production in regards to the harness. However, during the service tests of the NAF-69233 back pad kit the BuAer adopted the .38 Special with a shoulder harness as the new standard, and not the .45 Automatic.

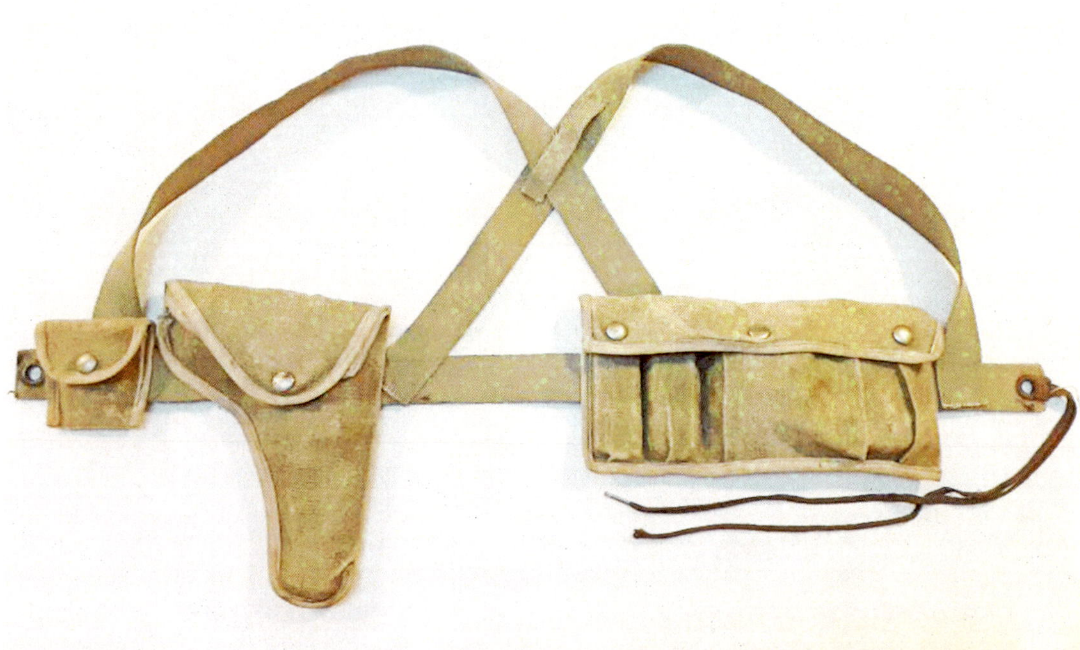

A surviving harness; under the right arm is pouch holding two magazines for automatic pistol, with additional pockets available, a lensatic compass pouch was provided near the holster. It is unclear if the contractor manufactured both the back-pad kit and pistol harness. Images of this holster in use all share these same characteristics. (Paul Faulkner Collection)

(Below) Members of VF-6 W/O Machinist Donald Runyon, Ens. Joseph Shoemaker and Ens. Wildon Rouse use .45 cal. 1911 automatic pistols aboard USS Enterprise, August 1942. Individual without pistol is Howard Packard AP1/c, who became fighter ace along with Runyon and Rouse. Each of the three wear locally manufactured shoulder holsters. (National Archives)

(Left) Personnel of VF-6 while serving aboard the USS Enterprise, c.1942. These aviators wear a mix of both M-1916 holsters and locally fabricated shoulder holsters. (National Archives)

Carrier and land-based aviation groups had access to local materials for repairs and maintenance. 36 inch wide rolls of duck and canvas were available to manufacture covers of aircraft engines, airplanes surfaces, and cockpits that were water-, oil-, and greaseproof. Most used were nonspecular sea blue waterproof cotton duck and 12 oz. aircraft grey Army duck. Another type was a processed cotton fabric, aircraft grey, for the repair of parachute packs.

Each squadron had an assigned parachute rigger responsible for the for the upkeep and maintenance of parachutes and lifesaving equipment. Additionally, each carrier and land base supporting air activity had a parachute loft and repair facility having the ability to cater the specific need of aircrews. Upon issuance of BuOrd letter G1-43, fleet requests streamed into the BuAer for pistol issuance to aircrew members. The .38 and the shoulder holster were only available on paper; to temporarily supplement demands, automatic pistols were requisitioned when supply permitted. The other demand was for shoulder holsters for which supply was not available for squadrons, which led to squadrons making do with local procurements and local fabrication. In the booklet, So You Are Going to the South Pacific, prepared by ComAirPac in August 1943, it states that holsters were allowed to each Naval aviator and each crew member of a plane crew. It further elaborates that if shoulder holsters were not available, any individual should be able to make one for himself after securing the proper amount of leather and webbing.

(Below) Typical parachute loft scene, USS Rudyerd Bay, April 1945. Rigger repairs parachute harness. (National Archives)

(Below) VMF-222 pilot makes own repairs, adds name tag to flight suit in parachute repair tent on Bougainville Island, April 1944. (National Archives)

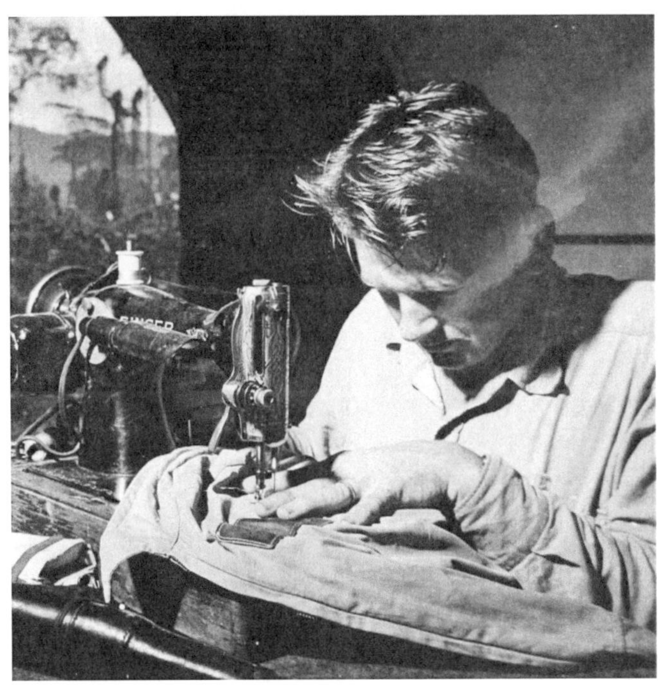

(Left) Pilot Lt. jg. Harold Johnston aboard the USS Saratoga with locally fabricated shoulder holster for .45 automatic pistol, likely made by Tamila, November 1943. (National Archives)

(Right) Lt. jg. William Totherow, who also served as Asst. Material Officer, aboard the USS Saratoga, with locally fabricated shoulder holster for .45 automatic pistol, November 1943. Holster near identical construction as worn by Tamila (below). (National Archives)

(Below) To accommodate squadron requests, parachute rigger Chester Tamila fabricated leather holsters in the parachute loft aboard USS Saratoga, February 1944. Each harness would be fitted to individual upon request. (National Archives)

(Above) W/0 Donald Runyon's locally fabricated shoulder holster. (National Archives)

(Right) Lt. George Polk, pilot qualified and CO of CUB-1 Aviation Unit at Tanambogo islet, Solomon Islands, wears shoulder holster for .45 1911 made from fabric, c.1943. Hunting knife secured opposite of holster. This holster and other shoulder holsters like that for Runyon would have been fabricated by personnel working in a ship- or land-based parachute loft. (National Archives)

(Below) Pilots of VF-10 aboard the USS Enterprise, February 1943. An assortment of different configurations of shoulder holsters can be observed as with the utilization of the M-1916 hip type. (National Archives)

Locally fabricated shoulder holsters worn by members of VT-10 aboard USS Enterprise, February 1943. (Above left) Ens. Robert Oscar. (Left) Elwin Rowell AMM1/c, Ens. Fred Hoover Jr. and Merril Hofer RM3/c. (Above) unidentified. These images illustrate an almost endless variety even amongst a single squadron. (National Archives)

(Above) Another pilot from VT-10, Lt. jg James Walter, February 1943. (National Archives)

(Above) PBY aircrew exit after mission, New Guinea, September 1944. One aircrew wears leather shoulder holster and other, a hip type made from fabric. (National Archives)

(Below) Pilots from VF-16, USS Lexington, April 1944. From left to right: Ens. Homer Brockmeyer, Lt. jg G.E. Hollingsworth, Lt. jg Duane Kenney, Ens. James Mullins and Ens. Edwin Ross. They display an assortment of styles of shoulder holsters. Ross uses standard holster for .38. (National Archives)

(Above) Pilot Lt. jg. Brownstein of VC-13 has a modified hip holster. The top flap is sewn to the pistol belt with an added strip of leather to secure the revolver in place, USS Core, August 1943. (Middle and right) TBF aircrew Harvey Donald AOM1/c and John Pancauge ARM2/c of VC-86 wear hip holsters modified to a shoulder type aboard the USS Bismarck Sea, January 1945. (National Archives)

(Below) Further illustration of modifying standard hip type holster to a shoulder configuration by two pilots of VF-16, c.1944. One harness has dual application of ammunition loops, across shoulder and chest strap. The other has pouch for securing extra ammunition attached to chest strap. (National Archives)

(Left) Local fabrication of shoulder holsters was also conducted for comfort and may not reflect unavailability of standard types. Here is a modified standard type should holster worn VB-5 pilot Lt. jg. Paul Johnson with shoulder strap removed, attached to left thigh of one-piece HBT coverall which in turn has added lower leg pockets. Lift-the-dot fasteners can be seen for easy removal for laundering. USS Yorktown, August 1944. (National Archives)

(Below) Ens. Robert Dooling of Air Group 19 has also removed the shoulder strap from the standard .38 Special harness, USS Lexington, February 1945. He has it secured to a trouser belt that also has the addition of ammunition loops applied. (National Archives)

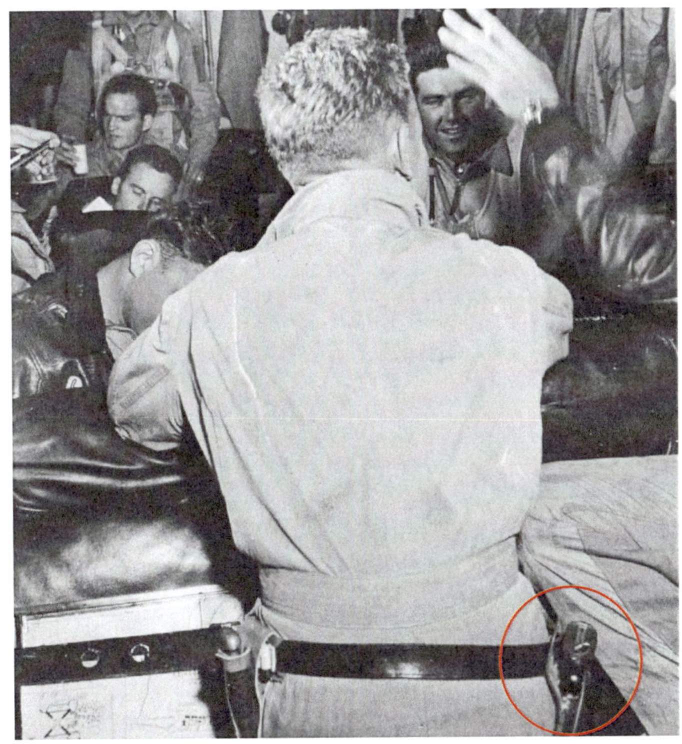

(Left) This standard .38 shoulder holster was modified with a wide fabric shoulder harness and leather chest strap with ammunition loops. The BuAer required the shoulder strap to be widened in lieu of the 3/4" on the M3 for comfort reasons, however, some still found it uncomfortable. (Paul Faulkner Collection)

(Below) Not all units had access to leather. Shoulder holsters were also made from locally available fabric. Daily journals are part of the squadron administrative duties, and one entry in the MAG-11 journal of 28 May 1943 for the parachute loft indicates personnel made .38 cal. shoulder holsters. Here are two examples, one for each .38 Special and .45 1911 automatic.

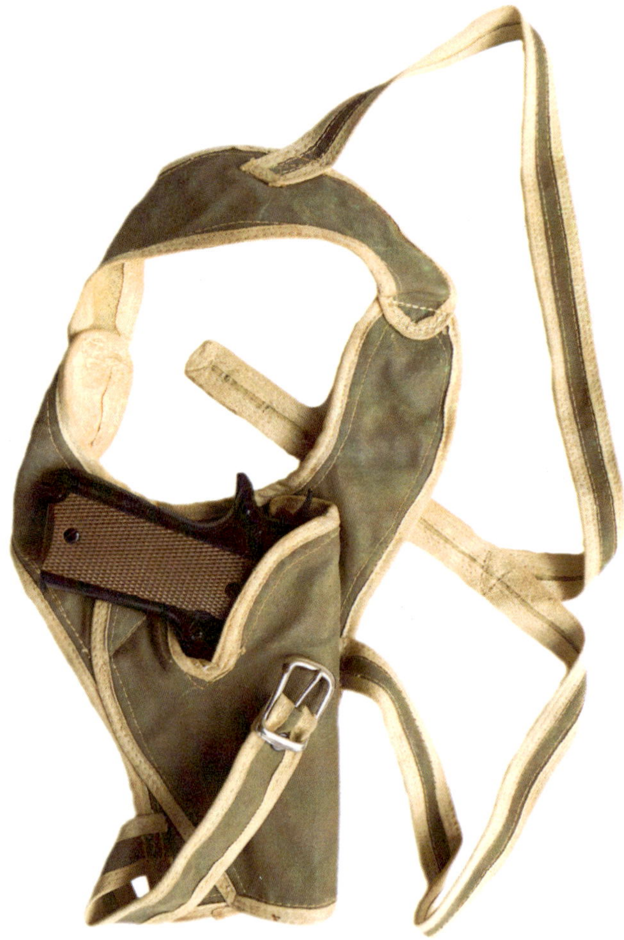

(Below, left and center) Major Theodore Olsen of VMF-313 wears a fabric vest incorporating holster for a .38 S&W Special revolver. (National Archives)

(Below) Unidentified TBF radioman from VT-5 uses a commercial or police service type holster for his Victory, USS Yorktown, August 1944. (National Archives)

(Above) TBF pilot of VC-5 shakes hands with fellow squadron mate after a successful attack on a Japanese battleship, Leyte, Philippines, November 1944. Ens. Charles Lee wears a private purchase type holster for his .38, revolver wrapped in cloth. (National Archives)

(Above) Privately purchased or locally procured hip holsters were obtained by Naval and Marine pilots from commercial sources. They could have been purchased by the individual, squadron, or group with funds in the US prior to overseas embarkation. VB-5 Pilot Lt. jg. T.J. Wiggins and his gunner Bennet Dees wear commercial type hip holsters, USS Yorktown, August 1944. (National Archives)

(Below) Unidentified aircrew of VT-10, USS Enterprise, c. March 1944. The gunner and radio operator wear commercial type pistol belts and holsters for their .38 Special. (National Archives)

(Below) Roy Williams of VT-20, USS Lexington, March 1945. He dons a private type hip holster for a .45 automatic. (National Archives)

(Below and right) Some shoulder holsters were manufactured at NAS Norfolk for the .38 revolver under drawing number 38H3. Two variations of WWII vintage have been identified. This Norfolk holster used by fleet personnel during WWII incorporating shoulder and chest strap. Note distinctive detail of gold embossed property mark US NAVY. (Mark Jones Collection)

(Below) 38H3 holster used by Lt. jg William E. Gehoe who flew Hellcats with VF-9 serving aboard the USS Essex from 1943 to 1944. These holsters were embossed MFG BY A&R DEPT. NAS NORFOLK VA, or Manufactured by the Assembly & Repair Department, Naval Air Station Norfolk, Virginia. (Paul Faulkner Collection)

(Below) Lt. Cmdr. Donald White of VT-9 wears 38H3 Norfolk holster. USS Essex, March 1944. The gold embossed US Navy can be see just in front of elbow. (National Archives)

(Above left) Fighter ace Lt. Edward Turner of VF-14 with Norfolk holster aboard the USS Wasp, October 1944. (National Archives)

(Above) Member of VF-14 donning Norfolk holster with bullet loops on chest strap, USS Wasp, August 1944. (National Archives)

(Left) TBF/TBM aircrew member Gilbert Johnson ARM2/c is wearing Norfolk holster while examining flak damage to his Avenger, Philippines, October 1944. Jackknife slug on oral inflation inlet valve of the pilot, center. (National Archives)

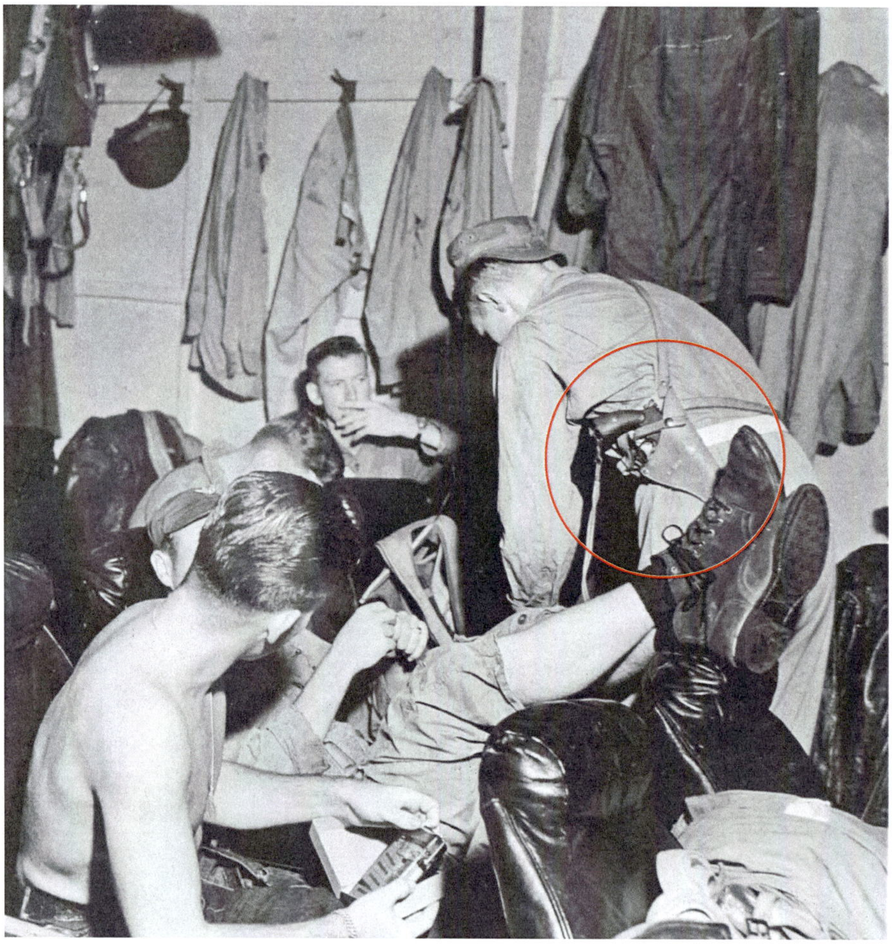

(Above and right) Two Norfolk 38H3 shoulder holsters in ready room of USS Ticonderoga, July 1945. The embossed gold US NAVY can be seen on the holster of individual leaning over and the back of a holster can be seen in a chair. (National Archives)

(Below) Second known example of the Norfolk 38H3 shoulder holster from WWII. Its manufacture was outsourced in 1945. The BuOrd distributed drawings for bids in May with responses by Boyt and the Enger-Kress Co.; Enger-Kress was awarded contract for 18,000 shoulder holsters in early June 1945. Construction closely resembles Norfolk model with addition of bullet loops on padded shoulder strap. The keeper strap holding the revolver in place was relocated to go over the back of the hammer instead of the trigger guard. ENGER-KRESS is the only marking embossed on holster's reverse. Deliveries of the new holsters commenced in July and were completed in November 1945. This example has extra elastic loops added to shoulder strap.

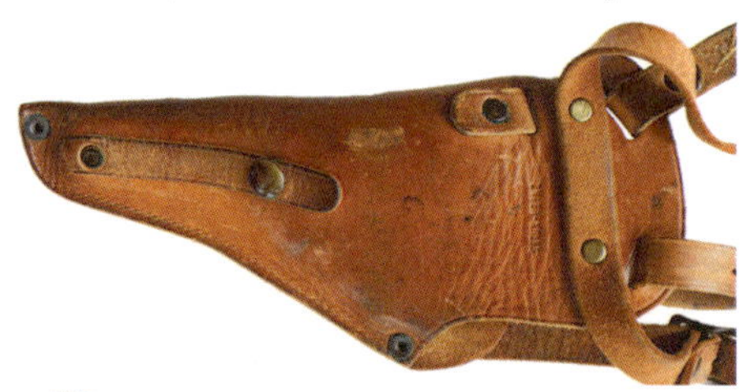

As previously mentioned, the .38 Special revolver was originally intended for the use of security forces within the confines of the USA. Commercial lead ball ammunition was procured for issuance. With the adoption of the revolver for aviation activities, the use of lead ball ammunition became problematic, because the Hague Convention prohibited the use of lead ammunition against the enemy.

To address the issue, BuOrd Circular Letter No. A103-43 was drafted and distributed on 27 September 1943. It stated the current Navy's stock of .38 caliber ammunition was unjacketed lead bullets and directed the use of unjacketed lead ammunition beyond the continental United States be prohibited. No additional ammunition was to be issued to aircraft aircrew departing overseas until the availability of a steel jacketed type. Further, personnel deployed overseas were ordered to dispose of existing stocks by dumping it overboard in deep water. Steel jacketed ammunition had already been placed on procurement and deliveries were projected later that month. Allotments were expedited to overseas supply points and aircrews preparing for overseas departure from coastal stations.

The diary of US Navy, Patrol Bombing Squadron 130 (VPB-130) lends a perspective on a reality in regards to sidearms and ammunition. The diary states: "When leaving the West Coast .38 caliber revolvers were issued for use as sidearms. Ammunition for them was drawn and shipped to our destination. The ammunition never arrived and it was impossible to draw .38 caliber ammunition anywhere in the Southwest Pacific, so far as could be determined. A few rounds were located here and there and as many .45 caliber pistols as possible were drawn. Ammunition for these was plentiful through Army supply. The .38 caliber revolver is an excellent survival gun. However, if it is to be issued for that purpose it is highly recommended that ammunition be made available in forward areas". VP/VPB-130 departed from Alameda, California 6 August 1944 to Hawaii then to their eventual final destination in the Philippines. The men with PV-1 Ventura's were shipped together with supporting supplies shipped separately. Supplies for carrier-based aircraft were required to be stored aboard, enough to support all requirements for duration of cruise. For the latter, if short, that would had been the deficiency of the of the Supply Officer.

(Below) Commercial lead ball ammunition manufactured in the same era. Initial allotments for aircrews were lead ball ammunition. US Army Ordnance procurements detail the procurement of lead ball .38 Police Special ammunition from both Remington Arms Co. and Western Cartridge Co., indicating in contract they were to be of commercial boxes. Examples of the type provided by Remington and Western Cartridge. Western boxes will include date code stamps."

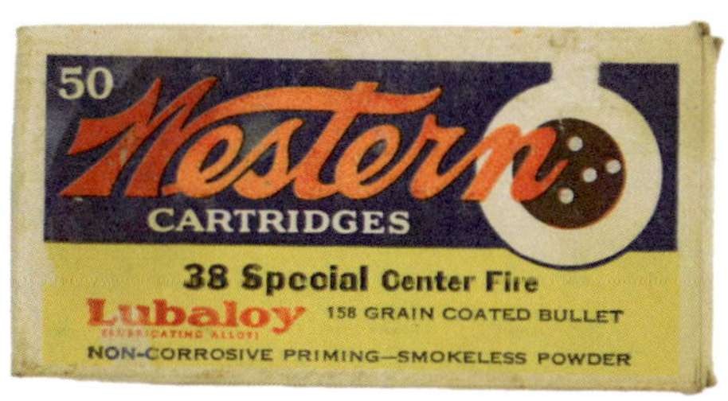

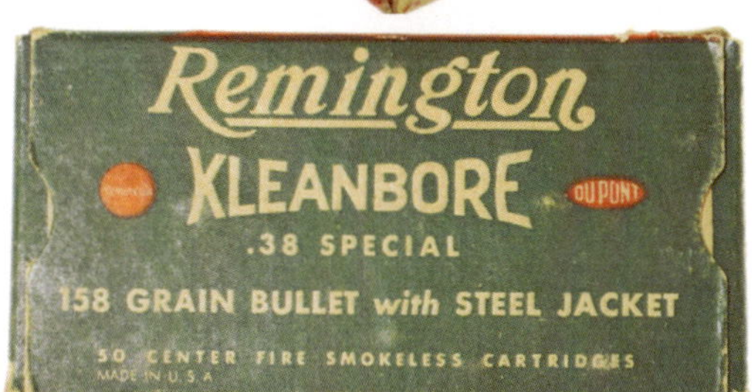

(Left and below) New steel-jacketed .38 cal. ammunition for use in forward area. Remington Arms Co., only known contractor, provided steel-jacketed ammunition in commercial box of 50 rounds. Lot number associated with government contracts was stamped on reverse with RA prefix to lot number. This ammunition was only manufactured during WWII.

(Left) comparative view between lead ball (left) and steel jacket (right) .38 cal. ammunition.

.38 special revolvers were issued with 100 rounds of ball ammunition but without means of carrying it, except the ammunition pouch 2-B-1500. Early examples of improvised methods for carrying extra rounds can be seen with loops sewn to trouser and pistol belts. Another method was to attach loops to the standard shoulder holster. No documentation surfaced discussing the origin of recommendation attaching ammunition loops to shoulder holsters but there was a variety of applications. Shoulder holsters were shipped from the factory without loops and they were added in service, whether at the assembly and repair facilities, a parachute loft aboard carriers, or land-based facilities.

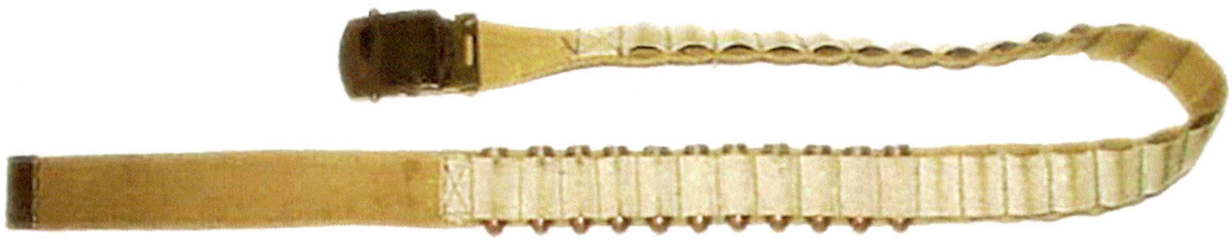

(Left) Web trouser belt adapted to carry .38 cal. ammunition. (Jerry Kolakowski collection)

(Below) Trouser belt with ammunition loops, Ens. Webb of VC-13 aboard the USS Core, August 1943. (National Archives)

(Below) SBD or TBM aircrew member aboard USS Yorktown uses modified trouser belt for ammunition, September 1943. (National Archives)

(Below and right) Modified trouser belt for .45 cal. ammunition. This commercial white leather type is similar to an officer's private purchase belt for white service dress uniforms. Elastic material used to repair flying goggles was used to apply loops. (Right) Ens. Charles Gibson of VS-10 wears leather trouser belt with loops for storing .45 cal. ammunition, USS Enterprise, February 1943. Gibson's sidearm is a .45 automatic in a shoulder holster. (National Archives)

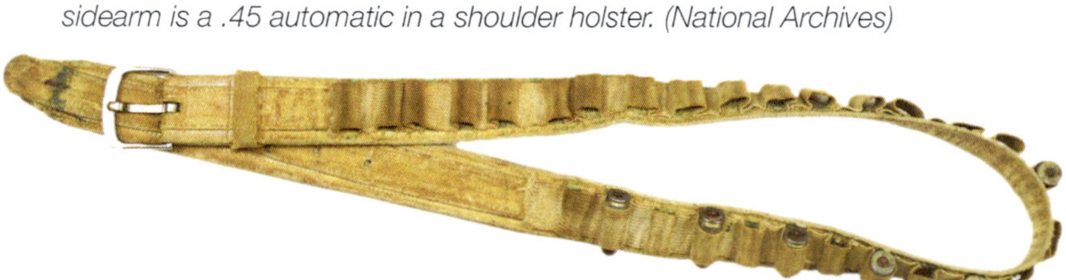

(Above) Pilots of VF-2 undergo briefing for air strikes, USS Hornet, September 1944. Ens. Ross Robinson has ammunition loops with .45 ACP rounds on strap of shoulder holster. (National Archives)

VF-8 pilot on the USS Bunker Hill in Ready Room, October 1944. Note custom rig for .45 auto hanging on wall with ammunition loops applied to belt. (National Archives)

Ens. D.C. Brandt of VF-2 has same leather slip with .45 ACP rounds on shoulder holster as Robinson. Probability of being tracer rounds. (National Archives)

(Above) VF-23 pilots Ens. Herbert Bonner, Lt. jg Charles Wiechardt Jr., Lt. Leslie Kerr, Ens. Henry Stasiuk, and Ens. Evan Adams wear modified pistol belts with loops for .38 cal. ammunition on flight deck of USS Langley, c.1945. Bonner and Weichardt have Vinylite covers for their revolvers. (National Archives)

(Below) Variety of applications for ammunition loops for the .38 caliber round on pistol belts.

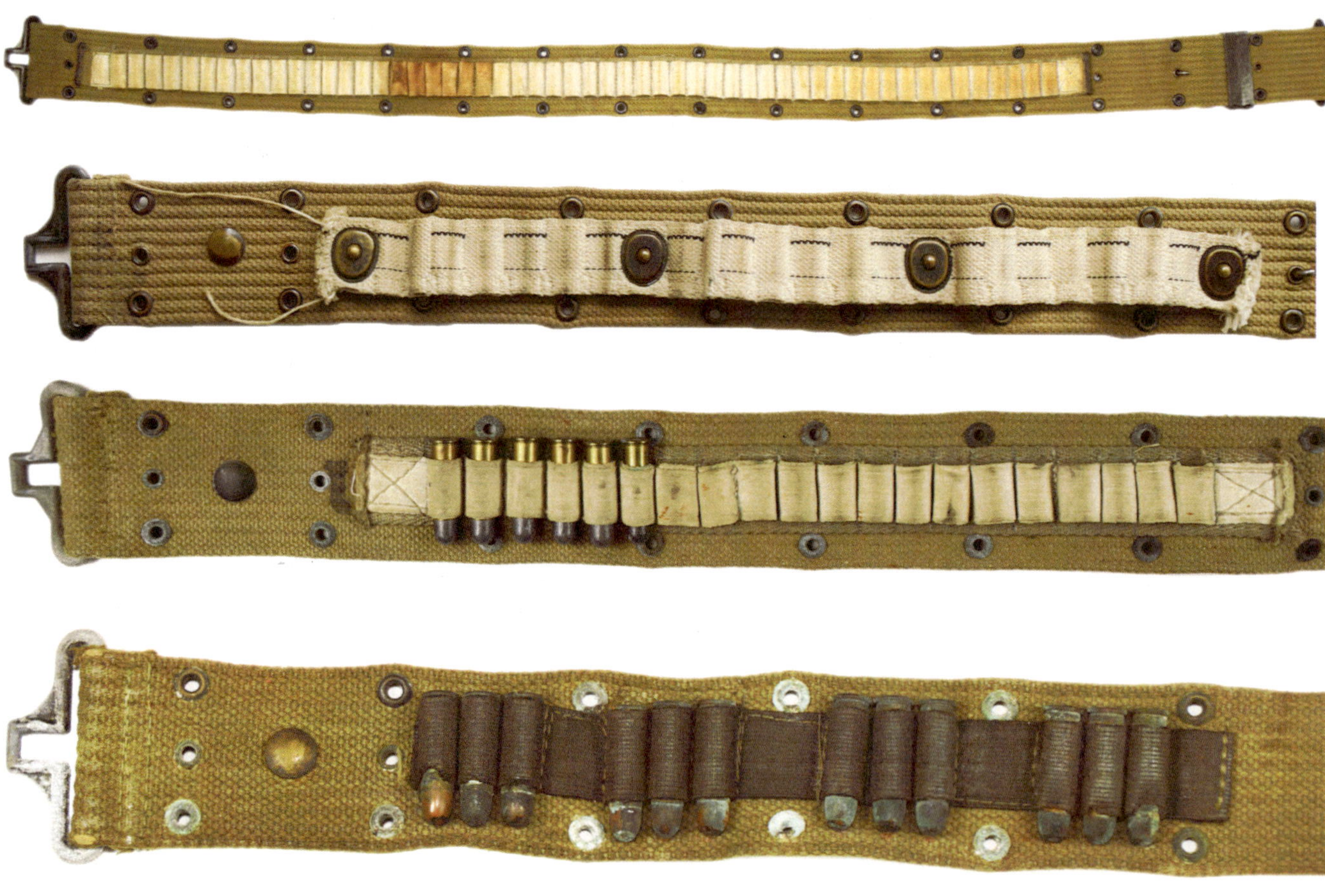

(Above left) Lt. jg. Fetter of VC-13 wears the standard allotment of the .38 revolver, shoulder holster, and 5" hunting knife as detailed in circular G1-43, with modified pistol belt for ammunition. (National Archives)

(Above middle) Pilot of VB-10 with hip holster, modified pistol belt and fighting knife Saipan, spring 1945. (National Archives)

(Above right) Member of VF-34 with hip holster and modified pistol belt for .38, c.1945. (National Archives)

(Below) VMF-312, Espiritu, c.1944. Pilot second from left has applied ammunition loops for .45 cal. rounds. (National Archives)

(Below) Examples of the application of ammunition loops on the standard .38 Special shoulder holster. A variety of materials was used: white and blue cotton webbing, elastic webbing, and, like one sample, yellow webbing used for life preserver repair. Standard ball ammunition, tracer rounds, or both were contained in the loops. There was a significant variable in the quantity of rounds carried, as evidenced here and in additional images.

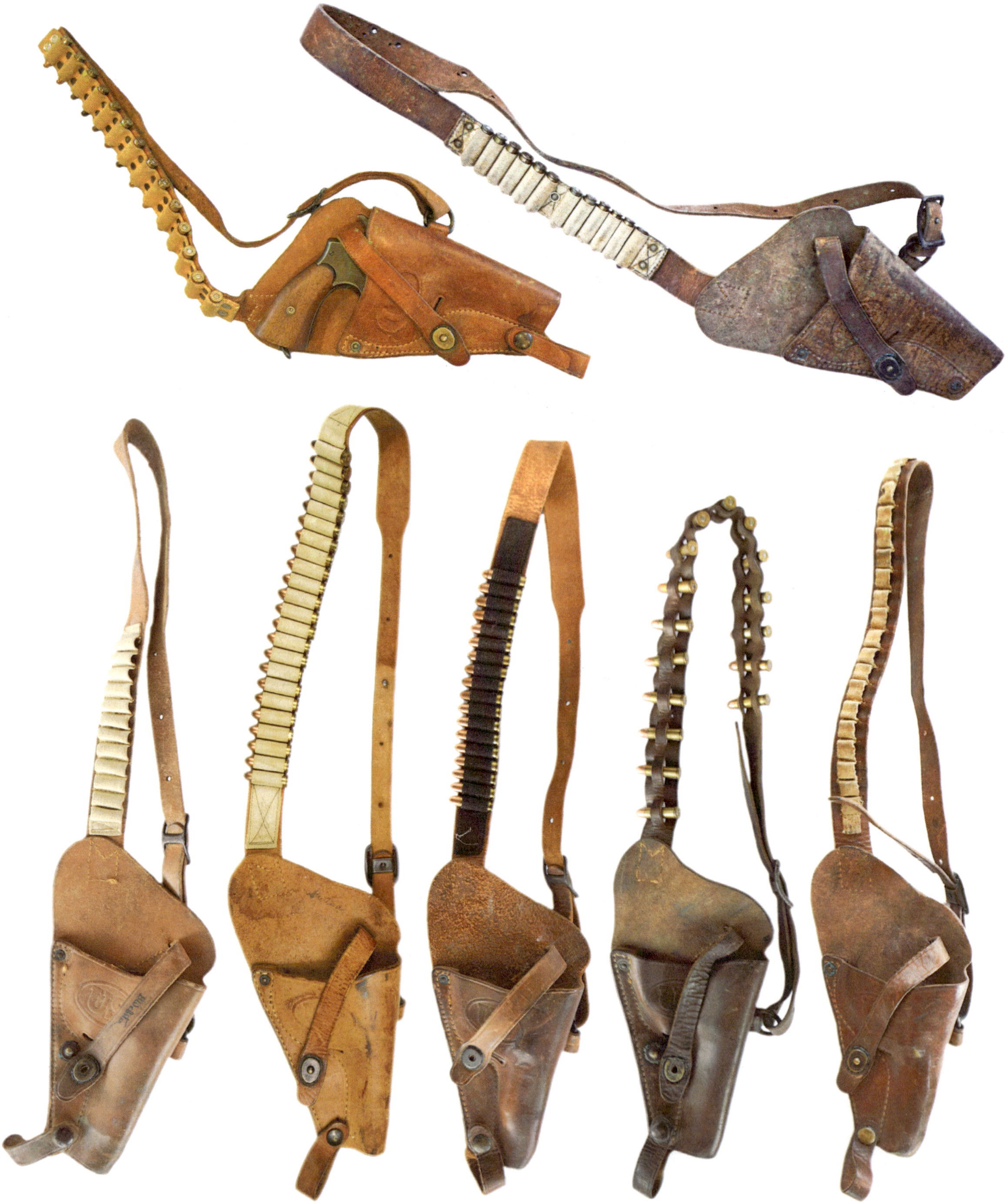

(Above) Ammunition loop variations on shoulder holster of Naval aviators. Enlisted men of Air Group 10 Milton Smythe ARM2/c and James Bennet ARM1/c aboard the USS Enterprise, March 1944. (National Archives)

(Below) Unidentified VB-5 SBD gunner aboard the USS Yorktown, September 1943. (National Archives)

(Below) One of the earliest examples of the loop application to the shoulder holster worn by TBF gunner and radio operators aboard the USS Yorktown, September 1943. (National Archives)

(Above) VT-47 TBF ball turret gunner Jack Snelson AOM2/c wears shoulder holster and trouser belt with added loops. (National Archives)

(Above) Lt. jg Thomas Wiggins of VF-5 wears shoulder holster around waist aboard the USS Yorktown, September 1943, This was more comfortable and convenient for some individuals. (National Archives)

(Below) Pilots of VF-27, c.1944. They wear a selection of the .38 shoulder holster with and without ammunition loops. The two pilots indicated have ammunition bandoleer. (National Archives)

(Above) Marksmanship was part of standard curriculum for aviation cadets at preflight schools. (Left) Cadets at St. Mary's on firing range with .45 cal. 1911 pistols, September 1942. (Above) Cadets at Del Monte preflight school becoming proficient with M1903 Springfield rifle, July 1943. (National Archives)

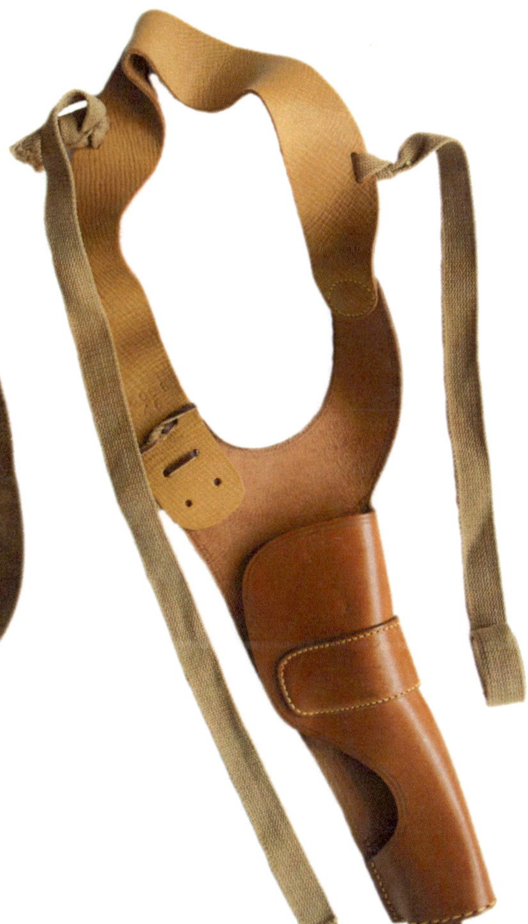

(Far left) Private purchase clip type shoulder holster for the .38 Special having added ammunition loops.

(Left) Private purchase clip type shoulder holster for a 45 automatic. Holster purchased by Lt. jg Bayard Webster for $4.50. Webster flew with VF-6 in 1943-1944 and with VBF-85 in 1945.

(Above) Lt. jg Schoby wears a tooled private type belt with added ammunition loops, USS Core, August 1943. Note waterproof wrap of revolver. (National Archives)

(Above)) Lt. jg Underman also wears a tooled private type belt with added ammunition loops, USS Core, August 1943. Note waterproof wrap of revolver. (National Archives)

(Below) VT-5 aircrew, USS Yorktown, August 1944. Left to right: Jack Hancock ARM2/c (killed In action), Ens. L.E. Benson and gunner W.J. Moak AMM2/c (shot down one of attacking planes at Truk). Moak wears private purchase western style pistol belt. (National Archives)

(Left) Unidentified rear-seat gunner of VT-11, March 1945. Another look at a private type belt adapted for the application of ammunition loops. (National Archives)

(Far below) Avenger torpedo aircrew, USS Rudyerd Bay, April 1945. Gunner, right, has a western style pistol belt rig with .38 cal. ammunition, no added loops to shoulder holster. He additionally uses variable density goggles. Radio operator, left, has them applied. Pilot, center, is utilizing the Vinylite pistol cover. (National Archives)

(Below) Example of an unattributed private type pistol belt. PAL RH-36 hunting knife with sheath sewn in between belt and tongue. Mounted type M-1910 canteen attached by paracord in lieu of belt hook. Emergency compass secured to forward ammunition loop. A length of paracord spooled at belt buckle.

A typical accoutrement used with the .38 revolver was the M-1917 pistol lanyard made of a woven shellacked cord and a metal clip. Adopted in 1917, it was used for both pistols and revolvers. It was superseded by the M1 in 1934 although it remained in service and used by the Naval and Marine aeronautical services. Use of the lanyard was recommended after reports of sidearm loss caused by the opening shock of the parachute and during initial struggles after emergency landings. (Above) Unidentified VT aircrew. (Left) Hellcat pilot of VF-82 aboard the USS Bennington, February 1945. Its application can be observed in images throughout this study.

M-1917 pistol and revolver lanyard.

(Above) All flight crews were to be issued a .38 revolver with a shoulder holster as of January 1943. However, due to lack of supply, issue was gradual. New items reached stateside central supply points, with units preparing for overseas departure first supplied with the new materiel. New items were expedited to Hawaii and later to Guam to resupply combat zone units. VF-14 pilots demonstrate typical configurations of small arms. (left) standard .45 with M-1916 holster, pistol belt and magazine pouch. (middle) Pilot wearing .38 with hip holster, which replaced the .45 in 1943. He uses trouser belt with ammunition loops. (right) Pilot with typical complement from 1944 through 1945 of .38 revolver, shoulder holster and pistol lanyard. Each is wearing configuration of personal preference. (National Archives)

(Right) William Thurston of VF-5 in cockpit of F6F Hellcat, USS Yorktown, September 1943. The sidearm for US Navy and US Marine aviators was a defensive and survival weapon, not intended for offensive use. (National Archives)

Marine aviators' sidearms were governed by different guidelines. Marine aviation cadets received standard 782 gear as recruits and in advanced flight training. They underwent short field training exercises under the premise of Marine first, Aviator second. At officer commissioning they received a .45 1911 automatic pistol as established under the T/O&E for amphibious forces when preparing for overseas departure. Marine aviation allowance was exempt under the initial distribution of G1-43 but was unofficially revised in February 1943 to comply with issuance of the .38 revolver. However, this transition did not start occurring until 1944.

G1-43 was superseded by BuOrd circular letter GV1-45 dated 24 February 1945, specifically addressing both Marine and Naval aircrew. After which date, the .38 Special was more prevalent in USMC use. Through the war, Marine units preferred the automatic to the revolver, though both were distributed.

(Below) Composite of small arms used by VMTB-232, c.1944. Left to right: .45 with M3 shoulder holster, M-1916 hip holster with .45, and standard aircrew complement of a .38 revolver with shoulder holster. (National Archives)

(Above and right) These images of USMC aircrew from VMB-413 illustrate the same type of assortment as seen with Naval aviators, c.1944. (Above) Four aircrew utilize M-1916, pistol belt and double magazine pouch. Two wear M3 shoulder holsters, one of each type shoulder strap width. (Above right) Bulk of crew with M3 shoulder holster with .45 auto. (Right) Officer on left has modified M3 with chest strap configuration, field modified to M7 pattern. (National Archives)

(Below) PBJ (B-25) aircrew of a VMB squadron in the SWPA, c.1944. These four aircrew have the .38 Special with shoulder holster. From 1944, issuance of Victory for USMC crews began while preparing for embarkation stateside. Attached to MAG-61, departed west coast in May of 1944, contributed to neutralization of Japanese defenses in Solomon Islands. (National Archives)

(Above) Pilots of VMF-213 wear an assortment of personal gear, Guadalcanal, c.1943. Modified pistol and trouser belts with ammunition. (National Archives)

(Left) Lt. Col. Frank Schwable, C.O. of VMF(N)-531 recently returning from patrol mission, Vella Lavella, November 1944. He carries .38 with shoulder holster and modified trouser type belt with ammunition. (National Archives)

(Left below) TBF aircrew members of VMTB-134 use standard allotment for Navy and Marine aviators of .38 revolver with shoulder holster, Peleliu, September 1944. (National Archives)

(Below) Unidentified VMB aircrew, c.1944. (National Archives)

(Left and below) Small arms, even in Marine units, would technically still be squadron property. Here, USMC contract M3 shoulder holster with added bullet loops, personalized by user with an EGA attachment . On back is factory embossed USMC/ Boyt/43 with property mark of VMSB-242. (Robin Bartel Collection)

(Right) VMF-213 pilot, Lt. Wallace Hallmeyer, USMC wears a locally procured leather shoulder holster for his .45 auto, Guadalcanal, June 1943. (National Archives)

(Below and below right) SBD radio-gunners of VMSB-244 on airstrip in Luzon, c.1945. Gunner on right wears locally fabricated vest. (Middle) M3 shoulder with lensatic compass pouch. (Left) M-1916 hip holster. (Below right) Filipino child wears flight gear of gunner. (USMC)

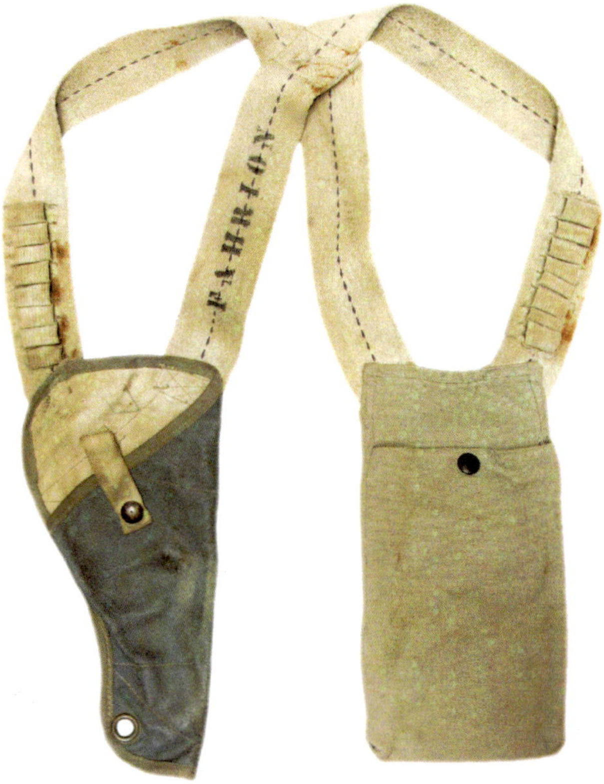

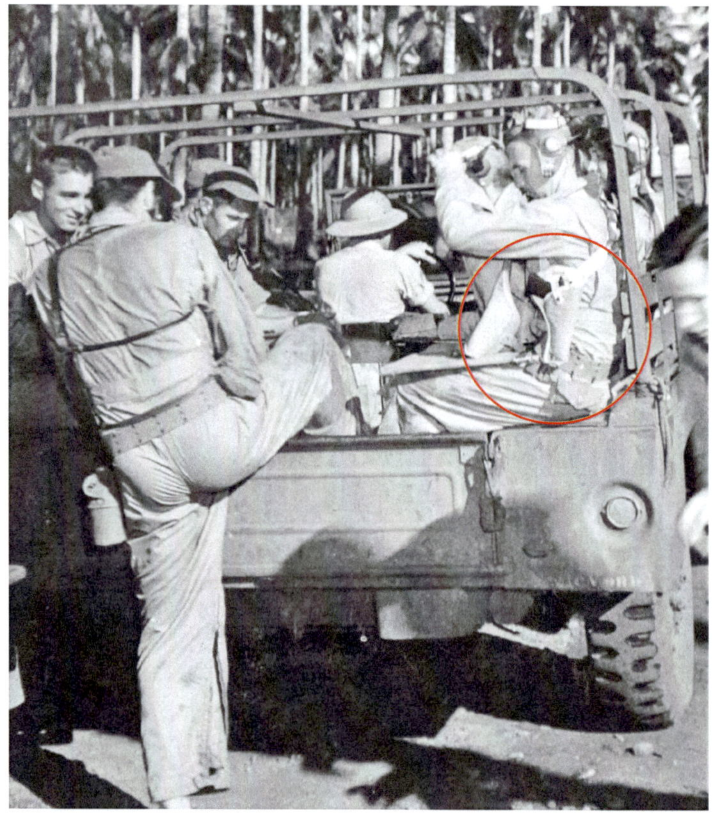

(Above) Locally fabricated shoulder holster for .45 1911, waterproof nonspecular sea blue holster with Army duck pouch. Shoulder harness made from parachute webbing. (Paul Faulkner collection)

(Above Right) VMF-223 pilots loading up to be transported to flight line, Guadalcanal, August 1943. Seated individual wearing locally fabricated shoulder holster for .45 1911, which shares many of the same features as the holster above. (USMC)

(Right) VMF-214 pilots illustrate an assortment of shoulder holsters. 1st Lt. Jack Petit wears a fabric shoulder holster at newly captured Munda Airfield, New Georgia Island, October 1943. (USMC)

(Above) Pilots of VMF-312, Espiritu Santo, c.1944. Pilot in right hand photo uses a spring clip type holster for his unidentified automatic pistol. (USMC)

Locally manufactured shoulder holster for .45 auto. Detachable holster modeled after commercial police types with added pocket for magazine. Chest strap repurposed lanyard from AN-R-2 one-man parachute raft.

USAAF issuance of small arms was similar to USMC's issuance: part of the T/O&E for field and flight officers. The US Army standard small arm was the Model 1911/1911A1 Pistol, Automatic, .45 cal. since before WWI.

Sidearms issuance to pilots outside of the continental United States began in the 1930s for hunting small game, personal protection and became typical when operating in isolated regions. Pistols were issued with a pistol belt, M-1916 holster, magazine pouch and two extra clips maintained by the squadron Ordnance Officer. Squadron PEOs would requisition the sidearms for distribution. Revised T/O&Es during the war authorized issuance to enlisted aircrew operating outside continental United States.

Standard allotment for USAAF personnel operating outside continental USA: 1911/1911A1 .45 auto, M-1916 hip holster, double magazine pouch, pistol belt and first pouch with packet. (Robin Bartel collection)

(Below) B-24 pilots Cpt. Robert Paullin and Lt. Patterson examine newly issued .45 1911 pistols in North Africa, August 1942 (National Archives)

(Above) Brig. Gen. Patrick Timberlake, C.O. 9th AF, and Col. Hugo Rush return from long range mission, North Africa, December 1942. Each carry basic armament with addition of M-1910 canteen. (National Archives)

(Above) Tallest man in the 9th AF, 2nd Lt. Carl Looker, with shortest, 2nd Lt. Edward ("Peewee") Rothkrug, North Africa, spring 1943. (National Archives)

(Below) Though small arms were reserved for overseas deployment, this local command thought prudent to issue small arms to its Pacific Coast patrol aircrew, 20 December 1941. (National Archives)

(Below) Pilots of the 11th Provisional Bomber Command undergo briefing before air strikes over Kiska, Alaska, c.1942. Two pilots wear standard M-1916 holsters. (National Archives)

Shoulder holsters for the .45 1911 had been proposed for standardization since before WWI, primarily from cavalry units. Upon entering WWII new service requirements were expressed by the USAAF, where the hip holster was a hindrance in cockpits and interfered with duties aboard an airplane. Both services reported that sometimes the hip holster would rip off from the pistol belt due to the opening shock of the parachute.

To address the issue, the US Army Ordnance Department developed a shoulder holster to meet requirements set by the USAAF, adopting a design on 12 October 1942 known as the Shoulder Holster, Model M3 for cal. .45 automatic pistols. Boyt Harness Co. was awarded the initial contracts the same month for what became officially known as the Holster, Pistol, Cal. .45, M3, available under stock number 74-H-149. The M3 was referenced under Dwg. No. 20-17-101 Specification X3, A140308 & B108108 with a 3/4 inch wide shoulder strap.

When this holster was field-tested by the US Navy in January 1943, it requested the shoulder strap be widened to 1-¼ inches for the .38 cal. model. This modification was made to the M3 under Revision-1, 15 February 1943. The USMC and US Navy made limited procurements of the M3 shoulder holster with an embossed US on the front and either an stamped USMC or USN on the back. One contract was awarded to Boyt Harness Co. in April 1944 for 4,000 by the BuOrd (USN) with an extension of 200 more in May under contract number NXSO-57033. On paper, the M3 would be superseded by the M7 shoulder holster in 1944.

(Left) First production M3 shoulder holster with 3/4" shoulder strap.

(Below) P-38 pilot of 459th FS, Capt. Maxwell Glenn wearing first production M3 shoulder holster in the CBI, June 1944. Belt strap was designed to secure to trouser belt. (National Archives)

(Right) Pilots of the 5th Air Force exhibit a variety of holsters, far right first generation M3 and seen is standard M-1916 hip. Pilot far left wears locally procured hip type with applied ammunition loops for .45 caliber, and middle is locally sourced shoulder type, c.1944.

(Below) Revision-1 M3 shoulder holster with 1-1/4" wide shoulder strap. Inset illustrates comparative view between the widths.

(Below) P-61 pilots of 10th AF, CBI wear Revision-1 holsters. (National Archives)

ComAirPac expressed officers and flight crews should be equipped with a pistol or revolver. Pilots in combat in the South Pacific state that the use of shoulder holsters for pistols were essential, because in about fifty percent of forced landings at sea or parachute jumps, a web belt becomes unfastened and the pistol was lost. The Army Ordnance type shoulder harness complete with knife, holster and medical supplies was considered ideal.

Extending on ComAirPac, the 5th AF published Regulation No. 133-95 28 December 1943. It prescribes; in order to safe-guard personnel in case of forced landings or parachute jumps, all combat crewman and passengers flying in the New Guinea area will be clothed and equipped with long sleeved shirt and long trousers -or summer flying suit -or one-piece HBT coverall and service shoes. Each person was to have a .45 cal. pistol, 21 rounds ammunition, shoulder holster, canteen, first aid packet and jungle emergency kit. There were many parallels between units in the forward area no matter the theater of operation or service. Like those early Naval air groups adapting use of a shoulder holster, the USAAF followed suit. Prior to the availability of the M3 shoulder holster, the prescribed allotment was as shown on page 56.

(Below) P-40 pilots of the 44th FS on New Georgia, Solomon Islands, August 1943. This squadron was stationed in Honolulu, Hawaii upon its attack on 7 December 1941 where it remained until December 1942, transferred to the SWPA. Pilots wear standard allotment with addition of canteen, as prescribed in T/O&E. Pilot far right wears a locally procured or fabricated shoulder holster. (National Archives)

(Left) 5th AF Douglas A-20 aircrew members 1st Lt. Raymond Wilkins and T/Sgt. William Ralls demonstrate typical complement of USAAF pilots and aircrew operating in the Asiatic-Pacific. The initial allotment was to officers then, if supplies were available, to noncommissioned crew. Like the US Navy, a small arm was recommended for all flight personnel to sustain life in isolated regions and for personal protection. As reported through the intelligence sections, "The Japanese shoot first and ask questions later." New Guinea, March 1943. (National Archives)

(Below) USAAF flight personnel also concluded that carrying a sidearm is more efficient with a shoulder holster. They too used locally procured and fabricated holsters, all three holsters observed here are of different pattern. Cpl. Richard Dodge, Sgt. Harry Rucker and 1st Lt. Robert Oestreicher of the 89th BS, 5th AF, Port Moresby, New Guinea, c.1943. (National Archives)

(Above) Early campaigning in the SWPA will illustrate a mix of standard sidearm allotment with modified M-1916 holster and locally procured shoulder types, as seen here: Aircrew members, 1st Lt. William Neel and T/Sgt. Edgar Crist of the 44th BS, New Guinea, c.1943. (National Archives)

(Above) Many pattern types can be observed, leather magazine pocket can be observed here. Unidentified officer of the 394th BS, Espiritu Santo, c.1943. (National Archives)

(Below) P-39 pilot 2nd Lt. Albert Rothhaar, New Guinea, c.1943. Leather magazine pocket and hunting knife under right arm. (National Archives)

(Below) P-38 pilot, Lt. John Roehm, Solomon Islands, c.1943. (National Archives)

(Above) Unidentified B-24 aircrew member of "Lucky Lucille", 65th BS, New Guinea, March 1944. (National Archives)

(Above) P-47 pilots of the 7th AF put the finishing touches on their Ready Shack/ Officers Club, Marianas, c.1944. 2nd Lt. Donald Kennedy uses his private M1873 single action Colt as a hammer. (National Archives)

(Below) Locally fabricated shoulder holsters for the .45 1911. (Left) Example of unknown origins. (Right) Manufactured in Auckland, New Zealand. USAAF, Navy and Marine commands had access to merchants and craftsmen residing in Australia and New Zealand with holsters like these purchased or commissioned by an individual or on unit levels.

(Left and below) Holster worn by B-24 ball turret gunner S/ Sgt. Howard Lee of the 371st BS. Note unique application of mission marks on back.

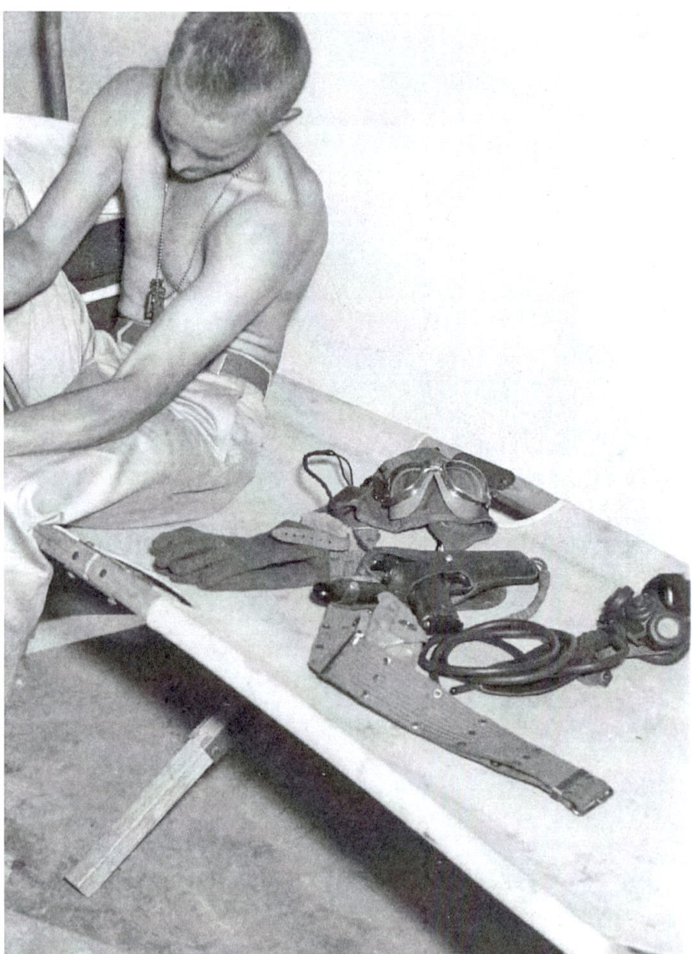

(Above) Unidentified B-24 aircrew, India, c.1943. Assortment of locally fabricated holsters for .45 1911. Units stationed in India and China had access to an abundance of local craftsman for their assembly.

(Left) Placed on his cot, private hip type holster.

(Below) Bombardier rests after mission, placed on night stand is private type shoulder holster. (USAF)

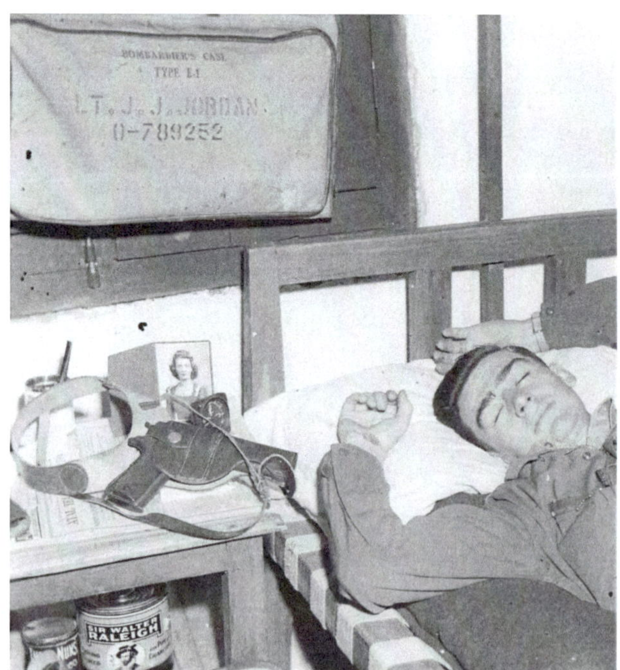

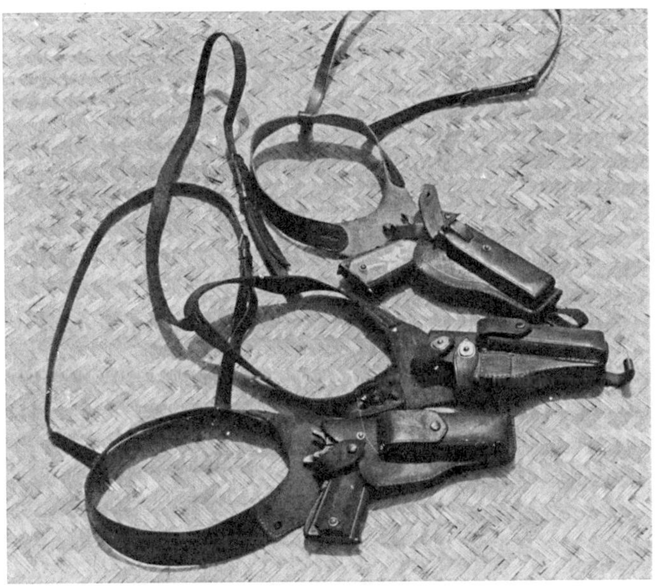

(Above) B-25 aircrew of the 341st BG interrogated by S-2 Intelligence Officer after mission, India, c.1943. Three air-crew wear standard complement with M-1916. Individual on right wears locally fabricated holster with ammunition loops. (National Archives)

(Left and below) Concepts of shoulder type holsters submitted to Materiel Command for evaluation of possible new standard to replace M3. Comfort and utility were emphasized, India, c. 1945. (USAF)

(Left) Unidentified 14th Air Force pilot stands on wing near visiting American artist Don Barclay while he was visiting China based forces, c.1943. He wears trouser belt with applied ammunition loops for his .45 auto. (National Archives)

(Right) S/Sgt. Harold Hadley watches a skirmish between Chinese Nationalist troops and local Yunnan Province militia a top 10th Air Force HQ, Kunming, China, c.1945. He has unidentified model of revolver with extra ammunition in loops attached to holster. (National Archives)

(Below) The addition of loops for ammunition storage to either holsters or belts was not as prevalent in the USAAF as the US Navy but can be observed. (Right) B-29 aircrew member Lt. Leslie Krech has loops on his M3 holster, China, c.1944. (Left) B-29 aircrew member Lt. M.D. Mantz of the 677th BS has loops applied to his M3 holster, China, c.1944. (National Archives)

(Left and below) Shoulder holsters were often worn around the waist by both aeronautical services, gunslinger style. This modified M3 shoulder holster had a snap hook added for easy removal. Rawhide thong was added as a leg tie attached to the belt strap, and bullet loops were added to the shoulder strap. (Left) Copilot of a C-47, Lt. R.S. Wilson wears a similar modified M3 holster, China, c.1945.

(Below left) Two B-29 aircrew members of the 676th BS, M/Sgt. George Vuletich and M/Sgt. Charlie Heyrend ,also use snap hook on the end of the shoulder strap. Heynard's M3 holster has additional modifications, including reinforcement with an aluminum plate secured to the tip of the holster. (Below) unattributed example of such a modification with ammunition loops added to shoulder strap. (National Archives)

Shoulder holsters were the preferred standard; however, some individuals preferred the hip type. Small arms differed between combat theaters. Units operating in the Pacific, India-Burma Theater and China Theater carried a sidearm as a mandatory standard as dictated by those theater commands; their use in the ETO was discouraged but occurred. The Japanese were not signatories to the Geneva Convention and did not abide by rules of warfare, so US airmen required some form of self-defense against aggression, but to also include wildlife. The primary reason for the sidearm was orientated towards that of survival, the ability to hunt game, later, to be used as a signaling device.

There were very few instances were the sidearm was ever used in self-defense, for those airmen forced down. The most common application was that as a gift to the native population. Those airmen forced down over the Philippines, China and Burma quickly encountered pro-American natives and Guerilla forces. Once in their custody, they would protect, feed and guide those distressed personnel, those airmen were actually a commodity. Commonly noted in evasion reports was how poorly equipped the Guerillas forces were, weapons include 19th century rifles, muzzle loaders and simply armed with machetes. A pistol was typically gifted to the leader or chieftain. For that of the Solomons and Papua/New Guinea, a small arm was way too big or value of a gift, in fact, it was stressed not to turn over your weapon to the natives in these areas. This is just a general overview of a much larger conversation that will be covered in-depth in a future volume.

For those operations in Europe, emphasis was on evasion and escape versus survival. Therefore, issuing of sidearms to personnel operating in Europe and the Mediterranean was discouraged by the intelligence sections responsible for E&E. However, sidearms were allocated as prescribed in T/O&E's regardless of operational theater. Upon ditching or bailout in occupied territory, airmen were instructed to immediately enter an evasion scenario and blend in. All military clothing was to be discarded in favor of civilian clothing, and ID was to be retained in case of capture. An individual in civilian clothes and carrying a sidearm or even a knife risked being treated as a spy or saboteur if caught behind German lines.

Lt. Byron Logie of the 458th BS makes note in his evasion narrative from Denmark that everyone in his unit had a pistol, but they were not allowed to carry them to the fact the Germans announced they were shooting armed airmen, as was relayed to him, claiming that they had been shooting German civilians. However, he wished he had a gun, and suggests carrying one in that area for the sole purpose of giving it to the helper. His experience proved that helpers were hesitant to guide evaders unless the helpers were armed. Logie's comments was shared by others who had evaded from Western Europe stating that they wished they had been carrying a pistol for the sole purpose of gifting it to the resistance, as weapons were badly needed. The actual distribution of the small arm was at the discretion of the Commanding Officer. S/Sgt. Carl Jennings of the 414th BS mentions that all men in their squadron had pistols but were not required to carry them. In contrast, in the 419th BS, only officers had pistols and were optional to carry.

Capt. Donald Manlove of the 746th BS discusses the small arm in his evasion narrative. The crew bailed out safely, only two were carrying .45 pistols, although all crewman in their unit were issued them. At one time carrying them had been optional, but recently the CO had issued an order to the effect that guns were not to be carried. However, all of the crew agrees that a gun should be carried as it could always be disposed of, if necessary.

Depending on the region of operations, sidearms were carried by airmen operating in Western Europe and the Mediterranean as an option. It then was banned to carry them based on intelligence of German reactions. When the 15th AF moved to bases in Italy, they established a policy that all airmen engaged on combat missions have on their possession a .45 cal. pistol while on operational flights, in the event they end up in Russian-controlled territory or the Balkans. A weapon fostered respect by people in those regions. Russians, Partisans of Yugoslavia, Hungarians, and Poles could not conceive of an unarmed soldier, aerial or otherwise. They looked upon them with disdain, impeding opportunities for gaining assistance. The larger and more conspicuous the gun, the greater the prestige in these areas. Aircrew therefore all donned sidearms under the 15th Air Forces policy, since their combat sorties were to Russian-controlled territories. As the war progressed, there was no longer vast German occupied territory and carrying of sidearms became more prevalent with in all Air Forces in Europe.

(Above) P-38 pilots have a meal and an interrogation after returning from a mission, North Africa, May 1943. Several of the pilots are carrying optional sidearms in shoulder holsters. (National Archives)

(Left and right) Shoulder holsters for .45 Auto and two magazines. (right) Holster used by 2nd Lt. Albert Duke of the 447th BS, 9th AF, North Africa.

(Above) 9th AF aircrew members under interrogation after bombing mission, Egypt, c.1943. Two crew members are utilizing 1917 .45 cal. Colt revolvers with M2 holsters. Aircrew member on left has half moon clip pouch. Note pistol on right is wrapped in cloth. (National Archives)

(Below) B-17 aircrew member 2nd Lt. Orlof Duker examining flak damage wearing locally fabricated shoulder holster with embossed designs, North Africa, c.1943. (National Archives)

(Below) VT-10 pilot Lt. jg. Jerry Rapp employs dual configuration of small arms. Shoulder holster carries 1917 Colt .45 revolver; on right hip is M-1916 holster for an automatic. It is not clear why he is using two sidearms. USS Enterprise, February 1943. (National Archives)

(Above) P-47 pilot Maj. Gilbert Talbot of the 345th FG wearing his .45 with an M3 holster around the hip, France, c.1944. (National Archives)

(Above) P-47 pilots 1st Lt. James McWhorter and Capt. Paul Van Cleef of the 9th AF wear sidearms while conducting operations from airfield in Germany, c.1945. McWhorter's M3 is around his waist; Van Cleef uses standard M-1916 hip type. (National Archives)

(Below) P-51 pilot Lt. H. Waggoner of the 352 FS, England, March 1945. (National Archives)

(Below) Unidentified Lieutenant of the 390th BG in England, c.1944. (National Archives)

(Below) .45 cal. ACP ammunition in 20 and 50-round boxes. 20-round boxes were typical until WWII, when 50-round packages were introduced. The 20-round boxes were standard for parachute emergency kits until the introduction of shot ammunition. Two 20-round boxes with red labels are prewar examples with head stamps dated in the 1930s. The 20-round box with orange label became standard in the inter-war years. Lastly, 50-round box. The .45 automatic pistol was issued with a clip and two spares in a magazine pouch, an allotment of 21 rounds per individual.

Small arms proficiency curriculum was instituted in USAAF preflight programs for flight cadets.

(Left) Cadets at Maxwell Field on the firing line shooting at targets 25 yards away.

(Below) Civilians were often conscripted to head the shooting ranges at cadet centers. Now sergeant, E.J. Mannie was vice president of the US Revolver Association and active member of Liberty Rifle and Pistol Club, San Antonio. During the war, he was an instructor in ground gunnery at the San Antonio Aviation Cadet Center.

By 1943, issuance of small arms to aeronautical services had become servicewide and in all world regions and climates. Arctic and jungle areas, with extreme moisture, sand, and saltwater corrosion, presented pistol maintenance problems. Salt water exposure, had the most immediate effect. Multiple Emergency Rescue Reports make mention that with in 24 to 36 hours their pistols or revolvers were badly corroded, they no longer functioned nor could they be salvaged. Even simply exposed to salt-ladened air, blooms would form on metal surfaces. Humid jungle environments were almost as damaging, many reported issues for those who Walked-Out. Those that trekked with natives, they had ingenious methods of applying animal fats and plant oils, pending the region one was in. Any ordnance manual makes it perfectly clear to always keep your firearm well oiled and cleaned, this fact would not had been lost on aircrew as part of the overview of preventive care.

In an extreme environment, an exposed oiled pistol would soon deteriorate. The first immediate step taken by the US Navy was to add a 3oz can of preservative oil to the back pad kit M-592, for continued treatment. In the instruction booklet for the M-592, it specifically states that the oil was provided for keeping the gun and other metallic parts in working order. As pointed out in several previous images, some pistols or revolvers were wrapped. Individuals took it upon themselves to take steps protecting their weapons with oil treated or untreated cloth. One Naval air group mentioned using segments of weather balloons to protect their revolvers.

In April 1943, the BuAer submitted an official request to the BuOrd to develop a waterproof bag or envelope for pistols to be issued with the shoulder holster. The BuOrd reached out to companies for possible designs, with samples for field testing becoming available by the end of the that year. The BuOrd had established a requirement of 100,000 waterproof pistol covers, awarding contracts to the National Carbon Co. by May 1944 under BuOrd Specification No. 1449, BuOrd Dwg No. 123110 titled Pistol Protective Cover for Cal. .38 Revolver and Cal. .45 Automatic Pistol. Quantities of the new waterproof covers were available by October 1944 under stock number 74-C-307 titled Cover, Protective, for Pistol or Revolver. The cover became available through both Army and Navy ordnance departments and it was incorporated into the USAAF Type C-1 vest as a standard accessory. Those covers for the vests were drawn from US Navy, BuOrd contracts. All further requirements by the USAAF were conducted by Purchase Orders on established fixed price contracts by BuOrd. The only known quantity for Army Ord. or USAAF can be extrapolated by C-1 vest production.

(Left) VF-32 pilot Lt. jg Richard May aboard the USS Langley, c.1944. Revolver protected in cloth. (National Archives)

(Below) Lt. jg Blake Moranville of VF-11, USS Hornet, January 1945. Moranville was shot down over Saigon 12 January 1945 where he and a group of other aviators spent time in a French military prison, a concentration camp, liberated and conscripted in the French Foreign Legion guerilla fighting Japanese until the party was finally flown to Kunming, China, 28 March 1945. (National Archives)

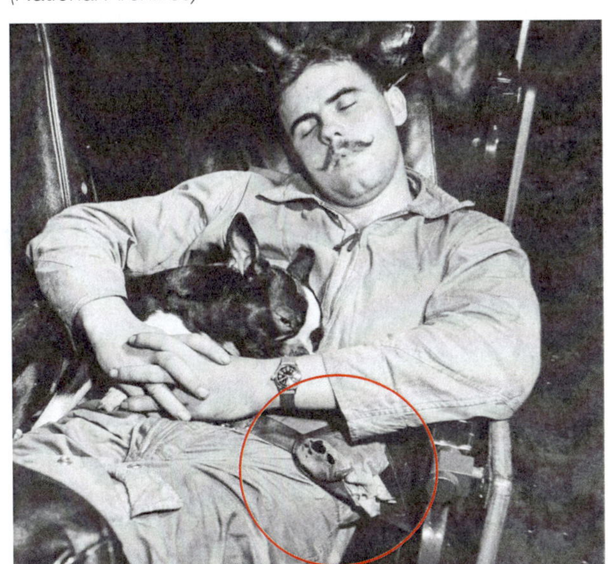

(Above) VF-16 pilot, USS Lexington, c.1944. His .38 appears to be in a typical black service sock. (National Archives)

(Above) Gunner, Oliver Tarr ARM2/c of VB-12 keeps his .45 pistol well protected, USS Saratoga, May 1944. (National Archives)

(Below) B-24 aircrew, Papua/New Guinea, c.1943. S/Sgt. Richard Faria keeps his .45 in some form of protective wrap. Other images from New guinea illustrate Marbles match safes attached to holsters, presumably for vials of oil. (National Archives)

(Below) Members of the 12th BG take recreation with a game of volleyball. One participant still wears his pistol rig with .45 wrapped in cloth. Sand was another enemy. (National Archives)

(Above) 404th Fighter-Bomber Group pilots inspect a Ju-88 at German based airfield, Germany, c.1945. Lt. Albert Greffenius wears his M3 around waist and cloth wrapped .45. (National Archive)

(Above) ATC pilots check-in from returning flight, India, November 1943. Protected .45 pistol. (National Archives)

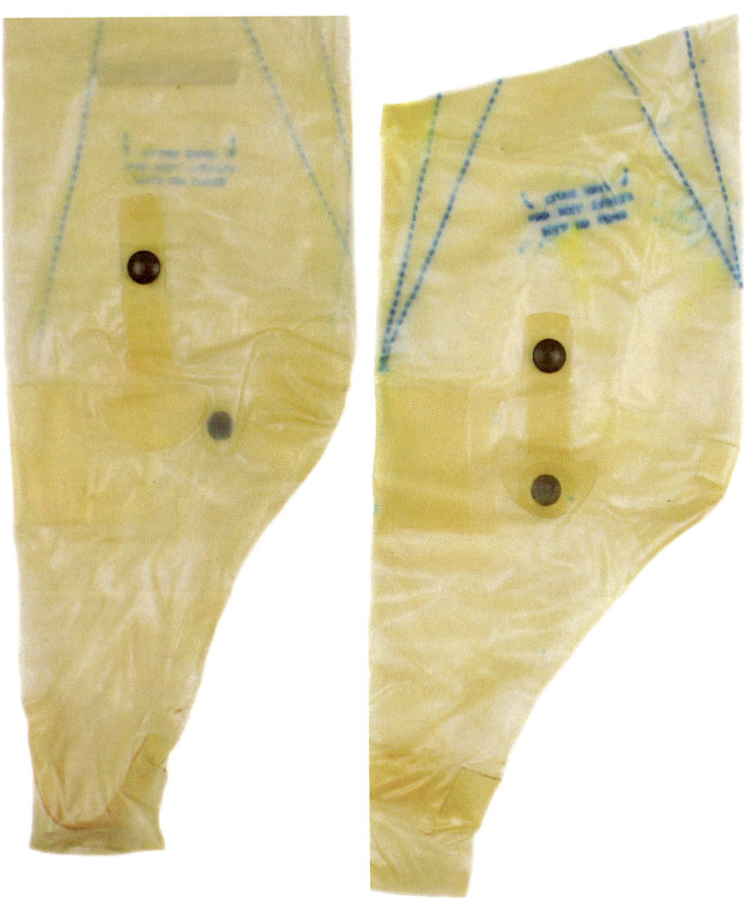

(Left) The National Carbon Co. Inc. waterproof pistol covers, cataloged as part number 339C. Cover with diagonal cut is from 1944 contract. In May 1945, BuOrd established an additional requirement for 500,000 more pistol covers. The cover with straight throat cut is from 1945 contract production.

(Below) Prototype, National Carbon Co.

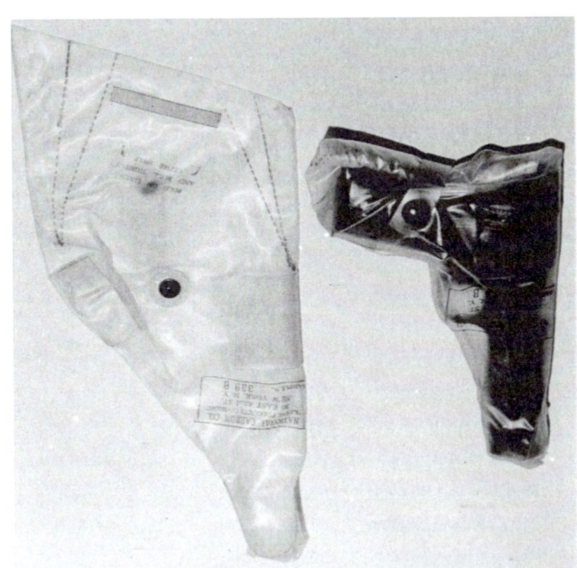

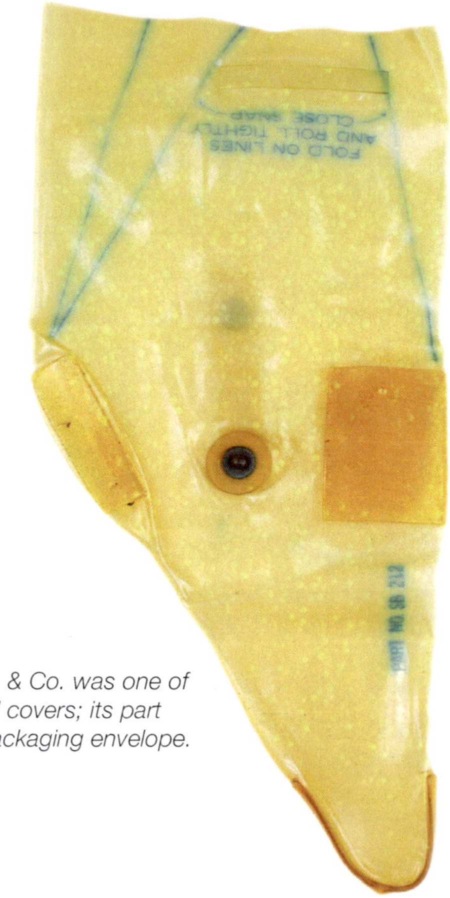

(Above and right) S. Buchsbaum & Co. was one of two contractors for Vinylite pistol covers; its part number was SB-212. Original packaging envelope. Production quantities unknown.

(Above) Experimental pistol cover with enclosed 1911. (National Archive)

(Below) Aircrew of VB-83, USS Essex, May 1945 with revolver/pistol cover (right) Pistol cover used by Lt. jg Ellis Meaker of VT-9, Okinawa, c.1945. (Inset) .38 revolver in waterproof pistol cover. (National Archives)

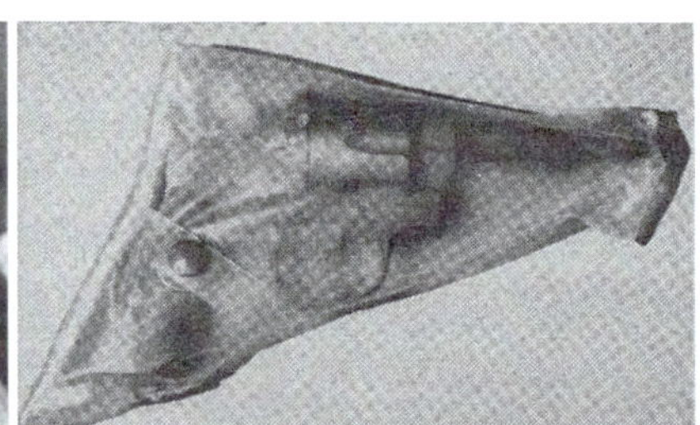

(Above) B-29 aircrew, India, c.1945. Once available, each issued pistol with holster was required one waterproof cover. Stock could be requisitioned for pistols already in service, USAAF and US Navy. (National Archives)

(Above) USAAF Brig. Gen. Thomas Power has a waterproof cover for his .45 auto. He discusses recent operations over Tokyo with Generals Curtis Le May and Lauris Norstad, Guam, March 1945. (National Archives)

Survivors; visual signals contributed greatly to the prevention of losing sight of them to finding them when they were lost. Signals used most successfully, in the order of their value, were: sea marker, smoke signals, signaling mirror and tracer ammunition, statistic based on US Navy analysis of Air Sea Rescue submitted in January 1946. The integration and adoption of tracer ammunition follows that of other personal signaling devices as detailed in Volume-2, Pyrotechnic Flares.

In 1939, the Army Ordnance Department adopted a .45 cal. tracer, designated Cal. .45, Tracer, M1. It was manufactured up to the spring of 1942. After field testing, the rounds failed to chamber and eject properly when used with the Thompson submachine gun, resulting in the cancellation of the requirement of .45 cal. tracer ammunition, May 1942. The primary manufacturer of the tracer was Frankford Arsenal, which produced their last lot in March of 1942. The Remington Arms Co. also produced the M1 tracer.

The BuOrd purchased these rejected lots at the request of the BuAer for distribution to flight personnel to be used as a distress signal since they were found functional with the .45 cal. pistol. At the time of the acquisition, January 1943, fleet activities were utilizing the .45 pistol as an unofficial standard sidearm with the M1 tracer expedited to the combat zone. Soon after, units preparing for overseas shipment were equipped with the .38 revolver.

It was not uncommon to lose most equipment during a ditching or a bailout, but an individual usually retained his sidearm. It was determined that, by having tracer ammunition, survivors retained the capability to signal. BuOrd Circular Letter No. A68-43 dated 22 July 1943 authorized the issuance of tracer rounds to all flying personnel equipped with a sidearm. The circular letter stated neither the .45 nor .38 cal. tracer ammunition were being manufactured. However, they were scheduled for production. The rejected M1 lots of tracer ammunition were to be issued to members of aircraft crews only.

Prior to issuance of circular A68-43, 100 rounds of ball cartridges were authorized allowance for each revolver. This was revised to add one tracer for two rounds, or ratio 70/30, still retaining 100 round allotment. Quantities of .38 tracer did not actually reach fleet activities until 1944. Tracers were distributed by the AESO or Ordnance Officer to individuals with sidearms and provided to units preparing for deployment. Use of tracer ammunition appears periodically in Air Sea Rescue or Emergency Rescue reports and proved highly effective. Recommendation was to load pistols with several rounds, especially during pre-dawn and night flights.

(Above) VF-4 pilots in ready room of USS Essex preparing for air strikes over Formosa. Two pilots do a final check of their revolvers before heading to flight deck, January 1945. Context, likely pre-dawn catapult launch. Standard Operating Procedure as directed by the CINCPOA for air-sea rescue, established October 1944; Carrier based operations at night, if the aircraft is in distress, the rescuers were directed to signal with green and the downed personnel with red by way of Very signal, tracer or flashlight. Probable these two pilots were preparing revolvers with tracer rounds. (National Archives)

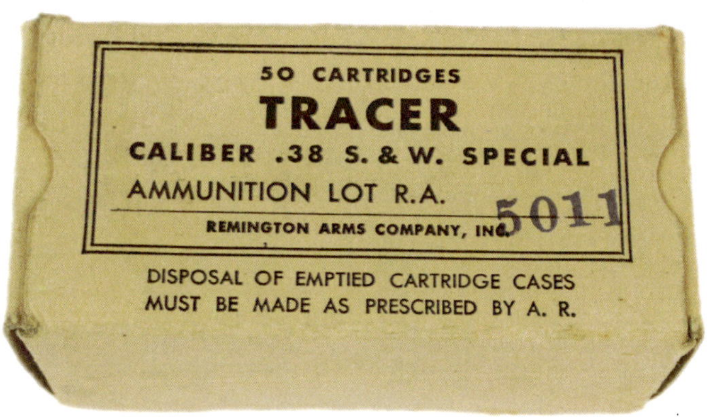

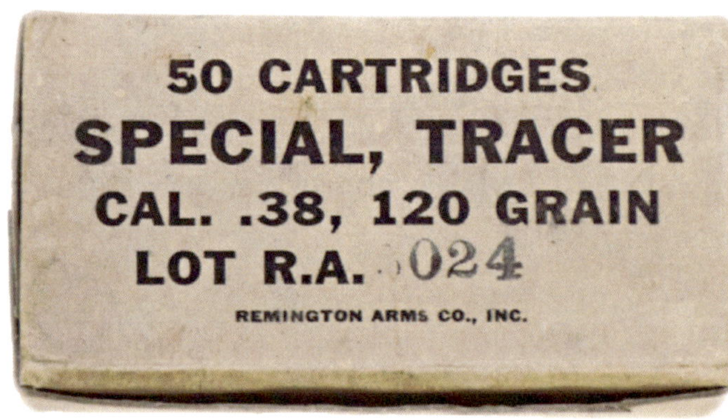

(Above) Boxed .38 cal. tracer ammunition manufactured by Remington Arms Co. (left) first production type. (Right) second production type, at some time in production the style of box was revised.

(Below) M1 .45 cal. tracer ammunition, packaged in 20-round cartons. (Left) first production carton, 1939 to 1941. In 1941, a change was made to the hardness of the brass casing. To identify this change, a star was added to head stamp of case. Lot Number 37 was accepted 1 December 1941. (Right) Last production carton, Lot Number in range for star head stamps.

The .45 cal. tracer scheduled for manufacturing was identified by BuOrd as Caliber .45 M1, but in fact that was erroneous on their behalf as it was no longer considered for or in production, they were just receiving previously rejected manufactured lots. As the .45 auto was phased from service and replaced by the .38 revolver, its use became limited in Naval aviation activities. However, it would continuously be referenced in use within USMC aviation activity. USAAF had interest in tracer ammunition, but no official allotment of tracer rounds for aircrew has been observed, leaving its distribution to local command authority. Use of the M1 tracer rounds can be found in accounts by the BuAer, USMC and USAAF during the New Guinea and Solomon campaigns through 1943. Tracer soon evolved to be recommended by intelligence sections responsible for evasion and escape and by Air Sea Rescue organizations into 1944. MIS-X, India, operating in the India-Burma Theater, recommended to the 10th Air Force that there is value to .45 tracer ammunition for ground-to-air signaling. In this letter, permission had been obtained to issue tracer ammunition to all flying personnel operating in North Burma, October 1944. They were issued a full clip (7 rounds) by the S-2 Intelligence Officer or PEO.

Since the spring of 1942, the US Army had no further interest in .45 cal. tracer. However, tracer development was revisited in the early part of 1944 by request of the Ordnance Officer of the Hawaiian Department. They requested .45 cal. tracer for use at the Jungle Training Center located at Kaneohe Bay, Hawaii. Frankford Arsenal loaded a modified M1 tracer but with a zinc-plated steal case that had been designated as Cartridge, Tracer, Cal. .45, T-30, 13 January 1944. Approximately 90,000 rounds were loaded and sent to Hawaii in cartons of 50 rounds. The T-30 was evaluated by various service testing boards concluding that there was no interest in .45 cal. tracer for combat use.

During this evaluation process, the T-30 ball was modified which significantly increased its performance. The T-30 continued to be packed in 50-round lots for training purposes from May 1944 to November 1945. In April 1945, manufacture ramped up for USAAF requirements to the amount of 8,518,000 rounds. These were packed in 20-round cartons then sealed in a foil lined tarred paper. At the time of the USAAF request, a new revised tracer was in development, designated Cartridge, Signal, Cal. .45, T73, 2 May 1945. It remained in development until the end of hostilities.

(Below) T-30 20-round carton packed for USAAF distribution. With such a late order date, it would be questionable if these 20-round cartons made it to the forward area prior to VJ-Day. Those .45 cal. tracer rounds used were of the M1 lots, and potentially requisitions of the T-30 50-round cartons assembled from 1944. Shot ammunition was integrated as a standard in parachute emergency kits and that of the Type C-1 vest. Ball ammunition was dropped in favor of it, followed by tracer for distress signaling. In design of the Type C-2 sustenance vest, one box of M15 shot and one box of T-30 tracer were to be included.

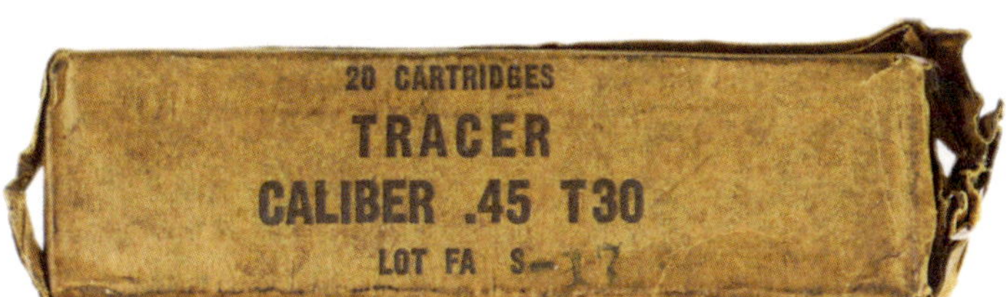

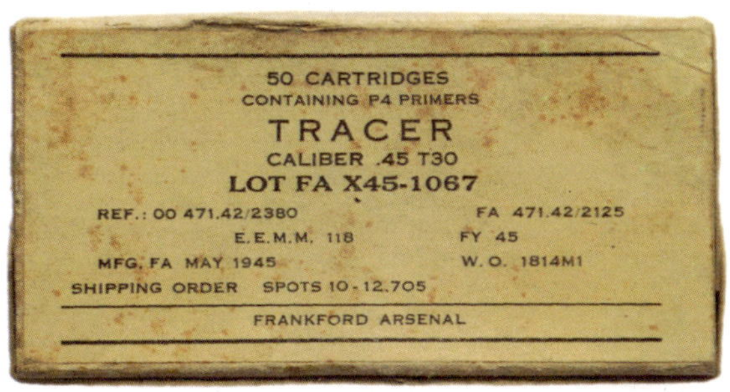

(Above) One example of the box for 50 cartridges of .45 .cal T-30 tracer ammunition.

(Right) Tracer rounds are identified by red paint on nose of round. Each had an approximate 200-300 yd. trace. .45 cal. (Left), .38 cal. (Right)

Ever since the implementation and development of cross-country emergency kits in the 1930s, the USAAC had retained an interest in methods of killing game for those stranded in isolated territory. The initial standard was to supply a .45 automatic pistol with 21 rounds of ball ammunition for emergency use. It was found that issuing the sidearm to the individual and removing it as a standard accessory from the emergency kit was optimal.

In the development of the parachute emergency kit, the addition of a .22 cal. pistol, specifically the Colt .22 cal. Woodsman model, in lieu of the .45 cal. was recommended. However, due to funding issues this was not done. It remained a topic of discussion and was highlighted again in 1940 for Alaskan sustenance kits, which required a rifle for hunting. Another attempt was made to integrate the .22 cal. pistol in the newly designed jungle parachute kit developed in 1941 but it was found redundant and was not adopted.

As the war progressed and survival and evasion reports began to accumulate, the USAAF established a requirement for a shot type cartridge based on experiences of starvation due to an inability to kill small birds and animals. In spring 1943 the Materiel Command experimented with the possible adoption of a Colt .22 cal. conversion kit for the .45 1911 but it was found unsatisfactory. A conversion kit was available on the commercial market.

In August 1943 the US Army Ordnance Department initiated development of a .45 cal. shot cartridge to meet USAAF requirements, and on 23 September approval of this was granted by the Ordnance Committee. An order was placed with Remington Arms Co. for the development and manufacturing of 10,000 cartridges titled Shot, Cal. .45, T23. These rounds were based on a commercial type patented in 1925 by Peters Cartridge Co., known as the Peters-Thompson Riot Cartridge, intended for use with Thompson submachine gun by police forces. Upon conclusion of field tests conducted at Aberdeen Proving Ground, the USAAF requested the T23 rounds be standardized for inclusion in parachute emergency kits packed 20 rounds per box. On 28 October 1943, the T23 shot cartridges were standardized as the Cartridge, Shot, Cal. .45, M12. A procurement of 10.5 million rounds packed in 20-round cartons wrapped in foil-lined heated sealed packages was made.

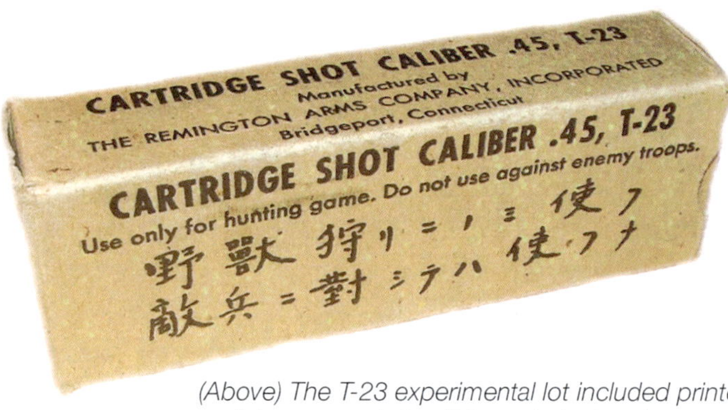

(Above) The T-23 experimental lot included printed instructions on the box in English, German and Japanese stating "Use only for hunting game/Do not use against enemy troops."

(Below) The M12 was fabricated almost identically to the Peters type; it included a waxed paper bullet with 130 pellets of No. 7½ chilled shot installed in a brass casing. Due to its length, the round had to be chambered by hand. For extraction, the magazine was removed and the casing dropped through the magazine slot in the handle. M12 shot ammunition packed 20 rounds per box. Instruction sheet was provided in lieu of printed warning. M12 was provided in compliance to T.O. 00-30-51-1 and in early procurements of the Type C-1 vest. A smooth bore barrel for the .45 was considered, but found logistically impractical.

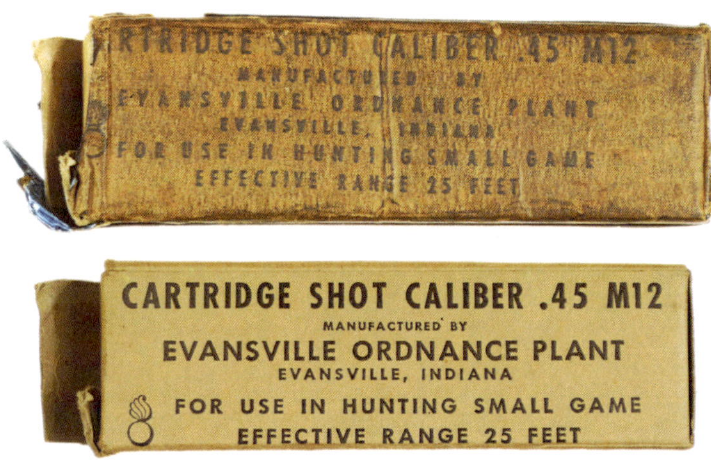

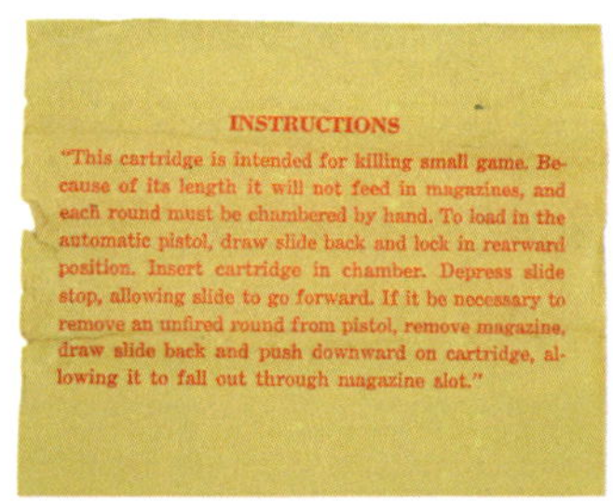
INSTRUCTIONS

"This cartridge is intended for killing small game. Because of its length it will not feed in magazines, and each round must be chambered by hand. To load in the automatic pistol, draw slide back and lock in rearward position. Insert cartridge in chamber. Depress slide stop, allowing slide to go forward. If it be necessary to remove an unfired round from pistol, remove magazine, draw slide back and push downward on cartridge, allowing it to fall out through magazine slot."

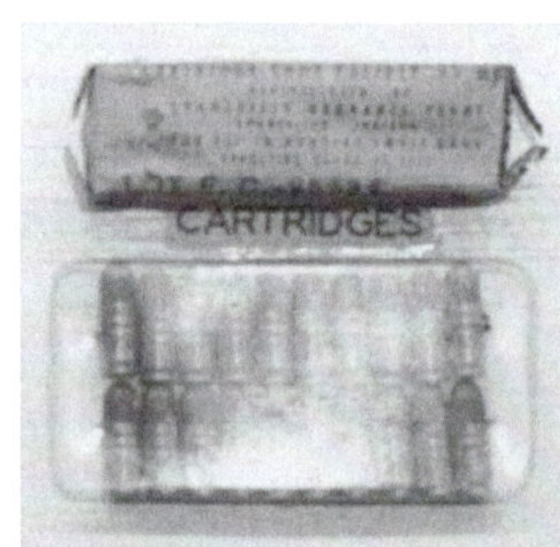

Prior to standardization of the M12, it had problems of swelling in humid conditions and failure of the paper capsule to completely open when fired. The Remington Arms Co. continued development and in November 1943 introduced an all-brass casing plugged with a wax wad and designating it Cartridge, Shot, Caliber .45, T29. This new round improved resistance to moisture and abuse and showed consistent signs of improved patterns after firing. On 30 December 1943 the T29 was adopted as standard, superseding the M12 and named Cartridge, Shot, Caliber .45, M15. Packaging material for the M15 was the same as the M12, and upon availability, two 20-round boxes were supplied in USAAF emergency parachute kits as per technical order, with one box becoming standard for the Type C-1 sustenance vest. The BuAer expressed a short interest in shot ammunition but made no further inquiries since by the time of standardization its emphasis was shifting toward rescue and away from survival proceeding into development of the hand held smoke signal and tracer ammunition.

(Left and below) Cartridge, Shot, Caliber .45, M15 packed in 20-round cartons. Shown, variants of outer wrapper provided by Remington.

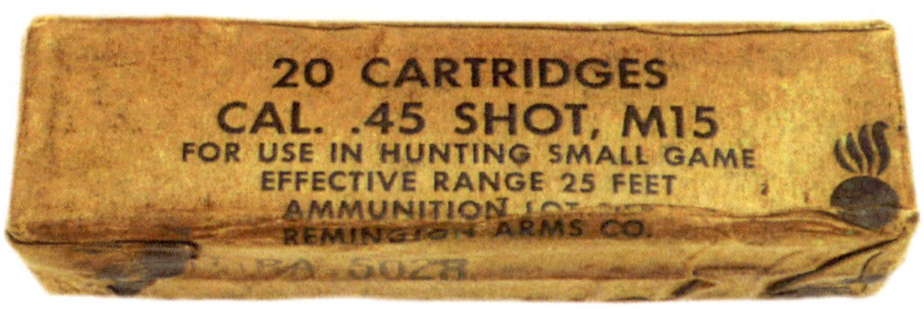

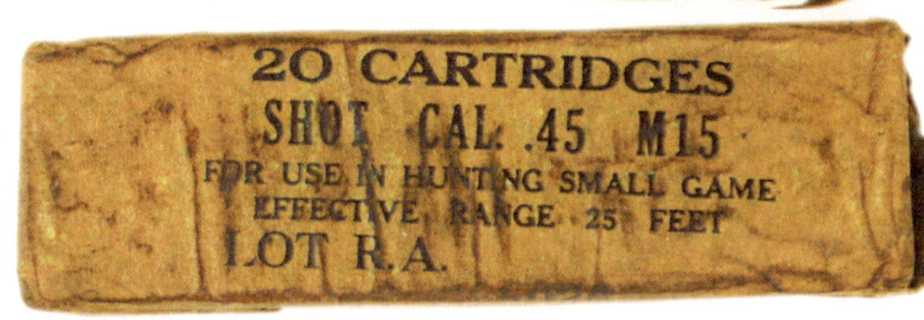

In Personal Equipment Bulletin No.2 distributed by HQ, FEAF, SWPA dated 10 November 1944 it addresses shot ammunition. It makes note that the Ordnance Officer now has quantities of the .45 buck shot, M12 for issue by requisitioning through squadron Ordnance Officers. It further made mention that the M15 had been requisitioned for that theater, but was not yet available for issue.

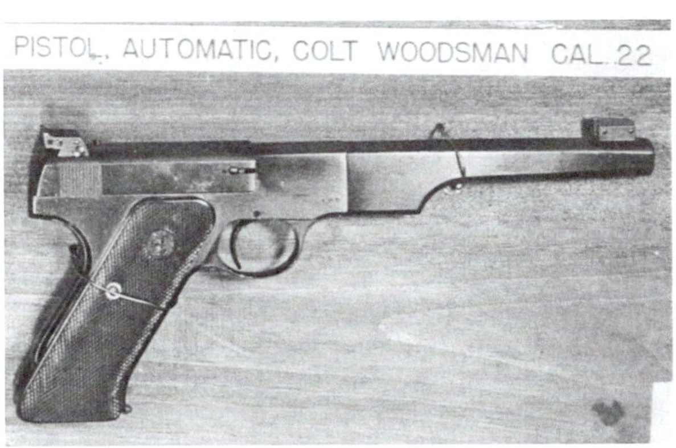

(Left and Below) As per request, a .22 cal. pistol was integrated into experimental jungle emergency kit, Type B-2. Type and model approved was Colt Woodsman. Ultimately it was found impractical financially and for accountability purposes. Small lots of this pistol were sent to Asia as requested by MIS-X and the 10th and 14th Air Forces for evaluation for use as a survival gun. Influence derived from Walk-Outs by downed airmen. Treks in the jungles of China and Burma could take up to 19 weeks, the ability to gather and harvest sustenance was of great importance. These sentiments were expressed through evasion reports stating that .45 ball ammunition was not practical for small game and a shot type would had been a welcomed addition. (National Archives)

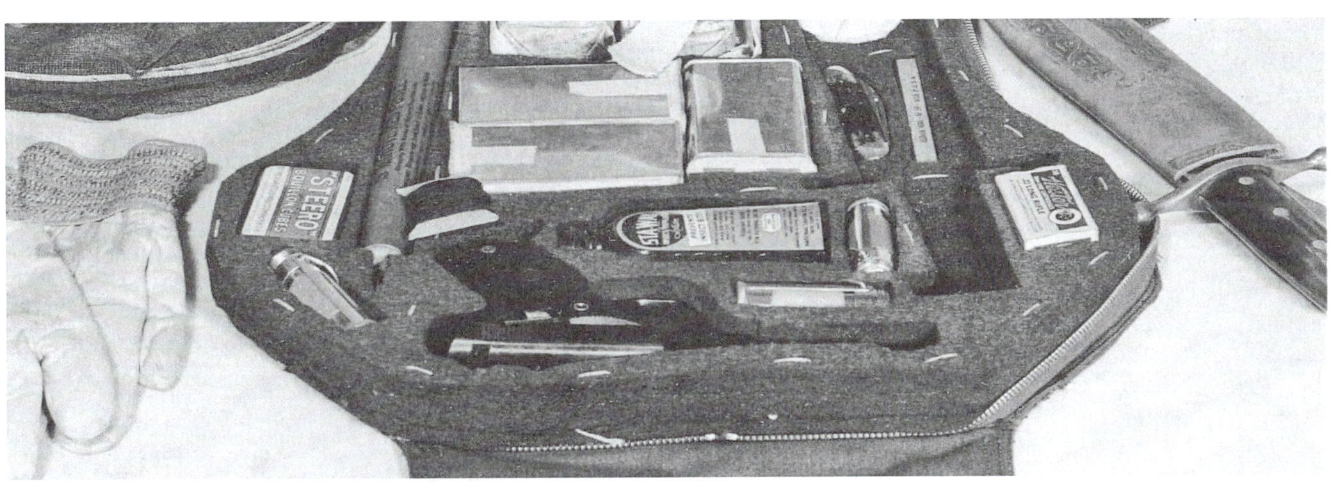

As development of a shot cartridge continued, all possible methods were investigated for killing game, including the .22 cal./.44 cal. (shot) over-under Marbles Game Getter Model 1921, and .22 cal./.410 gauge over-under model. These were compact enough for parachute emergency kits. The US Army Ordnance Department procured 100 15-inch barreled 22/410 Marbles Game Getters, with delivery expected in May 1944. This procurement's intended destination and purpose unknown but coincidently MIS-X, SWPA had requested a similar volume during this time. Model 1921 continued to be considered well into 1945 and influenced postwar development of the M4 and M6 survival rifles. 500 Game Getters were recorded as being sent to AGAS,China in the spring of 1944, procured through MIS-X as a survival rifle.

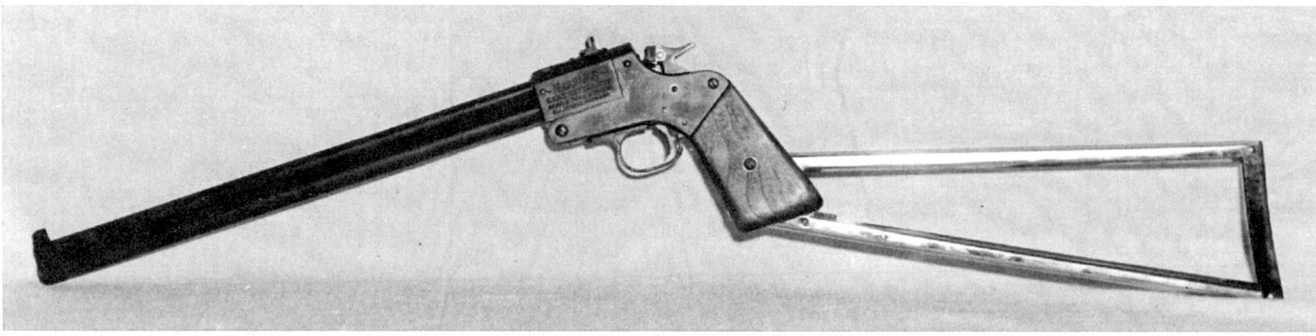

(Above) Marbles Game Getter 22/410 with folding stock. Interest in rifle was primarily expressed from arctic units but also sent to Asia for air droppable and cache rescue assemblies. (National Archives)

(Right) Game Getter was found to conveniently fit in the M3 shoulder holster. (National Archives)

(Below) Modified arctic parachute emergency kit that includes 15-inch barreled 22/410 Marbles Game Getter, May 1944. (National Archives)

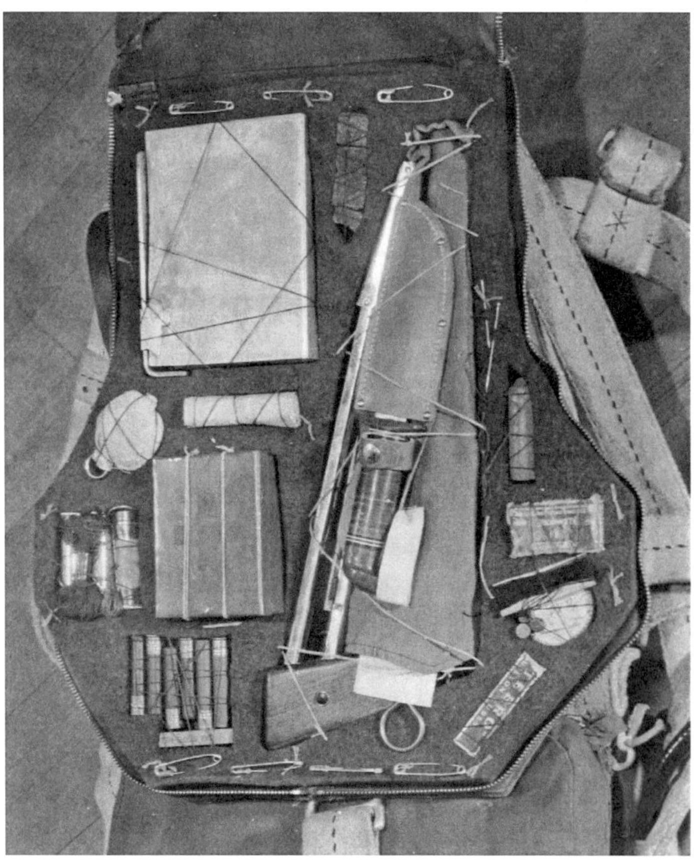

The Marbles Game Getter was always in consideration as a survival tool for hunting game. It was compact and versatile, having two calibers at the user's disposal; it was neither a pistol nor a rifle but something in the middle. However, it was slightly too large, consuming too much space in parachute kits, and not particularly suitable as a rifle for long range shooting, which were the two reasons it was never adopted.

With the development of the wilderness kits for Alaska, Ladd Field requested the inclusion of a .30-06 rifle for hunting purposes. After evaluation, the Materiel Division compromised on a rifle in between the .30-06 and the Marbles Game Getter. The rifle chosen was a six-pound over-under rifle-shotgun, Stevens Model 22-410, developed in 1939 and manufactured by J. Stevens Arms Co., a division of the Savage Arms Corporation. This rifle was capable of firing .22 cal. long rifle cartridges or high speed .22 cal. shots, and .410 gauge shot or slugs from a 24-inch over-under barrel. This rifle allowed users to hunt small game with the .22 cal., fowl with the .410 gauge shot, and large game with the .410-gauge slug. Initial procurements were executed directly by Wright Field with the earliest record in March 1941 for 50 rifles.

Upon the adoption of sustenance kits that included the Stevens 22-410, the rifle was listed as Contractor Furnished Equipment, procured directly by the contractor; the first procurement was for the Type E-2 sustenance kit. On 26 December 1942 it was ordered that the rifle and the appropriate ammunition be reclassed as Government Furnished Equipment. Army Ordnance records indicate procurements for 2,390 rifles in December 1942, 1,500 rifles in October 1943, 859 rifles in June 1944, and an additional 12,760 rifles that same month. Since the Ordnance Department had no specification at the time of the directive covering either the ammunition or the combination gun, commercial small arms were directed to be procured under US Army Specification No. 52-0-1. For the Stevens 22-410, the US Army Ordnance Department eventually assigned Specification No. AXS-1047 26 October 1943, and Dwg. No. 52-0-1B, 11 November 1943. The Stevens could be requisitioned via Ordnance Dept.. stock number M-21 under the nomenclature Gun, Combination, over and under .22 cal. and .410 gauge. The rifle was listed in Ordnance Supply Catalog SNL B-2. No printed examples of this SNL have been observed.

The Stevens 22-410 was marketed to the public beginning in March 1940. It retailed at $16.25, whereas the US Government was purchasing them at $10.64. The distinctive feature marketed at the time was the innovative application of a cellulose thermoplastic stock and fore end under the trademark Tenite, registered in 1932 by Eastman Chemical Company. It was also available utilizing a traditional wooden stock and fore end. Procurements by the US Government and by contractors of sustenance kits were of the Tenite stock option. The 22-410 remained in production, relatively unchanged, until 1950 when it was redesignated as the Savage Model 24.

The rifles were void of any serial numbers, making it nearly impossible to date their manufacture. However, one distinctive detail allows identification of a date range: incorporated into the design was a single hammer for both barrels, adjusted by a selector switch on the side. Stevens, under the assignee of the Savage Arms Corp., applied for a patent for the selector switch. This was filed in April 1940. From initial production until the award of patent on 14 October 1941, the frame was stamped PATENT PENDING on the side near the selector switch. Frames manufactured after October 1941 were stamped PATENT 2259397. Another helpful detail is that in 1946 the fore end design was changed, the late fore end can be secured to an early production rifle thus making it unreliable in establishing a manufacture date.

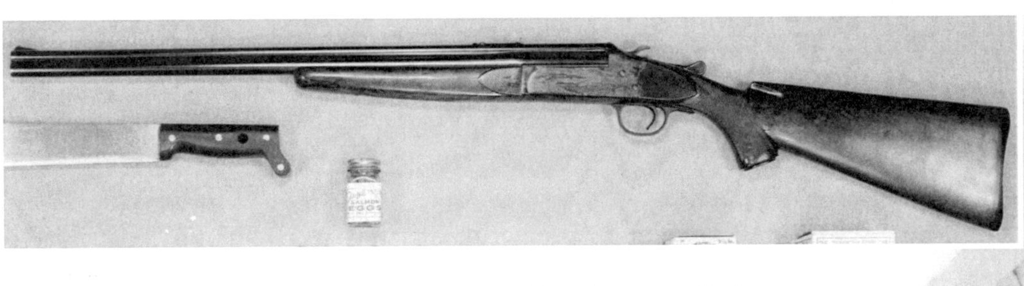

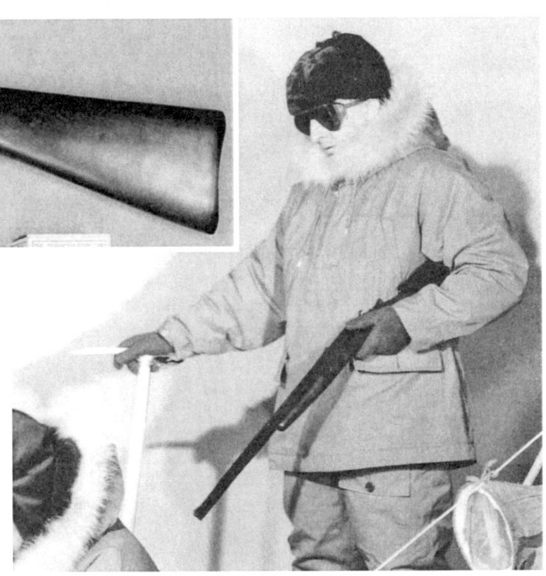

(Above and right) Two wartime images of Stevens 22-410, one on display board for the Type E-2 kit and another by a volunteer demonstrating equipment for the Type E-12 sustenance kit. Army Ordnance purchased at least 17,509 rifles, 50 in 1941 by Wright Field and 1,000 by Cappel MacDonald for E-2 assemblies in 1942. (National Archives)

The Stevens Model 22-410 over-under rifle included Tenite stocks for USAAF sustenance kits. The rifle weighed 6lbs. with a length of 40 inches. It could be broken down by unlocking the barrel and removing it. In December 1942, Army Ordnance took over procurement of the rifles, then supplied to contractors for assembly of sustenance kits before shipment to central supply points. Technical Order 13-10-1 dated 20 February 1949 details use and maintenance of the Type E-18 sustenance kit, stating that supplementary kits were no longer required and would be returned to stock and broken down, with all components returned to their stock classes or agencies. Rifles from obsolete kits were then returned to the Ordnance Department. With these rifles in inventory, an accountability system was instituted. Void of any factory serial number, serial numbers were added to rifles at least by one Ordnance Depot. These hand-stamped numbers can be found on the lower bottom of the frame. Prefix 49 (referencing 1949 date directive/fiscal year), followed by a 4-digit serial number and property mark USAF can be observed. Those existing examples have a sequential serial numbering system.

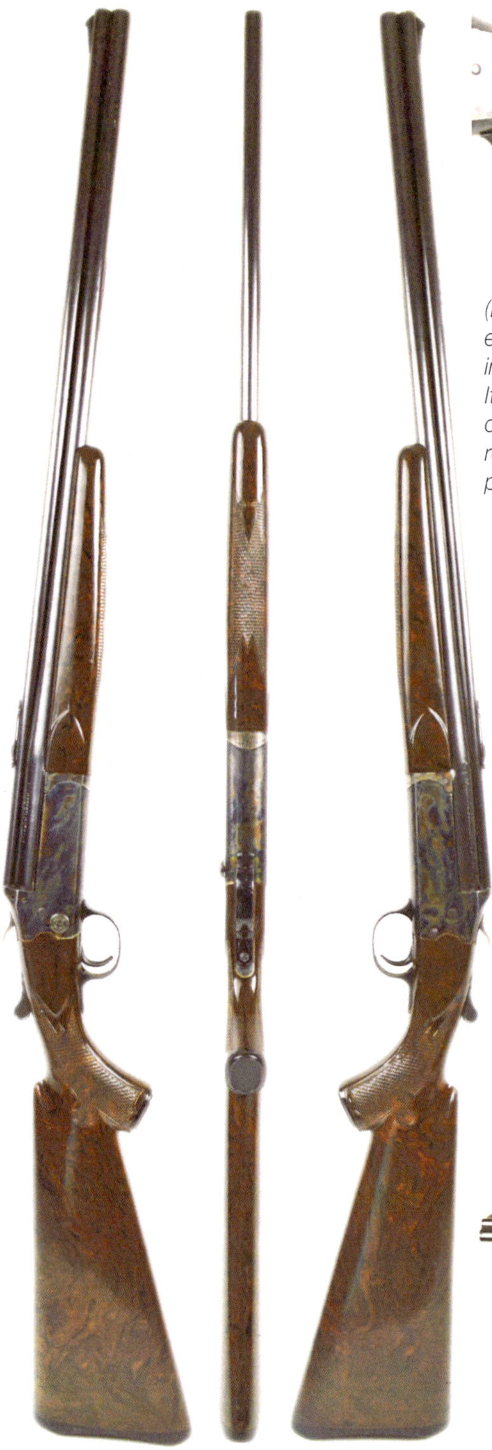

(Left and above) Stevens 22-410 as procured by USAAF. The Tenite stock and fore-end has burled finish with checkered panels. The wood version has a smooth finish instead of checkering. Tenite was chosen by the USAAF because of its durability. It was crack- and chip-proof as well as waterproof. It withstood moisture, extreme cold and heat without swelling, shrinking, or warping. This rifle would meet the rigid requirements in all climate zones without significant deterioration. Savage Arms applied for patents on the Tenite stock on 19 October 1939 and 21 November 1939.

(Above) Tenite fore-ends. (Bottom row) Original design with pot belly. Small molded PATENT PENDING on bottom near the metal ferrule. (Top row) Revised and streamlined profile fore-end introduced in 1946. Note difference in checkered panel between the two.

(Above) Case-hardened frame has marbled blued finish with selector switch located on the right side. PATENT PENDING or PATENT 2259397 was stamped near the switch.

Standard ammunition allotment for the Stevens rifle included in sustenance kits was four 50-round boxes of .22 cal. long rifle, three 25-round boxes of .410 gauge No. 7½ chilled shot, and five boxes of .410 gauge slug (5 rounds per box) with master carton. Ammunition was provided in commercial packaging. Three companies provided shot ammunition: Winchester Repeating Arms Co., Western Cartridge Co., and Remington Arms Co. Each box containing the shot and slugs consisted of 25 2½-inch cartridges.

(Right) A cleaning kit for the Stevens rifle was first applied to the revised Type E-8 sustenance kit, following kits include the E-10, E-14 and E-18. Cleaning assembly was to include a 3oz. can of oil, 50 cleaning patches, thong and .45 cal. bore brush. Cans of oil observed were either the early tin plated or OD standard ordnance types as with the 30-06 thong and .45 cal. bore brush.

(Left) Winchester supplied its Super Speed brand with lot number R10771/2C. Stamped US PROPERTY. Super-X chilled shot provided by Western was in yellow box with blue band at the bottom. Remington chilled shot ammunition box with US PROPERTY stamp.

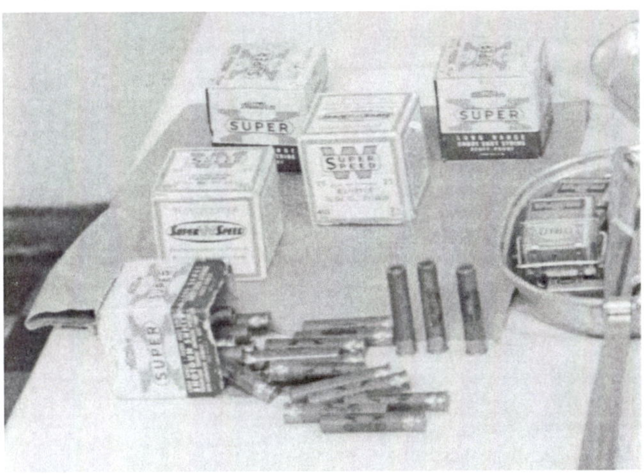

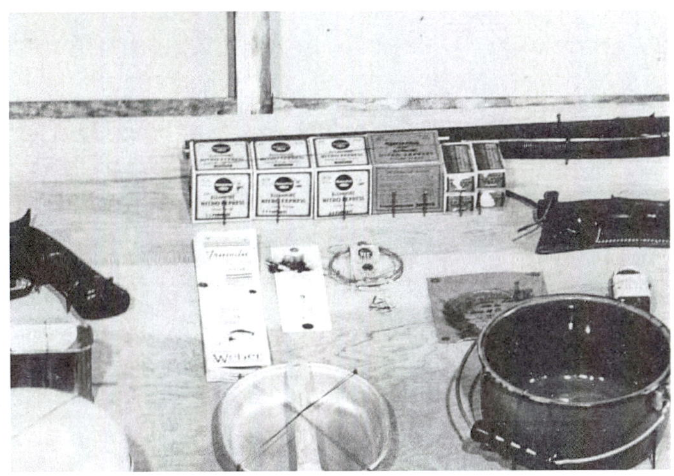

(Below) five round boxes of .410 slugs were contained in a master carton of 25 shells. Winchester and Remington are only two known contractors to provide slug ammunition. Remington is only company to stamp the cartons with US PROPERTY. Others retained their standard commercial packaging. Each of the five boxes were provided in a master carton.

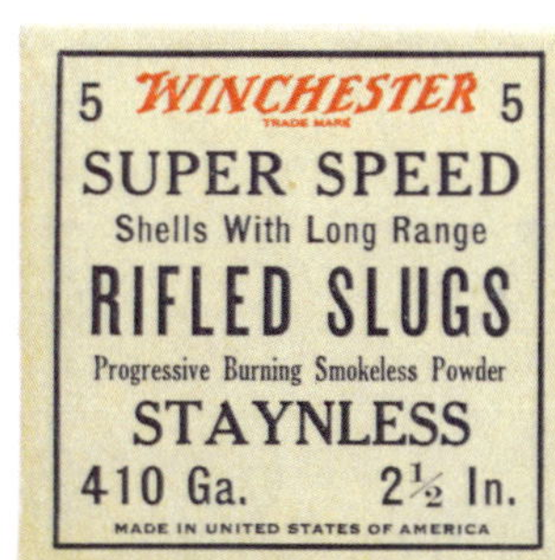

Twenty-two cal. long rifle ammunition was abundantly available in Army and Navy stocks. Both services and the USMC used .22 cal. rifles and pistols for training. All the ammunition was provided in companies' commercial packaging. Western and Remington supplied .22 ammunition for sustenance kits. The Remington box is referred to as the "dog-bone" pattern. Those supplied via Army Ordnance procurements were stamped with an Army Lot Number, often abbreviated ALN or AL No. Western supplied its Super-X brand, which came both with and without the ALN.

(Left) Western and Winchester used a lot number coding system that recycled every 20 years. The Western box presented here is from the WWII era; the 67 is the production line with the "VH8" the significant dating feature: V=Year, H=Month, 8=Day of the month. A double digit was also used, with numbers reversed to indicate the day of the month. This lot indicates the 8th day of August 1944. This coding system applied to all calibers of ammunition. Same method is applied to the .38 caliber ammunition.

(Below) .22 cal. ammunition procured by Army Ord. with Army Lot Numbers. Peters Cartridge Division, Remington Arms Co., and prewar pattern of the Western .22 long rifle cartridges. Peters has the ALN stamped on the back, Western has the ALN on inside of end flap. Ammunition requisitioned from standard stock for replenishment and new assemblies of emergency kits had the ALN.

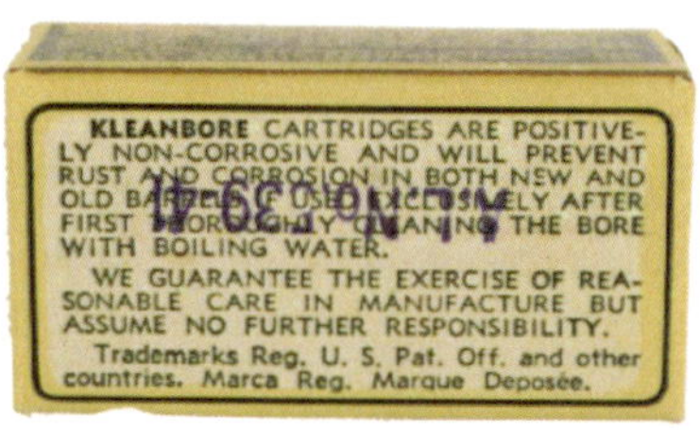

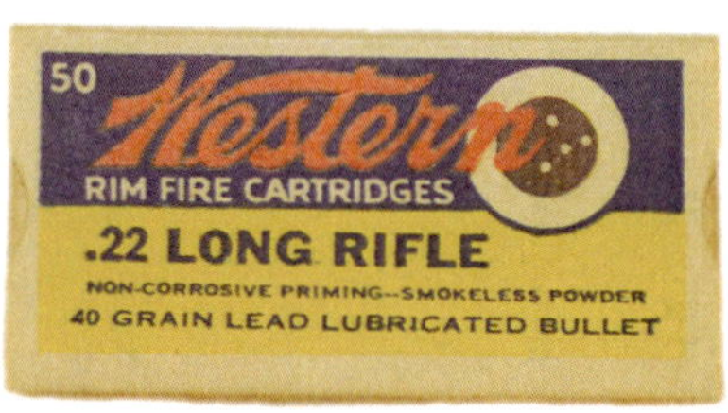

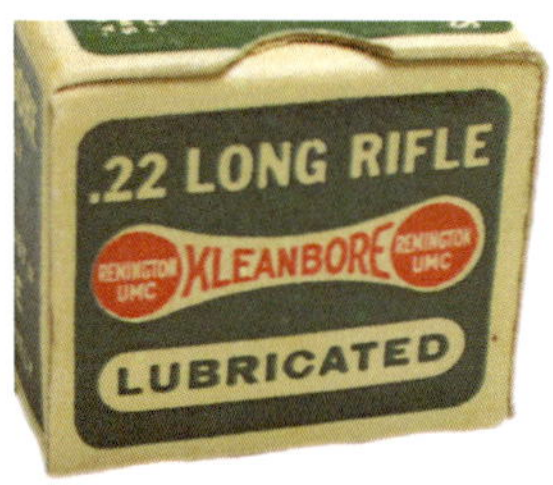

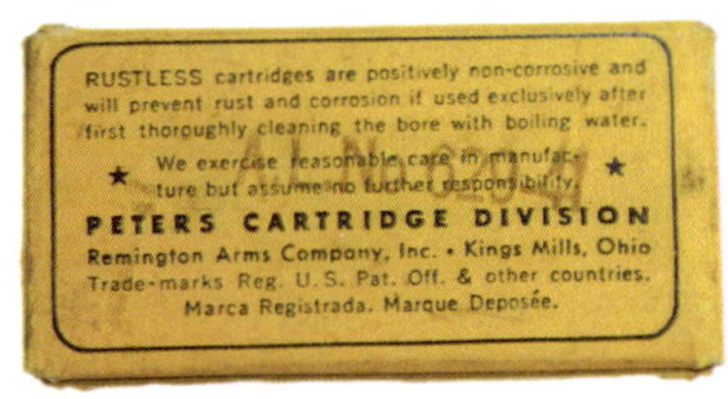

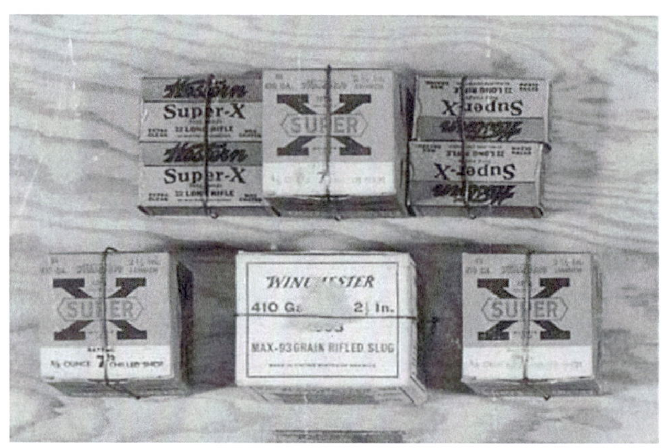

(Left) In March 1945, .410 chilled shot ammunition was directed to be packed in sealed metal tins, 30 to a can. Here, examples from Remington and Western. By this date slug ammunition was no longer provided as they violated the Hague Convention. Sustenance kits were then revised to include five boxes of 50 rounds of .22 cal. and seven boxes (25 per box) of .410 gauge chilled shot; later revised to include three 30 round tins once they became available. (Charlie Flick Collection)

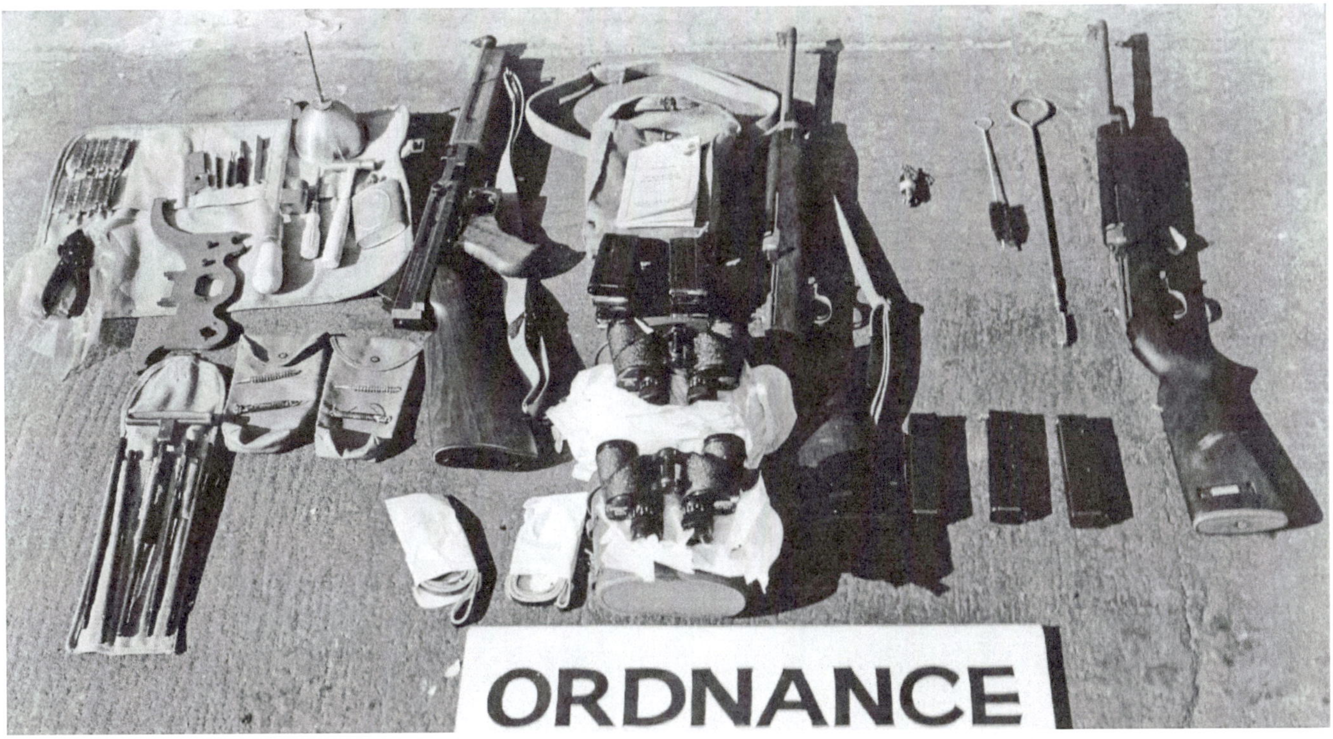

(Above) Assortment of ordnance provided in USAAF B-25 medium bomber for cross-country flights. Per T/O&E for bombardment squadrons, M1 carbines and .45 cal. Thompson submachine guns (later revised to the M3 submachine gun) were allotted for operations outside the continental US. Typical load for missions for Asiatic-Pacific. (National Archives)

(Right) By late 1944, cold weather units indicated a preference for the M1 carbine over the Stevens rifle. They thought the carbine was both better suited for protection against the enemy and provided better firepower for large animals. Cold weather sustenance kits were updated by local commands with either the M1 or M1A1 carbine and supplemented with M15 .45 cal. shot ammunition to replace the Stevens. In post-war development, the M1A1 carbine became the new standard, replacing the Stevens rifle in sustenance kits. E-12 sustenance kit updated with an M1 carbine. (National Archives)

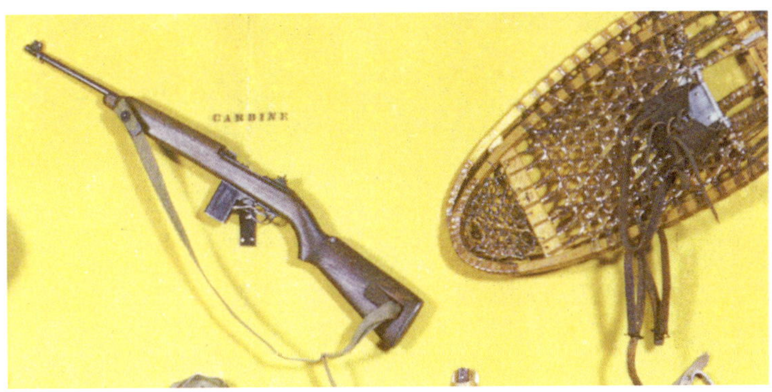

(Left) US Navy PV-1 Ventura copilot Lt. jg Phillip Brady of VB-139 cradles a Thompson submachine gun. On 9 July 1943, the BuOrd released Circular Letter GV19-43 detailing that one Thompson submachine gun was authorized for issuance to each airplane of classes VB, VT, VP, and VR operating outside continental United States. 1,200 rounds were allotted for each gun. It was deemed necessary to carry secondary sidearms in the combat zone. In a report from VP-91 dated April 1943 it states "There should be a Tommy gun in every plane. When a gun jams the Japanese notice it right away, and come right in. We tried to carry a Tommy gun with tracer ammunition in it. One of the men, while they were repairing a gun, would stand in the position of the .50 caliber gun and fire with this Tommy gun. The Japanese wouldn't know the difference. They'd keep clear as they have great respect for that .50 caliber. The kind of planes we ran into would keep clear of a Tommy gun just as if they would of .50 caliber." Evidence lends this courtesy to other patrol and carrier based aircraft as well. (National Archives)

(Above) Thompson submachine, compact enough for storage in all aircraft, provided effective firepower in the event of forced landing and enemy encounters. Many operational squadrons in and around island chains carried these weapons for self-defense on both offensive and patrol aircraft. (Left) Thompson being handed to turret gunner Cpl. Raymond Brown of unknown USMC VMTB squadron in the Pacific. (Right) Crew member of VD-5, January 1945. (National Archives)

(Below) USMC grasshopper pilot Capt. H. Grady poses with his L-1 Piper aircraft after returning from an observation mission during which his plane was damage by anti-aircraft fire, Saipan, June 1944. Accompanying him on his flights was a Thompson. (Left) TBM aircrew member of VC-98 poses with aircraft complement of Thompson, c.1944. (National Archives)

KNIVES & MACHETES

Historically, edged implements were integrated in the US military for a multitude of purposes, the styles of which varied according to the branches of service and their departments' requirements at the time. The topic is a much larger conversation than what will be highlighted here. For purposes of this study, hunting knives, pocket knives, and machetes will be presented as emergency and survival aeronautical tools.

Most often, knives in military use are considered fighting or combat knives, implying they are for front line troops which gives, in part, a false impression. Leading up to WWII, US aeronautical services had a different point of view on the role of a knife: there was no real consideration of their use for fliers until the development of emergency kits. The USAAC first considered the integration of a sheath or hunting knife as a standard complement in its cross-country emergency kit. This was revised to include a machete to address the needs for units downed over jungle terrain.

The Navy had a different approach. Knives of all types had been an integral part of nautical history, especially in the days of wooden sailing vessels, but their relevance in the modern era declined, although they remained in the inventory. Jackknives were the principal knife available to sailors but were not mandatory. Eventual wartime conditions demanded that all shipboard personnel be supplied with a life preserver, life preserver light, whistle and a jackknife.

US Navy units operating in tropical territories found it desirable to provide a hunting knife and a machete, but the decision to do so was a local one. Pocketknives or jackknives were eventually included as a standard complement for both services. All three types of edged implements quickly evolved to become a vital lifesaving tool during WWII and were utilized in all regions of the world.

Prior to WWII, a knife or machete was provided as a useful tool; in wartime such a tool became a survival requirement. In an emergency or survival situation, a knife was a morale booster as well as a means to sustain life until rescue. In warm climate regions, a knife or machete proved useful as a housekeeping tool, whether aboard a life raft or on land. Life in the tropics could either kill quickly or provide everything an individual needed to live indefinitely. Water could be harvested by vines, vegetation could provide shelter, cooking utensils and protective clothing, all accomplished with the assistance of a simple tool.

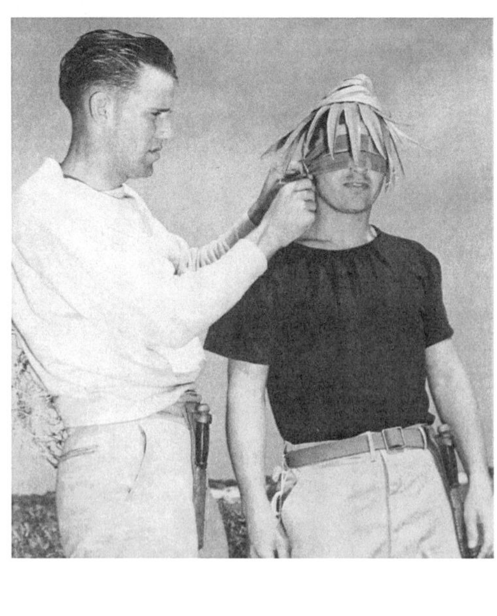

Through 1942, US Navy reports of accidents and after-action engagements indicated that many bodies were found trapped by debris. In some instances, victims were found with their feet entangled, which led to drowning. It became obvious that a knife of any type would have allowed service personnel to cut themselves free, saving lives. The Navy Department directed that a knife be carried on the person, ready for immediate use to clear away lines, nets, tarpaulins, remove clothes, and to cut away life floats in an emergency. It became the policy that a sheath or jackknife be carried by all personnel in the forward area.

A sheath knife was viewed to be more useful than a jackknife as a utility tool. Toward the end of 1942, the BuOrd began the procurement and negotiation of contracts for hunting knives with five-inch blades. Initial procurements were of readily available patterns on the commercial market pressed into service to meet wartime demand. The BuOrd made direct purchases and had the knives available in stock; additional purchases were executed for supply of advanced base use and turned over to ships' stores for resale or issued by the ship. The rapid procurement program introduced a wide variety of types of hunting knives into service. Eventually, specifications would be drafted formalizing patterns. The integration of edged implements coincided fairly equally amongst all the US military services.

In 1939, the BuAer adopted the jackknife as part of the standard complement for life rafts; it was the only knife in inventory entering WWII oriented to emergencies. One of the earliest accounts of the use of a sheath knife by aviators is an account that the skipper of VT-8, Lt. Cmdr. John Charles Waldron, made it mandatory to carry pistols and hunting knives by all squadron members. He had the foresight that both would be useful tools in emergency jungle landings.

Ens. George Gay, sole survivor of VT-8, recounted in a memoir about his experience in a life raft. Once he felt it safe to board the raft that day, he took inventory of the equipment. As he was processing his supplies Waldron's lecturing and planning crossed his mind, but of all that planning never once did they unpack and familiarize themselves with the rafts emergencies supplies. Those items being somewhat foreign without some measure of orientation. In part, Gay realized how valuable Waldron's foresight was with himself in that very predicament. 30 hours later, Gay was rescued by a PBY.

Survival booklets encouraged the use of knives and machetes among all service branches. In Castaway's Baedecker, the first survival booklet for servicewide distribution in the Pacific, printed in December 1942, the author, Kenneth Emory, stated that a machete was the most useful item of equipment, and that a hunting knife was a second necessity. The two articles were at the forefront of personal emergency equipment, aside from the life preserver and parachute. This complement of gear would assure that a pilot would have the best chance of survival the moment he hit water or ground.

In response to requests from the Pacific Fleet, the BuAer officially adopted a requirement that a small arm and sheath knife be issued on 21 January 1943, detailed by BuOrd Circular Letter G1-43. The directive was issued through BuOrd since both items were under their authority. In G1-43, it details the knife as Sheath Knife – Five (5) inch blade and sheath. Procurement of hunting knives was principally executed by the BuOrd; they could be requisitioned from Naval Supply Depots under stock number 74-K-58 and from ASO supply under stock number R41-K-640.

In the ASO Class-41 supply catalog the nomenclature used was Sheath Hunting Knife. The terms hunting knife, sheath knife, and utility knife were interchangeably used for five- and six-inch varieties. Seven-inch knives were typically referenced as fighting knives.

Prefixes to stock numbers represent the Federal Standard Stock Class and can be somewhat revealing. Class-41 was for Tools, Hand and Class-74 was listed as Individual-Equipment (field and landing forces). A hunting knife was both a tool and personal field equipment.

G1-43 established that a .38 revolver and a sheath knife would be issued 30 days prior to deployment. Following this circular, flying personnel were equipped with these two items prior to leaving the continental United States alleviating the burden on localized supply depots, particularly Hawaii. Knives were obtained in different ways: requisitioned formally through standard supply channels or obtained by local purchase, which many aviation units preferred.

Prior to deployment, it was the squadron and/or group commander, working in coordination with the squadron parachute rigger, later the equipment officer, who was responsible for securing all materials necessary to meet the needs of pilots and aircraft for the duration of duty; funds were allocated for such purchases.

Local purchase, with fewer specifications to meet, generated a variety of knives in service. Other methods of acquisition included private purchases by the individual either commercially or from the PX/ships' store, a collective pool of monies by the squadron members, or gifts from family and friends. Even upon requisitioning from standard supply channels, there was no guarantee that the lot would be the same type and pattern. There is therefore almost an endless variation of knives when viewing wartime images.

(Above) A hunting or sheath knife became a universal unofficial standard for early aviation activities. Members of VF-6 Willis, Cline, and Born clown around with knives aboard the USS Enterprise, June 1942. (National Archives)

(Left) Lt. Cmdr. John Charles Waldron, C.O. of VT-8, c.1942. Unusual combination of shoulder holster and M-1916. Hunting knife is a 6" Marbles Ideal.

(Below) Hunting knives being distributed to US Navy aircrew personnel prior to conducting survival training. After hunting knife became standard, aviation personnel would be trained in its proper use and care. (National Archives)

(Below) USS Wasp SBD pilot Lt. Robert Ware of VS-72 with hunting knife secured to his right lower leg. Ware had recently landed after shooting down a Japanese bomber, August 1942. (National Archives)

(Above and left) three torpedo-bomber pilots aboard USS Enterprise, February 1943, show variation in knife handle washer patterns. (Above left) Ens. George Wilhock, (above) Lt. jg Robert Nelson (left) Lt. James McConnaughay. Prior to distribution of G1-43, stateside Naval Air Bases and Air Stations purchased knives for expedited shipment to Hawaii and air groups preparing for embarkation, thence distributed with priority for units departing to combat zones. Knives from many commercial sources were intermixed. Case, Western, Remington, and Marbles knives were commonly purchased. (National Archives)

(Right and below) VF-16 pilot Lt. jg Youngberg wears a prewar commercially manufactured sheath knife, USS Lexington, November 1943. Youngberg dons an Original back pad kit. Colorful or decorative handles were made by many companies in the prewar era such as Case, Cattaraugus, and Remington. (Below) Pre-war type sheath knife by Western States Cutlery Co.

(Below) Many knives commercially manufactured prewar were pressed into service, as they were what was available in early procurements in 1942 and 1943 prior to the institution of standardized types. It is difficult to discern the make since manufacturers shared many handle and pommel designs. The hunting knife worn by VT-47 aircrew member Allen Maxwell ARM3/c is an example of a prewar commercial type. The handle here had a variety of colored washers. A second knife can be observed behind the shoulder holster; it appears to have a handle similar to the Western G46-8 or 6" blade. It was not uncommon to carry two knives in the event one should be lost. USS Bataan, May 1945. (National Archives)

(Below) Unique knife worn by Lt. Cmdr. Rodger Mehle of VF-28. Knives used in service came from a variety of sources beyond commercial manufacturers, including fabricated in facilities available to US Navy personnel, fabricated by hand through private parties in the US, and manufactured by foreign entities and sold commercially before the war. USS Monterey, June 1944. (National Archives)

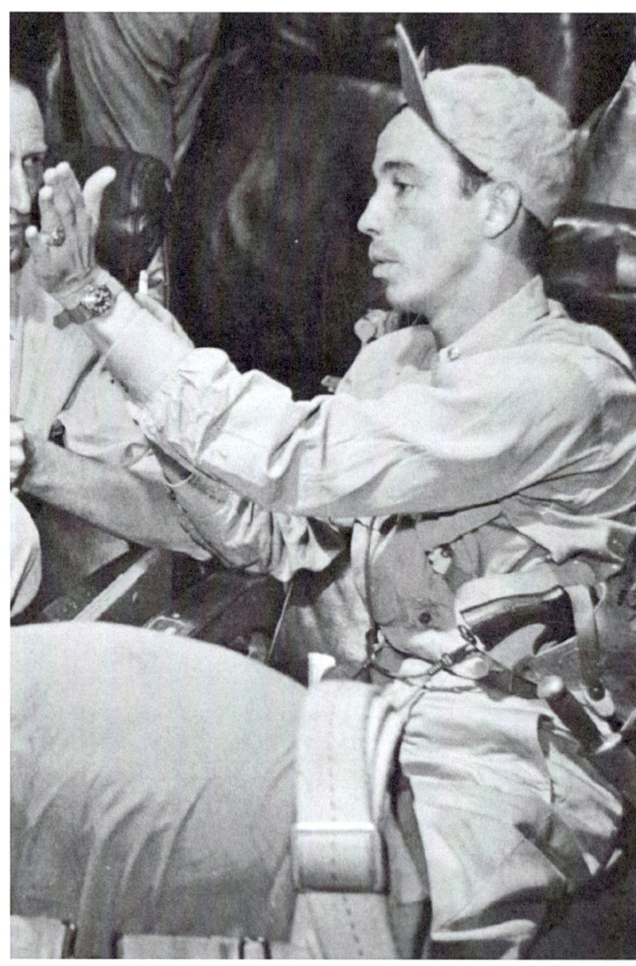

(Left and below) Initial procurements of sheath knives were of off-the-shelf models in many blade lengths, which still met the Navy Department's criteria for a utility knife. What was available to ships was also available for aviation activities. (Left) The knife worn by the sailor appears to have a 4" blade, USS New Jersey, December 1944. (Below) 4" blade sheath knife manufactured by Case belonging to Ens. Robert G. Evans. (Below left) Commercial 4-½" blade hunting knife manufactured by Kinfolks carried by Lt. jg Joe F. Bailey, who served aboard the USS Adhara. (National archives)

(Below) Sailor eating chow aboard LST in December 1943 wears 5" utility knife of a pattern manufactured by Union Cutlery Co. Union Cutlery was one of the companies in early negotiations with the BuOrd to supply hunting knives. KA-BAR is a Union Cutlery trademark but was also listed as a knife manufacturer for advance bases and ships stores for resale. (Below right) Pre- and interwar commercial KA-BAR hunting knives. Typical characteristics include a 5" blade with fuller, either a blued or bright finish, and a flat-ended wood pommel. (National Archives)

Marine pilots demonstrate more varieties of commercial pattern knives.

(Left) VMF-211 pilot Maj. Robert Harvey, Solomon Islands, November 1943 with unidentified make of knife.

(Below left) VMF-212 pilot J.G. Prestridge, Green Island, Solomon Islands c.1944.

(Below right) VMF-222 pilot 1st Lt. Ennis Walden, c.1944. Attaching knives to life preservers was typical, assuring quick access to it in an emergency as well as increasing comfort and convenience given cramped cockpits. The life preserver was a convenient location for attaching an array other emergency type equipment. (Gregory Pons collection) (National Archives)

(Right) Commercially acquired sheath knives came in many patterns. Sailor on the left has a distinct white-handled knife. Sailor, center, has Mark 1 utility knife and whistle at the ready. Sailor on the right has a jackknife 41-J-4, better known as E-Z opener, USS Kasaan Bay, August 1944. (National Archives)

(Below) Lt. R.B. McNees of VF-46 carries a short-bladed commercial sheath knife. (National Archives)

(Below right) Ens. Charles Hoffman of VT-14 has white-handled patterned sheath knife aboard the USS Wasp, August 1944. He also utilizes hip holster for .38. (National Archives)

(Above left) Unidentified Marine squadron member wearing knife with characteristics of being locally fabricated, often referred to as a "theater-made" knife. Another possibility is a modification of the standard 1219c2. (National Archives)

(Above) Pilot Capt. Harry Winfree Jr. of VMF-211 has one knife strapped to his lower right leg with another behind his M-1916 hip holster, Bougainville, February 1944. Both knives are a commercial type pattern. (National Archives)

(Left) Lt. jg John Hoeynuk of Air Group 20 has a US Army Case V-42 fighting knife attached to life preserver, March 1945. (National Archives)

(Right) The more unusual the knife, the more likely it was privately purchased. VMF-312 squadron member has knife with turned wooden handle and blade made from a Civil War era sword. (Below) Knife of the type in the image. (Robin Bartel Collection)

(Far left) The knuckle knife worn by a Marine squadron member of MAG-61 was likely theater-made. (USMC)

(Left) Hunting knife worn by unidentified pilot of VF-80 has ornate custom handle, c.1944. (National Archives)

(Above) Two unidentified pilots of VF-23 have M3 fighting knives with M6 pattern scabbards, c.1945.

(Right) This example belonged to Ensign Harold Orville Bosacki of VBF-1 who served aboard USS Bennington from July-October 1945 and flew Corsairs. This M6 pattern scabbard lacks M-1910 belt hook.

(Below) A Ship's Store located in the Forward Area where basic personal items could be purchased, and if stocked, hunting and jackknives, to also include sale of locally fabricated types.

(Above) Unidentified pilots of VF-34, USS Monterey, c.1945. Pilot on left wears a Knife Crafters.

(Above right) VMF-312 pilot with a Knife Crafters, c.1945. (National Archives)

(Left) Knife Crafters of Chestnut Hill, Philadelphia, PA was another manufacturer of knives converted from Civil War swords. Existing example a converted Civil War sword by Knife Crafters. The previous example is from unidentified manufacturer.

(Right) C.O. of VT-47, Lt. Cmdr. Ralph Mazza has a Knife Crafters as his personal knife, USS Bataan, May 1945. He additionally has waterproof pistol cover for .45 auto. (National Archives)

In 1942, survival equipment was in its infancy and applied by individual initiative, as illustrated by Commander Waldron. A hunting or sheath knife became a universal unofficial standard for early aviation activities. By the close of 1942, survival was being stressed to all aviation personnel entering the Asiatic-Pacific Theater of Operations. Hard lessons were learned but used to further educate that a quality knife will save a life in ditchings and bailouts. The most reported use of a knife, extracted from Evasion and Emergency Rescue Reports, was often used for cutting away parachute harnesses and chute lines.

By the summer of 1943, ComAirPac made it clear that all personnel going to bases in the tropics should realize that facilities and supplies were limited and that living is limited. Every effort should be made to procure objects as soon as possible after learning that one's squadron or unit was being sent to a combat area. Both officers and enlisted men should equip themselves either a five or six-inch hunting knife. Boy Scout knives and small Boy Scout axes would prove useful and squadrons should be sure they have adequate supplies. ComAirPac also makes note that hunting knives were part of squadron allowance for pilots and aircrews; not issued to others who must obtain them from commercial sources. Machete type knives must also be obtained from commercial sources.

In February 1944, a Naval liaison Officer assigned to MIS-X,SWPA flew from Brisbane to Port Moresby for intelligence briefings to VP-34. While there, he went on a bush trip with Australian intelligence officers and Papuan natives receiving an education on how to live off the land. From his experiences during the field trip, the Navy issue sheath knife (five inch) was inadequate either for cutting thru heavy undergrowth or hacking thru to the heart of a palm, one of the most accessible, plentiful and nutritious of jungle foods in the SWPA. A larger knife, with a much stouter blade and handle was recommended.

Following the occupation of Guadalcanal and continued expansion of air operations in the SWPA, the machete or similar tool was promoted to work jungle foliage. Large hunting knives could fill the role as they were suited to chopping. The large-bladed knife was more common, but also a personal choice based on looks and the notion that bigger is better.

(Below) Knives were sometimes purchased on a squadron level either by funds allocated for preparation for overseas movement or kitties in which members participated. Pilots of VGF-29 wear large hunting knives aboard the USS Suwanee, January 1943, en route to the South Pacific. Individual images reveal the knife to be a KA-BAR Trailblazer model, 13-¼" long with 8" blade. (National Archives)

(Above) KA-BAR Trailblazer used during field training survival exercises at NAS Pensacola, c.1944.

(Left) Ens. Robert Stonum of VGF-29. (National Archives)

(Left and below) Throat of sheath is embossed with KA-BAR as worn by Ens. Robert Stonum. This example was personalized with E. Petersen, a stamp similar to personal property markings found on US Navy clothing and other articles. Carved on sheath face is AFRICA 1942. USS Suwannee, with VGF-29, supported North Africa landings in November 1942. (National Archives)

(Left) Lt. John Pfeiler of VF-28 wears a harness with emergency equipment, including two sheath knives, August 1943. In front pocket is small knife; attached and worn on back is a large hunting knife. Knife at front is for immediate emergency use and small tasks. Large knife was used like a machete for chopping brush and branches. At this time, access to small machetes was limited but some were obtained by select Naval and Marine aviation personnel. The M-592 back-pad kit with the V-44 machete was not available until the close of 1943, leaving units to improvise. (National Archives)

(Below) Knife worn by Pfeiler and in the two images below is a Marbles Ideal Hunting Knife No. 45, introduced in 1899 by Marbles Arms & Manufacturing Co. and available in a variety of styles. It was produced with blade lengths of 4-1/4, 5, 6, 7, and 8". It had stag or leather handles, stag or zinc pommels, and full or half guards. Knives here are 8" with leather handles and stag pommel. (Below left) Jim Swope of VF-11 wears a No. 45 attached to his life preserver waist strap, USS Hornet, January 1945. (Below) Member of VGF-29 Ens. William Ambrosio wears a No. 45 in lieu of the KA-BAR, USS Suwannee, January 1943. (National Archives)

(Above) Large-bladed hunting knife worn by Ens. Greene with characteristics of an E.G. Waterman, NAS Melbourne, October 1944. (National Archives)

(Above) TBF pilots of VC-26, Lt. Rhodam McElroy and Lt. Robert Taylor, report after an attack on Munda Point, New Georgia Is. from their base of operations on Guadalcanal, April 1943. They both wear large hunting knives with zinc cast pommels, make unknown. (National Archives)

(Below) Western G46-8 hunting knife with 8" blade ID'd to Navy or Marine person M.J. Marr.

(Below) Pilots of VS-10, Lt. jg Thomas Ramsay and Lt. jg John Richey, each have a Marbles hunting knife as their personal utility tool, USS Enterprise, February 1943. The No. 45 worn by Ramsay on his shoulder holster is either 6" or 7". Rickey is wearing an 8" Ideal. Both have zinc cast pommel. (National Archives)

With the dramatic increase in demand for knives to equip ships and aviation personnel, procurement was rampant, unorganized, and unable to meet established quality criteria. As a result, the wide variety of knives included inferior types that could not sustain rigorous use and the climate conditions. To formalize and consolidate procurement, in 1943 the BuOrd drafted two drawings and procurement specifications detailing minimum requirements for a five-inch Utility Knife and a seven-inch Fighting Knife. This consolidated issuance of knives to Naval personnel to two patterns and resulted in competitive bidding, establishing a predictable fixed price. This also increased production efficiency, since, in concept, all successful bidders could use multiple suppliers of parts, and with multiple contractors making the same item, deliveries to supply points wouldn't be affected by individual plant delays.

The drawings for the two patterns of knives were referenced as Utility Knife, Mark 1 Dwg. No. 394832 for the five-inch, and Fighting Utility Knife, Mark 2 Dwg. No. 394831 for the seven-inch. Interestingly, the Mark 2 preceded the Mark 1 numerically by drawing and procurement specification. The Mark 2 was adopted as standard 13 April 1943, but it is not clear when the Mark 1 was adopted, although evidence suggests the same time frame. These drawings established dimensions and materials, and requirements of Rockwell hardness testing and corrosion resistance.

The Mark 1 was to have a five-inch blade and a leather handle with pommel. Many knives met those criteria; seven companies are known to have been contractors for this style of knife for the US Navy. The initial procurement of the five-inch knife, prior to the release of the drawing, was referenced as Knife, Hunting, with leather handle. Drawing No. 394832 redesignated the official nomenclature to a government classification of Knife, Utility, Mark 1. Of the variety of knives observed, many can be defined as having been procured as the Mark 1 from the BuOrd; others were purchased for advanced base use and resale through the Bureau of Supplies & Accounts under authority of the BuOrd and by supply depots or other Naval bases. Some contractors were making knives under contract for the Mark 1, and in turn selling lots to other government entities. All of this makes it difficult to determine what was an official Mark 1 knife and what is not. However, there are subtle details that help make this distinction.

When reviewing the five-inch hunting knife it is practical to start with the Marbles No. 45, which became the standard by which all other like styles were measured. The adoption of the No. 45 came much earlier for the USAAC, but it was observed in a BuAer jungle emergency seat kit developed for use in the Panama Canal Zone in 1937.

(Below) No. 45 Marbles with 5" blade and half guard included in jungle seat kit from Panama, c.1937.

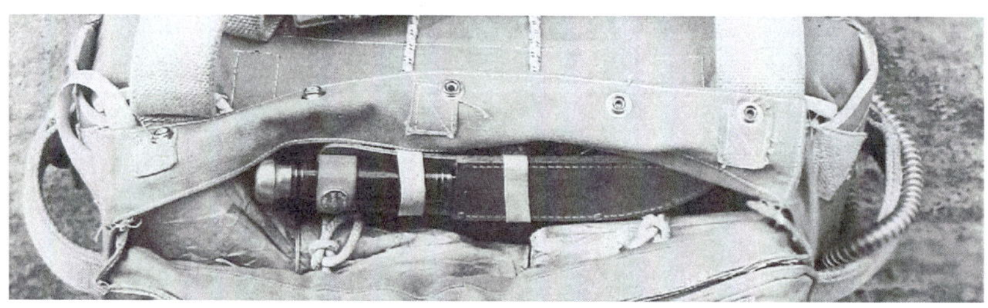

(Left) Carrier-based pilot Ens. Robert S. Nelson greeted in Hawaii by Adm. Charles Lockwood after rescue by the submarine USS Tang off Truk Island 29-30 April 1944. His personal knife is a 5" Marbles Ideal.

(Below) 5" Marbles Ideal attributed to Richard Hoppins MM2/c. He served aboard the USS De Haven, Fletcher Class destroyer.

(Left) VF-17 pilot Ens. B.D. Popp wears a Marbles Ideal secured to his sea marker, Bougainville Island, February 1944. As of 2 December 1942, 55,000 No. 45 Ideal hunting knives were on procurement from the Naval Air Base Station, Port Hueneme, California. From here, circulation of the knives could be requisitioned by units preparing for overseas movement and shipboard personnel. (National Archives)

(Below) Flight deck officer aboard the USS Bunker Hill directs a SB2C Helldiver; attached to his trouser belt is the 5" Marbles Ideal hunting knife, June 1944. (National Archives)

(Far left) VT-10 pilot Lt. MacDonald Thompson uses a Marbles No. 45; the scabbard was personalized with a hand-applied USN property mark and his name, USS Enterprise, February 1943. Pouch secured over his right hip is a locally fabricated emergency kit containing first aid, fishing and ration items; pouches were fabricated for pilots of the entire air group. (National Archives)

(Left) Example of a 6" Marbles Ideal with similar applied USN marking used by Lt. jg Kilmer Schmuck Bortz of VB-13.

The five-inch hunting knife became the preferred standard for general purpose and utility. Interdepartmental BuAer correspondence dated October 1943 discussed suitable knives for use by aviators. The BuOrd stated it had standardized the Robeson Shuredge hunting knife No. 20, and it was to be issued to all Naval pilots by the BuOrd. No such suitable knife had been procured for aviation activities, however; this was one individual's opinion about the deficiencies of the No. 20 and similar styles in circulation at the time. Moreover, utility sheath knife was used as the description. Recommended, was adoption of an aluminum-handled hunting knife made by R.J. Richtig, models No. 22 and No. 12; these knives were being sold at $2.00 and $1.75 each. It was felt they had more desirable features not incorporated in the No. 20 or other styles. By this and other accounts, The Robeson Cutlery Co. Shuredge No. 20 was the foundation for what would become known as the Mark 1 utility hunting knife. Simultaneously and shortly thereafter, other manufacturers would produce similar styles with some having their own unique manufacturer distinctions.

(Left and below) Robeson Shuredge No. 20 hunting knife had a polished zinc cast pommel pinned to the tang, but in production it changed to a wood pommel as detailed by BuOrd drawing. These were procured as hunting knives and were void of USN property markings on ricasso and sheath. (Left) Lt. R.W.L. Ross, USNR was a VD-1 pilot with 40 combat sorties when this image was taken in February 1944. He wears flying clothing and equipment assembly typical of Naval and Marine aviators, including a Robeson No. 20 hunting knife.

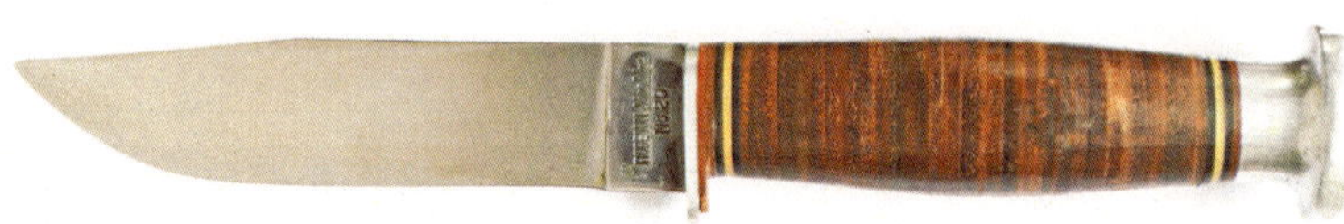

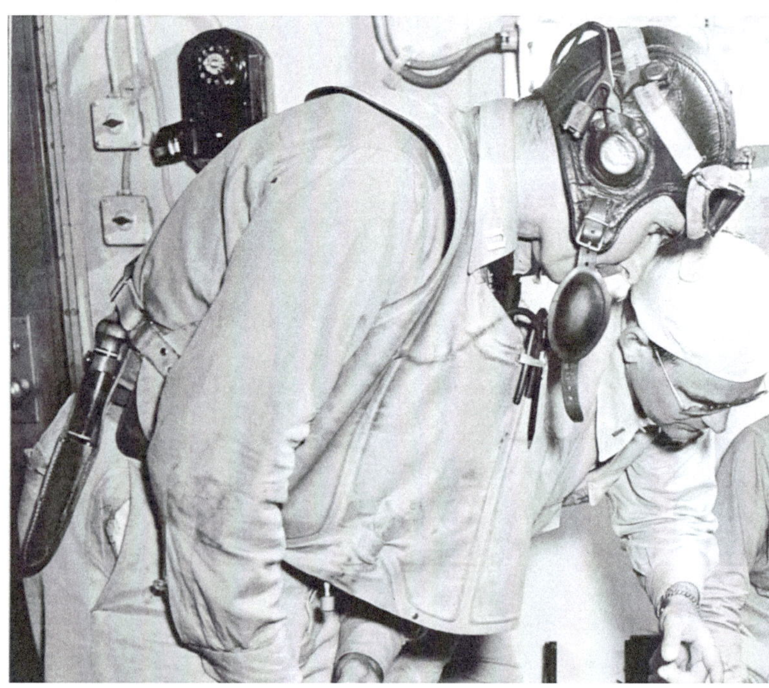

(Right) Unidentified pilot of VC-22 wears a Robeson No. 20 with wood pommel aboard the USS Independence, May 1943. Sheath knives were part of the standard allotment for aviation personnel. They were also issued to all Naval personnel operating in the forward area. Procurement from Robeson was initiated in December 1942 and another March 1943 for accumulative total of 185,000. (National Archives)

(Below) An example of the next generation as designated Knife, Utility, Mark 1. While formalizing a new standard utility knife, pommel was revised to be secured by a split locking nut, a feature adopted from Marbles. Robeson manufactured its early generation hunting knives for the US Navy with a brightly finished flat ground blade, half guard and wood pommel, retaining most commercial features. Split locking nuts provided by Marbles were brass; due to restrictions, other manufacturers used steel. Robeson added lead filler. (right) Comparison of the Marbles and Robeson pommels. The ricasso was stamped only on one side with the manufacturer's brand.

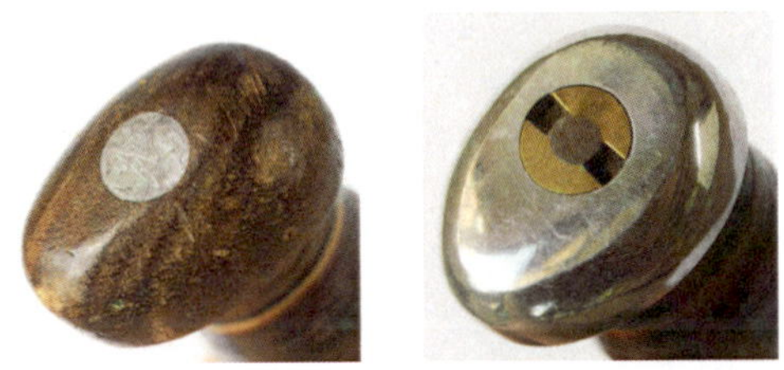

(Below) Dwg. 394832, distributed spring 1943, required property markings. USN property markings were then applied by manufacturers into the existing assembly process on the ricasso opposite their brands. BuOrd also authorized property markings on scabbards. This transition occurred through the summer of 1943. Most Robeson Mark 1s lacked scabbard markings, however. (below) Robeson No. 20, Mark 1s with USN stamps on ricasso and variants of colored handle washers. The lead filler in the pommel remained.

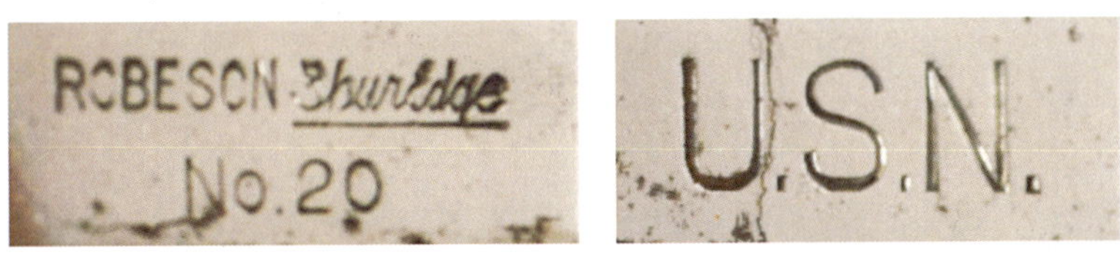

(Below) Robeson manufactured a full guard version. Evidence suggests these were made concurrently with the half guard type. One example lacks USN property mark on the ricasso like those in the early manufacturing process. The other has the USN mark applied as well as property mark on scabbard. Each has flat ground bright blade with wood pommel.

(Above) Lt. jg Tom killefer, and Ace, of VF-17 has Robeson No.20 secured to his M-566 sea marker, Oct.1943 to Mar. 1944.

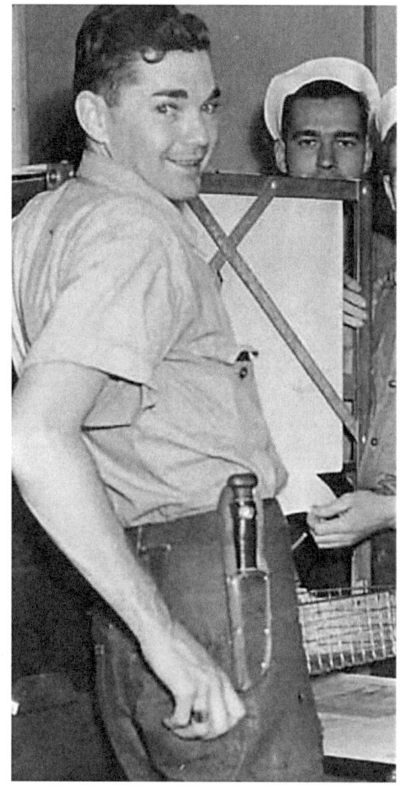

(Above) Unidentified sailor exhibits Robeson No. 20.

(Below) Robeson full guard worn by a sailor aboard the USS Kasaan Bay, August 1944.

(Below) Pilot Ens. Blanks of VC-13 wears a No. 20 aboard USS Core, August 1943. Sheaths not marked as property of USN, which was typical of knives delivered from Robeson. (National Archives)

Robeson manufactured the No. 20 under US Navy contract with a full guard, half guard, and a wood pommel. In spring 1944, the BuOrd requested that manufacturers of the Mark 1 switch to a zinc cast pommel like that on knives provided by PAL Blade Co. Robeson and Geneva Forge appear to have been the only companies to comply with the request. Manufacturing switched to the last generation Mark 1, with features of a blued flat ground blade, full guard, and leather handle with zinc cast pommel pinned to the tang. The property mark on the ricasso was also changed to USN/M.K.1. By the close of 1944, deliveries were beginning of knives with the new plastic scabbard. Robeson Cutlery Co. was reported to have manufactured at least 175,000 Mark 1 knives but this does not reflect the quantities for fiscal year 1945, when an additional estimated 100,000 were made. The application of Mark 1 or an abbreviation of it on the ricasso occurred in manufacturing and awarded contracts after March 1944, coinciding with a BuOrd directive changing nomenclature designation on all ordnance materials. Also, in November 1943 the BuOrd revised the requirements for corrosion resistance affecting all manufacturing of the Mark 1. After that contractors applied a parkerized finish to the blades; however, deliveries of bright blades continued until those were exhausted, likely in 1944.

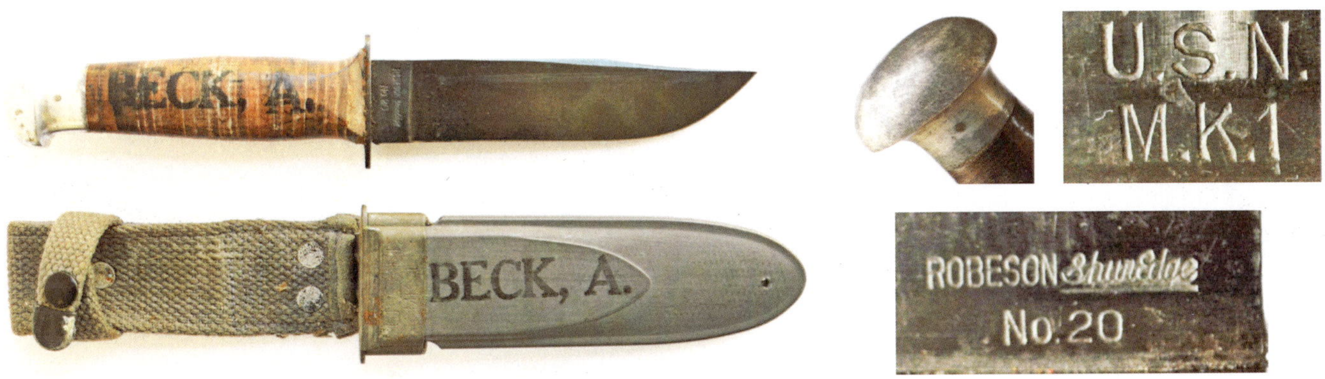

(Above) Late production Robeson, Mark 1. Zinc cast pommel and treated blade.

(Left) The bulk of negotiations and invitations to bid on the new utility knife occurred from January 1943 through the spring of that year. Contracts were awarded to the companies recognized as having produced legitimate Mark 1 knives. The bulk of production occurred after release of Dwg. 394832 with those knives conforming to the standards. A 5" hunting knife, similar to the Robeson No. 20, was produced by PAL Blade Co. In 1940, the company purchased cutlery tooling equipment from Remington Arms Co. and established a manufacturing facility with it. For sheath knives, PAL retained the pattern for the knife known as the Remington Hunter. One model in the Remington Hunter line was the RH-34, which had a 4"flat ground bight blade and a variety of pommel styles. This pattern was a Boy Scout knife; the BSA crest was etched on the blade and/or embossed on the scabbard, on many types observed. The RH-34 is best recognized as being a commercial pattern purchased as a utility hunting knife by US Government entities. These are some examples of the PAL Blade Co. RH-34.

When PAL was invited to bid on the new utility knife, it revised the RH-34 to a five-inch blade, conforming to US Navy requirements. PAL designated it as the RH-35,with the "5" signifying blade length. The PAL RH-35 produced under initial BuOrd contracts shares characteristics with the Robeson No. 20: flat ground blade, half guard, leather handle, and a wooden pommel. However, the RH-35 blade was parkerized and the pommel secured by a visible split steel locking nut. Under the original requirements, corrosion resistance was specified, and bright blades met those requirements. Production of the PAL Mark 1 (RH-35) occurred after Robeson and Camillus manufactured their versions, in the summer of 1943 and close to the corrosion-resistance change. The property mark was applied to the ricasso. Production by PAL was estimated at 550,000.

(Left) Typical 1943-1944 production by the PAL Blade Co.

(Below) This knife has the same characteristics but with the branding PAL BLADE CO./MADE IN USA. The significance of this alternate branding is unknown.

(Left and below) Sheaths were manufactured by a variety of companies that supplied knife manufacturers. Scabbards were not proprietary. Here, a scabbard ink-stamped for the US Navy, as many clothing and equipment items were.

All manufacturers of the Mark 1, or five-inch hunting knife, saw a progression of changes. Companies in early negotiations maintained their commercial features, such as bright blades and lack of property markings. The application of property markings on blades and scabbards began in the summer of 1943. By spring 1944, PAL had switched to the application of a pinned zinc cast pommel. This transition occurred after the release of the revision to Dwg. 394832 in November 1943. Two companies were contracted for the molding process: Mount Vernon Die Casting Co. and New England Die Casting Co. The MARK 1 stamp on the ricasso occurred late in the war or post-March 1944. The more common applications are USN/MARK 1 and MARK 1/PAL.

(Above) Last production type by Pal Balde Co. to include zinc cast pommel. (Right) variants in ricasso markings.

(Right and below) Aircrew of VB-20, USS Enterprise, March 1945. (Right) Ens. Ralph Collins. (Below) Edward Ruopp ARM1/c. PAL Mark 1 worn by each member. (National Archives)

Union Cutlery Co., manufacturer of the KA-BAR, developed strong early ties for production of military knives. The company has become recognized for production of the seven-inch fighting utility knife 1219C2 and Mark 2. They produced the Mark 1 and similar like types. Union Cutlery's five-inch hunting knives were also of a blued flat ground, produced with and without fullers, with and without the property mark USN stamped on the ricasso.

In breaking down the evolution of the KA-BAR, the progression can be best observed in the pommel. The first generation had a thin pommel round peen. That was followed by a thick pommel with round peen, and then a thick pommel with rectangular pinned tang. Both the thin and thick pommel types had flat and fullered blades. USN property marks stamped on the ricasso have been observed only with the round peened types, not with the rectangular tang.

Union Cutlery negotiated with the US Navy as early as October 1942 for the supply of utility knives; those purchased from that date until the early spring of 1943 were considered hunting knives. The examples with USN on the ricasso are legitimate Mark 1s as procured by the BuOrd, and their five and six-inch hunting knives were procured by the US Navy (BuS&A) and US Army for PX and advanced base use.

(Above) The two primary patterns of the KA-BAR 5" hunting knife, with and without fuller, each having a flat round peened pommel. Neither has USN property mark on ricasso, consistent with early purchases from late 1942 into 1943 as a utility hunting knife.

(Left and above) USN property marked Ka-Bar with flat ground blued blade.

USN property marked Ka-Bar with blued fuller blade.

Production of the five-inch utility knife for the US Navy was limited since the company's manufacturing capability was oriented to fabricating the seven-inch 1219C2 and Mark 2. Union Cutlery had no contracts for the Mark 1 after 1943, having produced upward of 40,000, at minimum,, in addition to those made under the designation of "hunting knife."

(Below left) Lt. jg Keaton VC-13 aboard the USS Core, August 1943. (Below right) Ens. Hudeck of VC-13 wears a 5" Ka-Bar hunting knife attached to shoulder holster strap, USS Core, August 1943. (National Archives)

Camillus Cutlery's story is like Robeson's. Camillus entered negotiations for the manufacture of the five-inch hunting knife early, presenting its pattern before March 1943 and receiving a contract in April. The Camillus version is similar to Marbles No. 45 and was model No. 5679 L76, known to have been procured by the USAAF. After revision of the blade it was designated model No. 5683 L75. The new model was saber ground, included a full guard, leather handle, and Resinox pommel with split steel locking nut. Three procurements were executed: the first in 1943 for 135,000, next in 1944 for an additional 220,000, and finally a procurement in 1945 for 195,000 at $0.85 each.

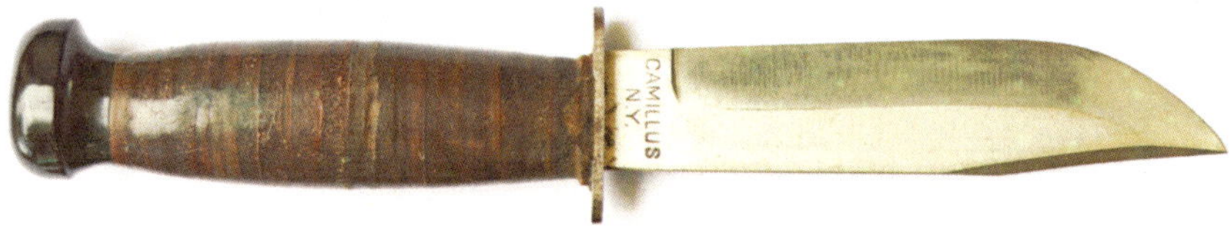

(Left and above) First generation Camillus Hunting Knife. Bright blade, no property mark on ricasso and scabbard, steel split locking nut.

(Above and right) By the summer of 1943, Camillus' design was revised to include the USN property mark on the ricasso with the bright finished blade. Camillus production utilized Resinox pommels.

(Above and left) After November 1943, Camillus treated its blades with a parkerized finish to meet new corrosion resistant requirements.

(Below) Sailor with two Camillus 5" knives, across waist and left hand. The one in left hand has a bright blade.

(Above) Flight deck of the USS Manila Bay, January 1945. Camillus Mark 1 or Hunting Knife on sailor's trouser belt. (National Archives)

(Above) Flight deck officer of USS Saratoga, All American football player and Hall of Fame coach of the University of Southern California, Lt. Julie Bescoes, lifts wounded TBF ball turret gunner Kenneth Bratton AOM, November 1943. Bescoes uses a Camillus Mark 1. (National Archives)

VB-5 gunner LeRoy Olsen ARM2/c has standard complement for pilots and aircrew, August 1944. Issued with the life preserver were two sea marker packets, a .38 revolver with shoulder holster, and sheath knife. Knife is a Camillus Mark 1. The revolver has M1917 lanyard added. (National Archives)

Pilot: Ens C.R. Van Buren, and gunner: J.B. Woodbright AOM2/c of VT-5, USS Yorktown, August 1944. Each wear Camillus Mark 1 hunting knives. (National Archives)

(Above) Contracts awarded after March 1944 required the ricasso to be stamped MARK 1. Camillus marked one side of the ricasso USN/MARK 1. Two variations existed, referenced as a small and large USN, shown here. New contracts were awarded in July 1944, but due to outstanding deliveries of earlier procurements, assembly on revised blades began toward the end of 1944. By this time, restrictions on brass had lifted, allowing it to be used in fabricating split locking nut. Camillus reported the first 14,000 on its new contract used a steel nut; remainder used brass.

Henry Boker & Co. also made a five-inch Mark 1 hunting knife. Boker's knives had a blued and bright flat ground blade, leather handle, full guard, and Resinox pommel with lead filler. Three variations include sterile ricasso, H. BOKER & CO/USA stamped on one side only, and H. BOKER & CO/USA on one side with USN on the other. Boker appears to have been awarded only one contract for a quantity of 80,000. Production mirrors Robesons and Camillus in that early manufacturing produced bright blades lacking property mark, while later production was of blued blades with USN property mark. This suggests manufacture extended into 1944.

(Above) Boker Mark 1 identified to G.A. Kreitzer GM2/c, USS Indiana, Division-3.

(Right) VF-20 pilot Lt. jg Donald Reeder, March 1945, wearing Mark 1 with similar characteristics to a Boker with elongated pommel and lead filler. (National Archives)

Geneva Forge Inc., a subsidiary of the Sta-Brite Products Corp., was another manufacturer of the Mark 1 utility knife. Sta-Brite entered negotiations in early 1943. They were not invited to bid but approached the BuOrd, eventually earning two contract awards totaling 140,000 units at a cost of $0.91 each. Like Robeson's, Sta-Brite's initial production had a flat ground bright blade, leather handle, half guard, and wooden pommel with lead filler. This first generation had the stamp of USN on one side of the ricasso only and a full sterile ricasso. The knife changed to include the company name stamped to one side of the ricasso and USN on the other, replacing the wooden pommel with a polished zinc die cast like that of PAL. In December 1943, Geneva Forge was awarded a second contract that dictated new requirements of blade treatment and the designation of MARK 1 on the ricasso was added after March 1944.

(Left and above) Some principle examples of the Geneva Forge Mark 1.

The Mark 1 utility knives illustrated thus far conform to a set of guidelines and are similar, but have subtle differences. Specifications are better thought of as guidelines, with contractors allowed variance outside of technical precision instruments. As long as the items met the minimum, approval for the deviations was usually granted. Companies merely submitted the proposed deviation, which was often approved by the relevant bureau.

Many companies promoted deviations that resulted in costs savings. Colonial Knife Co. was one company that stood out for a clear distinction on its Mark 1. In the spring of 1943, Colonial Knife had proposed the handle be of a molded plastic, a form of cellulose acetate. Cellulose acetate, commonly used in making camera film, could be manipulated to make a variety of plastics. It had become a popular material during the war to replace natural materials such as the handles or scales of pocketknives. Companies also experimented with it for ration and personal aid kit containers. As the wonder material of its day, the BuOrd requested it undergo a series of stringent tests to assure its durability as a handle for knives. After successful tests, the proposed molded handle was approved, followed by a contract in May 1943 for 50,000 knives. Follow-up contracts were released with a recorded total production by Colonial of at least 175,000.

The Mark 1 manufactured by Colonial had a prototype and four variants. The prototypes were unmarked and included a tan, almost translucent molded handle. After approval and contract award, the production handles were molded in black. To meet demand, production was expedited and no markings were stamped on the ricasso. These early generation knives had both bright and parkerized blades. Later Colonial Mark 1s had a ricasso stamped with COLONIAL/PROV. R.I. on one side and USN on the other. Aside from the lack of markings on the ricasso, there is a distinguishing feature between the pommels of the early production handles: each had a mushroom shape. A lead filler was later integrated into the pommel.

(Left and above) Early generation Colonial Mark 1, with and without lead filler.

(Left) Colonial changed its design to become similar to other Mark 1 designs, principally in the pommel shape. With more material, the need for lead filler was eliminated. This example was carried by Martin PBM pilot, Ens. Donald Price of Patrol Squadron 21, VPB-21.

(Below) Colonial Mark 1 being used by US Navy personnel. (Right) Crewman aboard a carrier fuel and arm F6F Hellcat for strikes against Manila Bay in the Philippines, January 1945. (Left) Landing craft crewman holds plasma bottle of a wounded marine removed from Okinawa; his hunting knife is late pattern Colonial with plastic scabbard, c.1945. (National Archives)

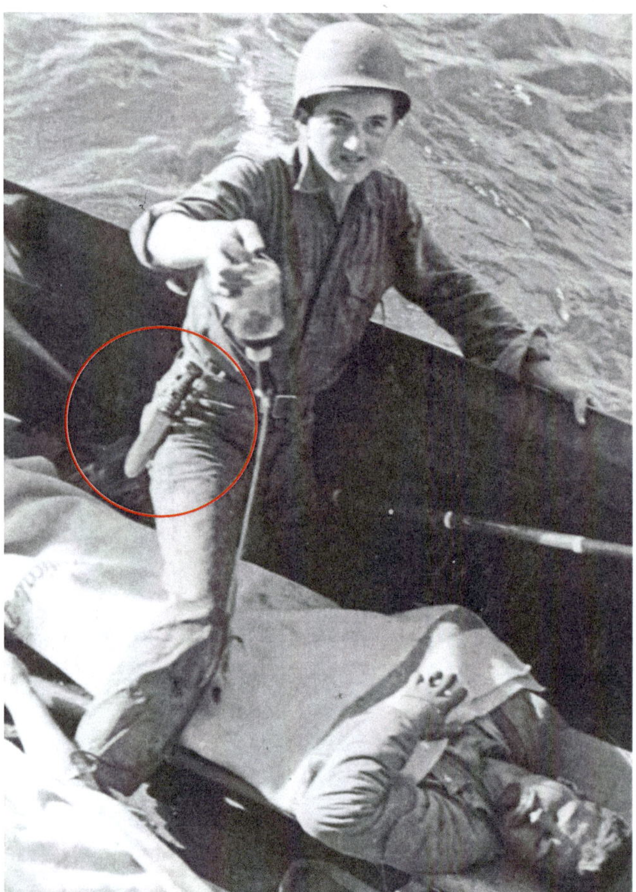

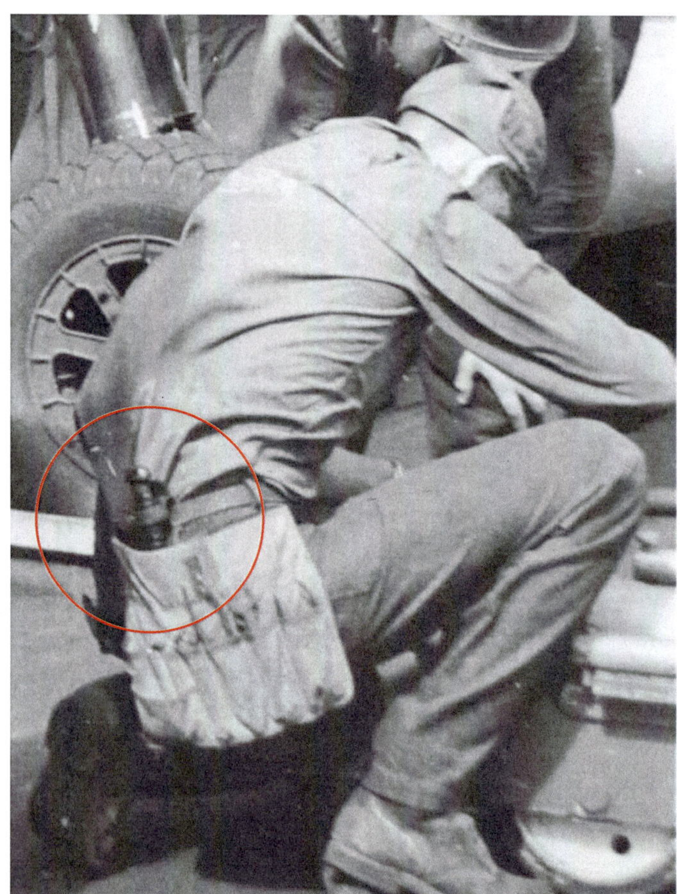

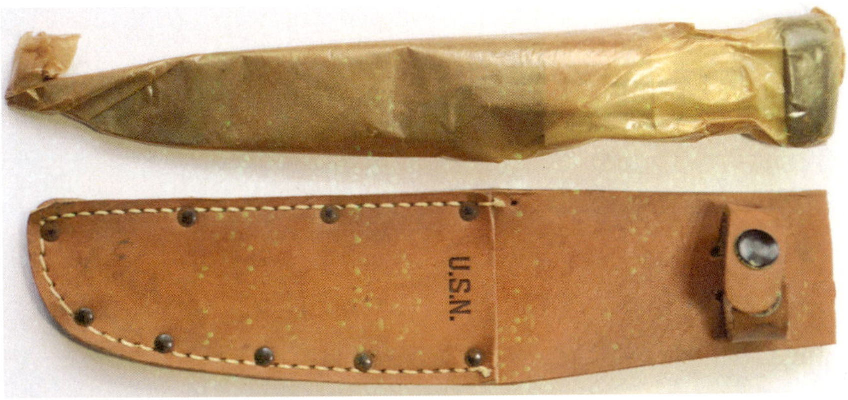

(Left and above) Late production Colonial Mark 1, as issued sealed in wax paper and provided with instruction sheet. Markings and pommel design on late production types. Significance to note, issuance with leather scabbard.

Western States Cutlery Co. produced a knife similar to the Mark 1, but there is no evidence of it being procured under that pretense. Western made knives to sell to the US Armed Forces, which it did in great numbers. The Western No. G46-5 and No. G46-6 are highlighted here because they were used by the BuAer and USAAF. The G46-5 was a five-inch hunting knife with a fullered blade with bright and blued finishes. Guards were available in half or full size in brass, steel and Tenite, and pommels were made of Tenite and zinc. All these features were available in a variety of combinations.

(Below) Two Western variants with steel and brass guards in half and full size, polished zinc die cast pommels with and without lanyard hole, and subtle variation in colored handle washers. Both have polished saber ground blade, variants can be found with blued blades. WESTERN/BOULDER COLO./PAT. No.1967479 on one side of the ricasso. These were often referred to as a "Baby Sharks," a collectors' term derived from Western's designation of "Shark" for its 6" model. No Westerns' have been found factory property marked USN.

(Right) Volunteer during evaluation of the integration of the side arm and sheath knife for Naval aviation personnel, December 1942. Knife is Western G46-5 with Tenite pommel and guard with a high handle strap. Scabbards had alternating features of high and low retaining straps. (National Archives)

(Above) Western produced a knife similar to the Mark 1 KA-BAR, designated as model No. L71. In advertising, Western listed it as SEEBEE–No. L71. It had a flat ground blued blade, half steel guard, leather washer handle, and flat steel pommel. Western often coined trade names for its knives: "Shark," "Bushman," and "Commando" existed in addition to SEEBEE. It is considered a like type.

(Right) Ground crew member of VMF-215 refueling Major R.G. Owens's corsair, October 1943. He wears a knife similar in appearance to a G46-5 with a steel or brass guard. (USMC)

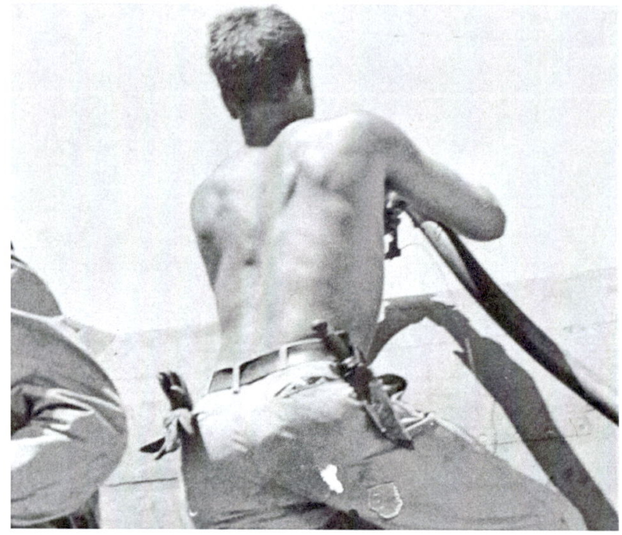

The Navy rapidly integrated the five-inch hunting knife into use in all theaters of war. Its principal function was a lifesaving tool. The average cost for the Mark 1 ranged from $0.85 to $0.91 each. Adding the cost of the leather sheath, which ranged from $0.23-$0.25, the average total cost was approximately $1.15. Under the designation of Mark 1, stock number 74-K-95-150, was assigned for requisitioning the Mark I knife and sheath.

(Left) Ens. Robert Geldart of VOF-1 aboard USS Tulagi, August 1944.

(Bottom left) Ens. Arthur Johnson of VF-5 aboard USS Yorktown, August 1943.

(Below) VF-66 pilot, July 1945, whose knife has a similar appearance to late production Robeson No. 20. (National Archives).

(Below) A variety of leather sheaths provided for the Mark 1 utility knife. These were purchased at $0.23-$0.25 each, depending on the number of rivets. Known manufacturers for leather sheaths provided for US Government contracts include Nitke Bros.., Brauer Bros.. MFG, Boyt Harness Co., A.L. Siegel Co., and Mosser Leather. (Left) two sheaths manufactured by Mosser Leather; one variation had KEEP THIS KNIFE WELL OILED embossed on the face. (Far right) two sheaths made by Boyt, with and without USN embossed at the throat. Property marking of leather sheaths was approved by BuOrd in spring of 1943.

(Below) Sheath manufactured by Brauer Bros.. Mfg. can be found with and without USN embossing (left). (right) Sheath manufactured by A.L. Siegel Co. This model also produced with and without USN.

(Below) Sheath knives were carried many ways: on the pistol belt, trouser belt, flight suit belt, shoulder holster strap, life preserver waist strap, sea marker tie, and by lanyards from around the neck and life vest. Two pilots of VF-16 aboard the USS Lexington dress down after air strikes. (Left) Individual removing 5" knife from flight suit waist belt. (Right) LCDR. R.H. Isley, enjoying coffee, wears an RH-36 on his belt and camouflaged Type B-4 life preserver, July 1943. (National Archives)

By the close of 1943, emphasis within the Navy Department had shifted toward a variety of larger knives. The BuOrd began purchasing six-inch knives for advanced base and ship's stores referenced as Knives, Utility and Fighting, available under stock number 74-K-96. The BuAer had also transitioned to the procurement and issuance of six-inch hunting knives.

The transition from five- to six-inch knives occurred due to a change in operational requirements. Naval shipboard personnel spent more time ashore and in greater numbers in the combat zone, which established a need to carry a close-in fighting knife for defense. The five-inch knife lacked some of the versatility needed, which the larger types provided. The BuAer also felt these larger knives brought a significant benefit to survival and emergency situations. The six-inch knife thus became the unofficial standard for aviation activities in March 1944 with the release of the new preliminary ASO supply catalog, which replaced the five-inch type. The nomenclature in the catalog was Sheath Hunting Knife, stock number R41-K-640. Its description: Blade is approximately six inches long – furnished with leather handle, double guard, approximately 4-1/2 inches in length from guard to end of the handle – end of handle to be molded Bakelite – sheath to be acid-free leather with handle strap and belt loop. This description supports the notion that Western had become a prominent knife producer, using Bakelite or Tenite pommels. The description was revised in December 1944 to be more generic: Blade is six inches long – furnished with leather handle, double guard, and leather sheath. BuOrd circular GV1-45 superseded G1-43 and indicated that one sheath knife of five or seven inches be authorized for issue to aviation personnel.

The BuOrd listed manufacturers of the six-inch utility and fighting knife as Case & Sons, Cattaraugus Cutlery, Robeson, Union Cutlery and Kinfolks, Inc. Others likely existed. (Left, top to bottom): Union Cutlery, Case, Cattaraugus, Case and Kinfolks 6" utility and fighting knives.

(Below) The Robeson was issued to a Naval person identified to K.P. Lord. He personalized the scabbard with a Navy pennant decal and a decal from his station, US Naval Net Depot, Melville, R.I.

The procurement of knives was under the authority of the BuOrd, as the BuAer regularly noted. However, some exceptions, as previously mentioned, occurred. For example, the Aviation Supply Office of Philadelphia awarded two direct procurements to knife contractors. The first of these was to PAL Blade Co. in June 1943 under contract 288-14513 with the generic listing of Sheath Knives; the second was to Western Sates Cutlery Co. in June 1944 under contract 288-24171, again for Sheath Knives. It is unknown exactly what sheath knives were placed under procurement by the ASO, but many wartime images illustrate the use of the PAL RH-36 and the Western G46-6.

(Left and above) Early generation PAL RH-36 with embossed scabbard. Early knife had bright blade with fuller, full guard and zinc diecast pommel. This example has black plastic washers at guard and pommel.

(Right and below) The most commonly seen 6" utility knife in aviation was the PAL RH-36. Here, each has leather sheath with embossed oval PAL.

(Right) RH-36 secured to life preserver of VT-14 pilot Ens. E.F. Hoffman aboard USS Wasp, August 1944.

(Below) Pilot Ens. W.H. Larson of VF-16 wears PAL on flight suit waist belt, USS Lexington, November 1943. (National Archives)

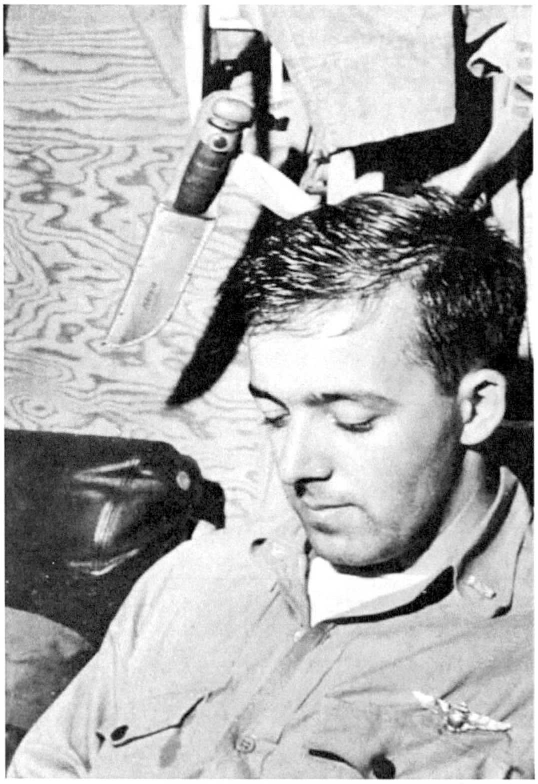

(Above) PAL embossed sheath hanging in ready room of USS Yorktown, June 1944. (Right) VT-5 TBF pilots in the ready room of the USS Yorktown, September 1943. Pilot on left has PAL RH-36 with embossed sheath. (National Archives)

(Left) VT-20 gunner Bert Hutsebus AOM2/c, March 1945.

(Below) TBF pilot of VT-11 Lt. M.L. Tegge, USS Hornet, March 1945. Both wear RH-36. Tegge wears personal emergency kit next to knife. (National Archives)

(Left) PAL revised the RH-36 twice to be similar to the Mark 2. The first revision was to the scabbard pattern; later, the blade was parkerized. Examples of those two alterations. Other notable variations include colored washers in handle.

(Below) VT-47 gunner John Thomas AOM2/c wears a revised RH-36, USS Bataan, May 1945.

(Below left) VF-16 squadron member wears revised RH-36 secured to the riser strap of his parachute harness, USS Lexington, c.1944. (National Archives)

Western Cutlery referenced their pattern of G46-6 as the Shark, often called the "Shark knife." The G46-5, a smaller version, has been dubbed the "Baby Shark." It is not clear exactly when Western applied this designation, but it references history. Native peoples proficient at skin diving carried blades they called "shark knives". Influenced by observing nature, fabricators had made these knives resemble a shark's tooth. They had a double edge tapering to a fine point, providing excellent penetration and slashing.

Sharks, however, were a threat to survival in some accounts. Wartime sailors referred to their knives as "shark knives" because, if they found themselves overboard, the knife would be their only defense against sharks.

Training pamphlets mentioned knives for defense against shark attacks. When defending against a shark, they recommended, "kicking or stiff arming them in the nose," followed by "knifing them in the gills." One pamphlet suggested, "Use a knife at close quarters in a showdown." The pamphlet Shark Sense, says, "In case of shark attack in the water, swim out of the line of his charge, grab a pectoral fin as he goes by, and ride with him as long as you can holding your breath. If you are fortunate enough to be armed with any kind of knife, cut the shark's belly open."

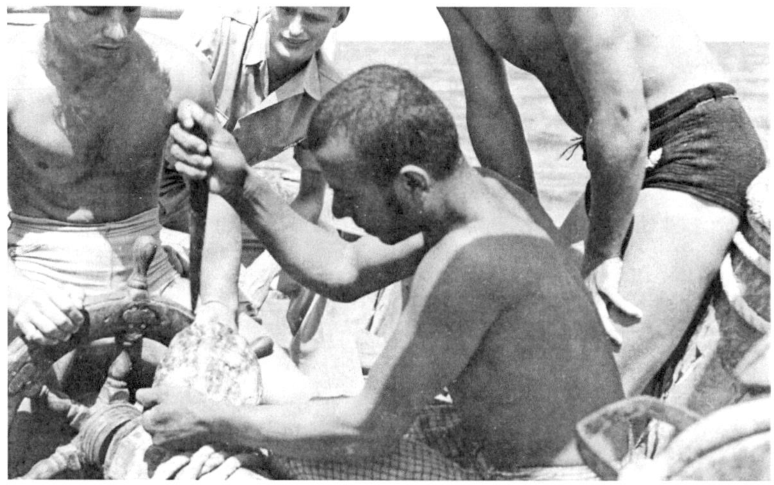

(Right) Native pearl diver opens an oyster off Bahrein (now spelled Bahrain) Is., Persian Gulf, c.1944. Caption described it as a shark knife, and the blade characteristics match a shark's tooth. The G46-6 does not share this pattern; the designation was a marketing label. (National Archives)

(Below) Early accounts by survivors stated they wished they had a knife to defend against sharks circling their life rafts. They often used oars to jab and slap at the water. Lt. jg Butler being groomed by four of his squadron mates using the Western G46-6, Ugi Island, Solomon Islands, November 1943. (National Archives)

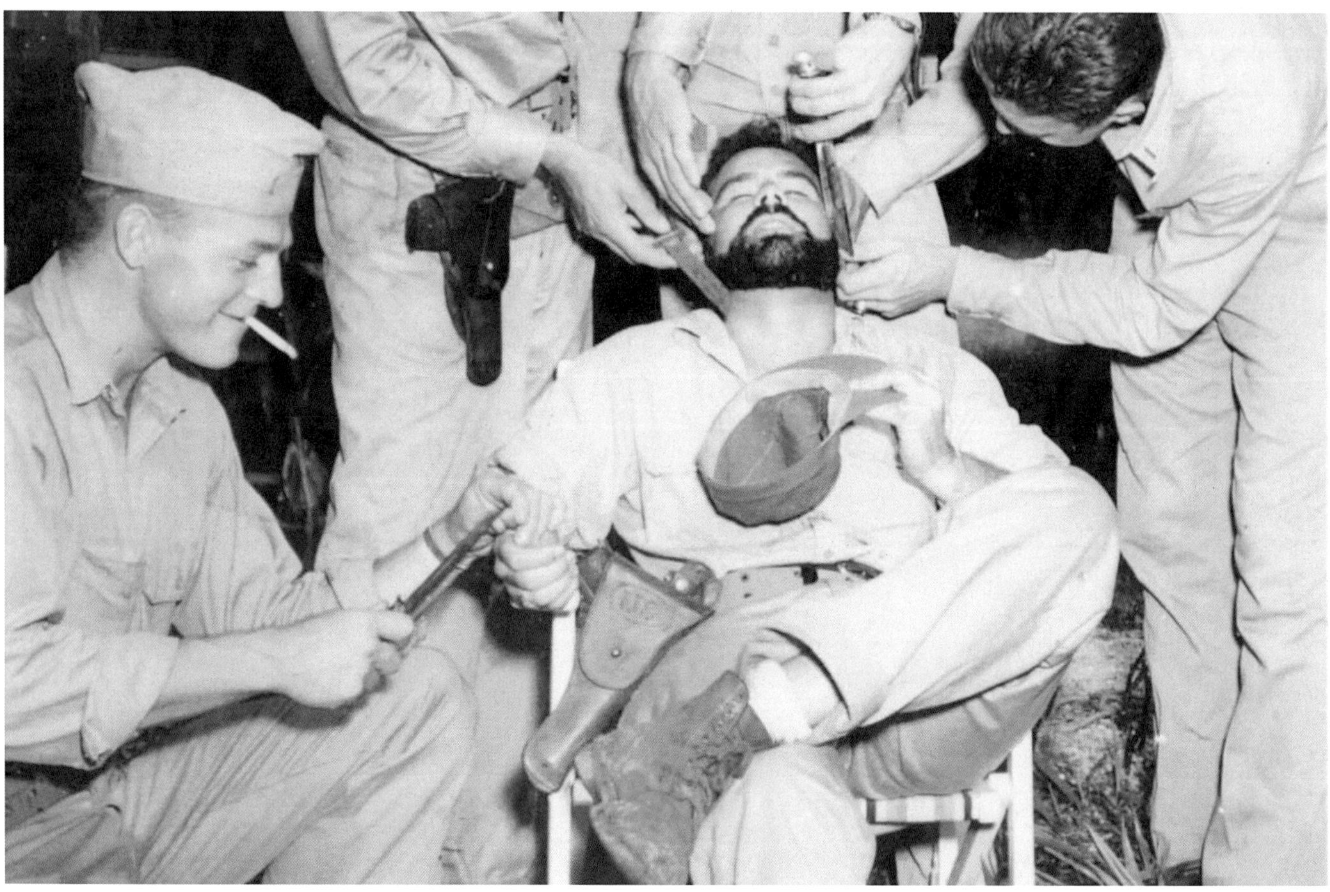

(Above and right) Variations of the G46-6. They have leather handles and 6" fullered blades, either bright or blued finish. Pommels were either Tenite or polished zinc, with or without lanyard hole. Guards were brass or steel. Sheaths are typical for this pattern with either bright or blued rivets. The profile, however, is similar to the silhouette of a great white shark.

(Left and below) Western G46-6, property marked in Naval fashion with A.N. Larsen stamped on the handle and leather sheath.

(Below) VF-5 pilots Lt. J.A. Granger and Lt. jg T.D. Crow carry Western G46-6 sheath knives aboard USS Yorktown, September 1943. Crow's Western has lanyard hole in pommel. Like the sidearm, there were reports of knife being lost during a struggle. The reports resulted in two recommendations: provide a knife lanyard or carry two knives. (National Archives)

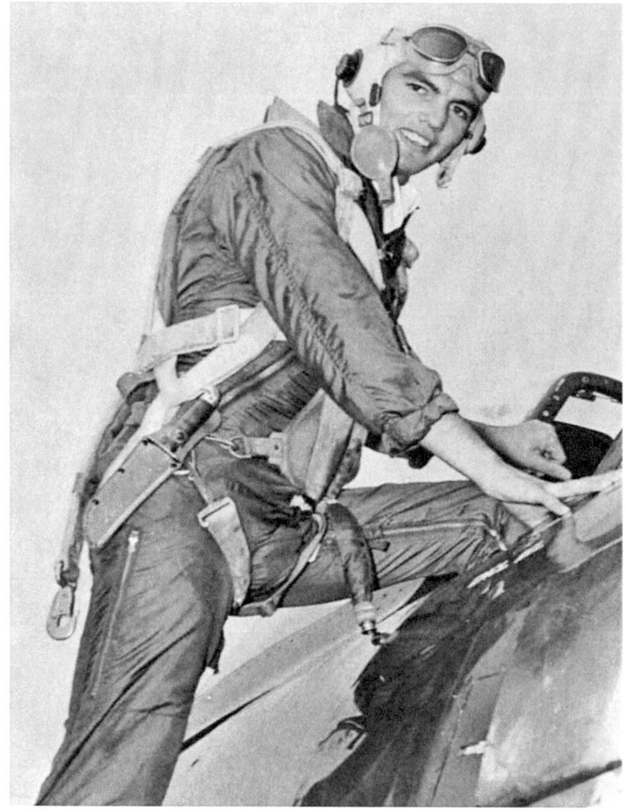

(Above left) Western's G46-6 and the PAL RH-36 became widely used by pilots and aircrew of the US Navy and Marines. Westerns were often used by Marine aviation units, including member of VMF-222, c.1944. (Courtesy Gregory Pons)

(Above) Ens. Malcolm Gottschalk of VF-66 has a Western with a sheath with blued rivets, July 1945. (National Archives)

(Left) VC-85 TBM aircrew member James Sullivan AOM2/c has a G46-6 secured to life preserver, March 1945. (National Archives)

(Below) VMF-215 squadron member with either a Western L76 or L77 secured horizontally across waist, c.1944. (National archives)

(Right) Adding pockets to flight suits or other garments was common, from pencil pockets on the upper sleeve, through patch pockets on the thighs and lower legs, to enlargement of factory applied pockets: whatever the individual needed. Here, pilots of Air Group 20 are debriefed by the ACIO after air strikes over Formosa, USS Lexington, January 1945. Highlighted pilot has flight suit refinements that include added left shoulder pocket and large lower left leg pocket. Such modifications were made by the squadron parachute rigger at an individual's request. (National Archives)

(Below) The pictures here were taken by ZP-32 in Spring 1945 and submitted to Chief, BuAer, for consideration of making permanent a hunting knife pocket on flight suits. Exhibits closer details of application methods. Leather sheaths sometimes modified with dome snap for extra security to suit as seen in image and artifact named to Lt. L. Bell. (National Archives)

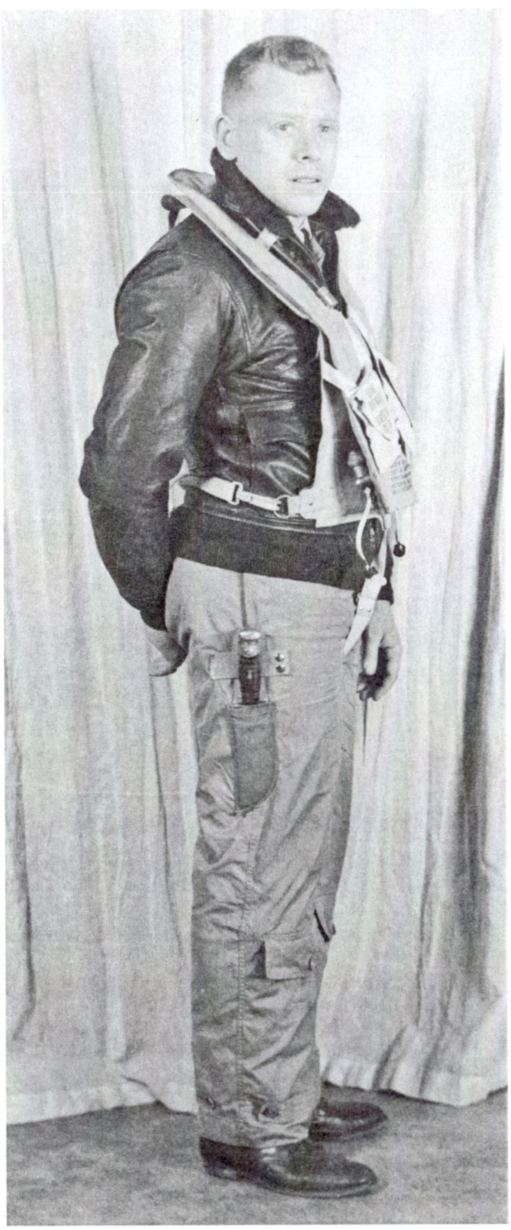

(Above) VB-2 pilot Lt. jg A.J. Norris has knife at lower leg position, he additionally has round chromium signal mirror slung from vest. Gunner is M.A. Patillo ARM1/c, USS Hornet, September 1944.

(Below) PAL RH-36 with four dome or glove snaps for attachment to flight suit.

(Right) Air Group 9 Commander Lt. Cmdr. Phillip Torrey, Jr. has a series of additional pockets applied to his flight suit, left leg secures hunting knife. Pocket on right leg is of an unidentified object. March 1944. (National Archives)

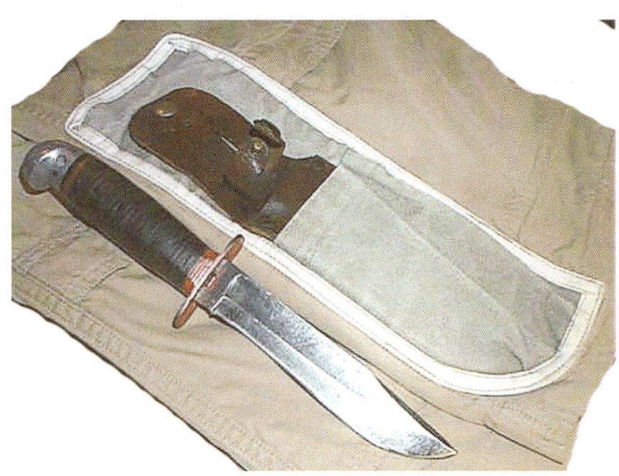

(Right) The advantage to the pocket is that the knife was clear of the user's flight clothing or life preserver, ensuring no interference with parachute descent or movement in the water. Whether standing or swimming, the knife handle is easily found. Pocket positioning was consistent, but variations occurred in how the knife was secured. This was achieved with aircraft grey duck fabric. Leather sheath was modified with the female glove snap and the male stud is secured on the flight suit.

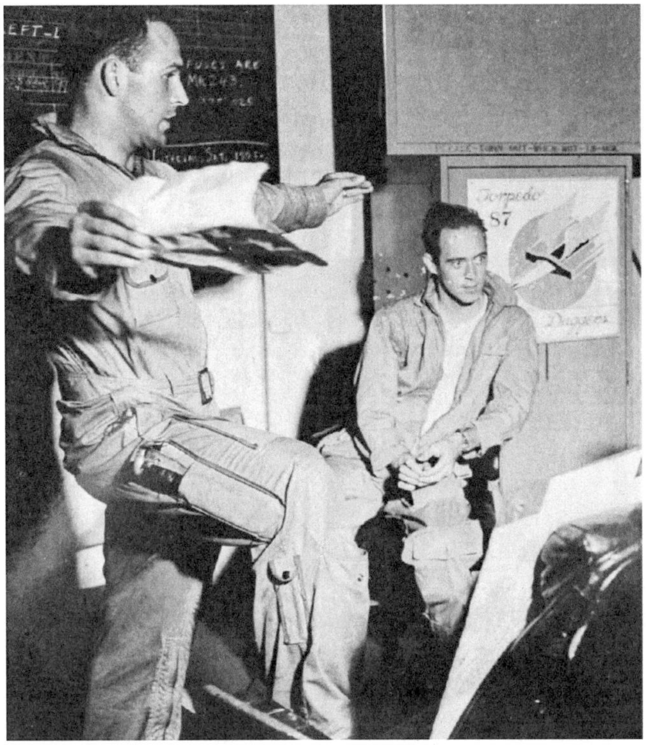

(Above) Pocket placement was typically on the thigh of the user's dominant hand. Here, additional thigh pocket positions. VT-87 pilot Ens. Giovanni Brega, USS Ticonderoga, July 1945. PAL RH-36. (National Archives)

(Above) VB-11 pilot Lt. jg Gallatin, March 1945. After evaluation, Chief of the BuAer opposed permanent pocket on flying suits. It would interfere with proper cleaning and obstruct access to pockets already provided, as new M-668 flying suit had zippered internal thigh pockets. (National Archives)

(Below) VF-17 pilot, Harold Bitzegaio, Bougainville, c.1944. He is wearing early pattern HBT coveralls with added shin pockets. The hunting knife is secured in the utility pocket provided on that pattern of suit. (National Archives)

(Below) VMF-212 Corsair pilot whose 1219C2 is secured by a canvas strap over the sheath behind the handle, Green Island, c.1944. (National Archives)

(Above) VT-2 TBM aircrew member prepares for a mission with 5" hunting knife appropriately sized pocket, USS Hornet, June 1944. (National Archives)

(Above right) LCDR. "Lambo" Lamberson, CO of VF-3, carries US Army M3 trench knife in leg pocket, USS Yorktown, c.1944. (National Archives)

(Right) Lower pocket on flight suit of Christofferson ARM2/c of VB-18, enlarged to accommodate additional items, USS Intrepid, November 1944. (National Archives)

(Below) Patch pocket attached to lower leg of BuAer contract AN-S-31 flight suit. Two additional pockets for pencil and jack-knife. Waterproof aircraft grey Army duck.

(Above) VB-10 pilot Ens. C.R. Mester has added shin pockets in addition to knife pocket applied to the lower legs of standard flight suit, USS Enterprise, February 1944. (National Archives)

(Above) Lower leg pocket on Ens. R.H. Baker of VF-6, USS Hancock, April 1945. Baker has Mark 2 fighting knife. (National Archives)

(Left and right) Knife pockets applied to nylon flight suits with nonspecular sea blue waterproof fabric. (Right) The Mark 2 is secured by a canvas strap through belt loop of sheath. (Left) Pocket on other suit fabricated for Mark 1 knife.

Pockets not permanently attached to flight suit. (Above left) Securing apparatus used by member of VMF-312 is a frog, like used for bayonets, then strapped to the lower leg. (Right) Member of VMF-212 with same installation but enclosed pocket for knife. (Center) leg scabbard for 7" fighting knife 1219C2 identified to 1st Lt. G.M. Holderer of VMF-215. (USMC)

With the increasing use of hunting knives among fleet personnel, reports emerged that leather sheaths deteriorated after prolonged use. This was partially due to initial procurements of substandard sheaths. With procurements of Mark 1 and Mark 2 knives, sheath quality improved. Those of lesser quality were reinforced with aluminum plates, which were readily available in aircraft assembly and repair facilities, both aboard ship and in land-based Carrier Air Service Units. Typically, the two wear points were the tip of the sheath and the throat, and the entire sheath sustained abuse.

Typical applications include only aluminum applied to the face, then secured by rivets; some had back reinforcing as well. Other methods include reinforcement at the tip or a band near the throat.

(Right and far right) Flight deck crew members aboard USS Lexington whose hunting knife has had aluminum panel applied over sheath, December 1944. (National Archives)

(Above) VC-13 aircrew members have reinforced knife sheaths, August 1943. (National Archives)

(Below) Sailor standing atop a PBY at NAS San Diego, June 1942. Reinforcement at sheath throat was personalized. (National Archives)

(Below) VF-11 pilot grabs a snack before conducting first air strikes over Japan, USS Hornet, February 1945. His sheath is reinforced at tip and throat. Knife appears to have been manufactured by Union Cutlery. (National Archives)

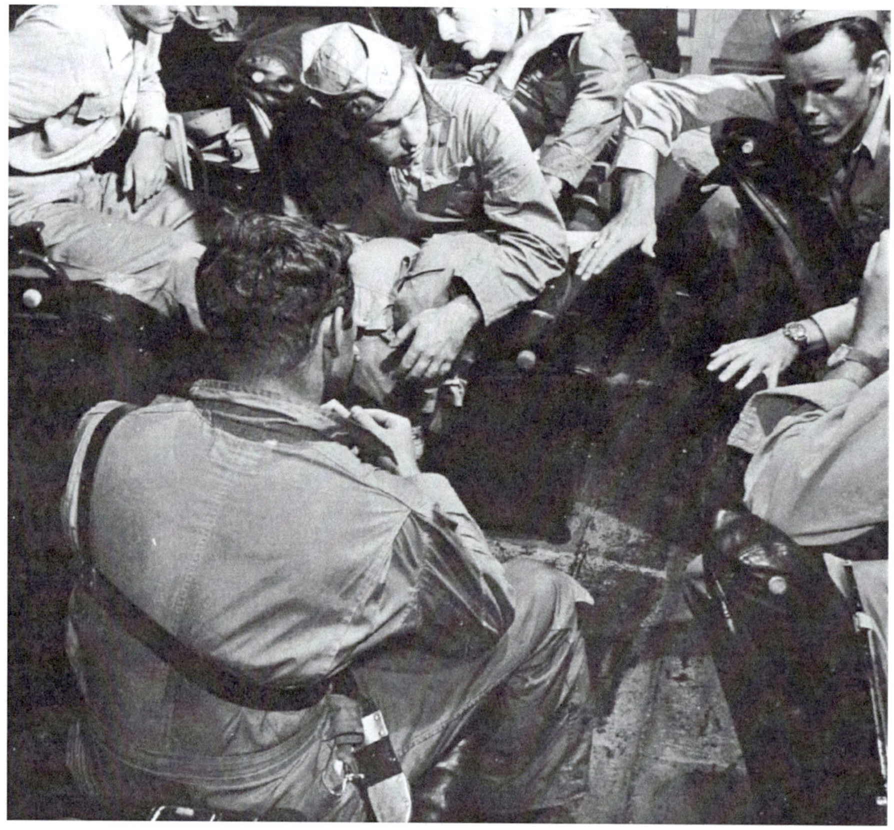

(Above) Reinforced sheath with single aluminum sheet applied to the face and riveted. Knife is a prewar Western and on the blade is stamped RUSSEL D. DAVIS. Such personalization was not uncommon for US Navy personnel. (Above right) VF-16 member has reinforced knife on tip and across throat visible over right hip, USS Lexington, November 1943.

(Below) More sheath reinforcements, one containing a PAL RH-36. The Western G46-6 has it at the tip of scabbard. Aluminum is inscribed to an aircrew member of VPB-18.

(Below) In December 1945 issue of Naval Aviation Newsletter, a BuAer official addressed safety concerns of the fleet. The article includes a drawing for the application of reinforcement across the sheath throat and tip, to be implemented locally.

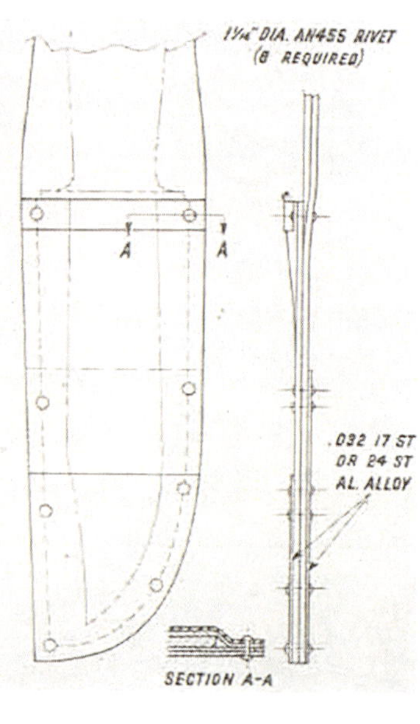

Deterioration of leather sheaths became an early issue servicewide, and, in response, the BuOrd began negotiation and procurement of replacement scabbards made from a fiber-reinforced Tenite-II cellulose acetate butyrate plastic. The first contract was awarded in November 1943 to Beckwith Manufacturing Company under contract NORD-4723 for 600,900 scabbards, equally split at 50% for the Mark 1 and 50% for Mark 2 knives at $0.85 each. This contract was to replace sheaths already in service and for knives currently on procurement. Though the contract was negotiated and approved in November, BuOrd did not approve specifications for the new plastic scabbards until the end of January 1944. BuOrd Dwg. No. 300340 was assigned to the Mark 2 scabbard and Dwg. No. 300341 for the Mark 1. Contracts were awarded for new production, but this time to a Beckwith subsidiary, Victory Plastics Co., in July and August of 1944. Contract NORD-6581 was for 165,000 Mark 2 scabbards, and NORD-6804 was for 235,000 Mark 1 scabbards. In March 1945, these two contracts were still in production.

The next contract was awarded in March 1945 as NORD-8114 for 825,000 scabbards, equally split at 50% for the Mark 1 and 50% for Mark 2 knives. Another contract, NORD-8676, was awarded the same month for 470,000 Mark 1 and Mark 2, equally split contract, scabbards but was soon cancelled. It saw a limited production run. The initial cost for the plastic scabbards was $0.85 each. Interestingly, Beckwith voluntarily refunded the US Government Treasury Department just under $31,000 due to savings in the pressed steel manufacturing of the throat between January and September 1944. The Treasury Department received the check in November. The price quoted thereafter was for $0.75 each on all future procurements. The metal throats were manufactured by New England Pressed Steel Co., which reported consistent delays. Knife manufacturers were notified that initial deliveries of the new plastic scabbards could be expected by September of 1944, but due to delays very few were received by December of that year.

Because plastic scabbard delivery and distribution came late in the war, evidence of their use is limited. With the projected availability of the new replacement plastic scabbards, stock number 74-S-98-50 was assigned for the Mark 1, and 74-S-98-68 for the Mark 2 scabbards. Knives returned with damaged sheaths were placed into stock and the leather sheath was discarded. Stock number 74-95-140 was assigned to the Mark 1, and 74-K-95-115 was assigned for the Mark 2, complete with plastic scabbard."

(Below) Ens. George Bailey and Lt. William Miller of VF-23 wear Mark 2 knives with plastic scabbards, USS Langley, c.1945.

(Above left) Unidentified PB4Y crew members of a Navy patrol squadron, crewman on right with Tenite scabbard, c.1945.

(Above) Ready room of the USS Langley, VF-23, c.1945. Mark 2 with Tenite scabbard can be observed in lap of pilot.

(Left) Pilot of VB-16, Saipan, c.1945. Wears a brand new issue Mark 2 with plastic scabbard, still has protective wrap around throat.

(Right) Examples of the plastic scabbard for the Mark 2 and Mark 1.

BuOrd Circular Letter G1-43 was revised 24 February 1945 as GV1-45, which prescribed the issuance of knives five to seven inches for Navy and Marine aviation activities, to include a six-inch type. By this date, however, larger bladed knives were preferred and requested. To simplify requisitioning, they were placed on allowance lists and could therefore become standard stock items at aviation supply depots. Aviation personnel regularly used the USMC 1219c2 and US Navy Mark 2. These seven-inch fighting utility knives originated within the USMC, developed to supplement the bayonet for front line combat troops, adopted 9 December 1942. The US Navy adoption came on 13 April 1943. Stock number 74-K-97 was assigned for requisitioning from either USMC or Naval depots. This was later revised, for the Mark 2 only, to stock number 74-K-95-120 when its plastic scabbard was introduced. The models are virtually indistinguishable, the Mark 2 being near a direct copy of the USMC pattern.

The USMC awarded contracts to Robeson, PAL, Camillus, and the Union Cutlery Co. (KA-BAR). First production Robesons and KA-BARs had blued blades, later transitioning to parkerized. PAL and Camillus provided only parkerized blades. Distinct features include seven-inch blades with a fuller, leather handle and flat steel pommels. Most examples were stamped USMC on the ricasso or guard, depending on their manufacture date.

(Right) Sampling of 1219c2 knives for the USMC; first, second and third generations. Knives delivered to the USMC were provided with leather scabbards only.

(Below right) VMF-312 Corsair pilots look over a target chart on an airfield in Okinawa. The pilot on the left has hunting knife; pilot on right pilot has a 1219c2. His rig includes locally fabricated holster, lensatic compass, and ammunition loops sewn to the pistol belt. (USMC)

(Below) VMF-312 member wears his 1219c2 on right hip. Okinawa, c.1945. (USMC)

(Above) 7" fighting utility knife 1219c2 was available from standard stock and issued as part of 782 gear to aviation cadets. Its use and wear were optional. Two unidentified USMC Corsair pilots on left wear 1219c2 knives. Third from left wears Camillus Mark 1. Believed to be VMF-322 on Okinawa c.1945.

(Below) VMTB-232 aircrew member wears 1219c2 over left hip along with a TL-122A flashlight, Solomon Islands, c.1944. (Courtesy Gregory Pons)

(Below) VMTB-454 pilot poses in front of a TBM while stationed stateside, c.1945. (Courtesy Gregory Pons)

The march through the Central Pacific began in 1943, starting with the Gilberts, then on to the Marshalls, and then the Caroline Islands. Hundreds of thousands of Naval personnel filtered into the forward operating areas at island bases, providing a multitude of support functions. Naval procurement of Mark 2 fighting utility knives occurred after the USMC's purchases, but sufficient quantities were available by Spring 1944 to equip Naval activities. With the large procurements of the Mark 1 and Mark 2, ship and shore personnel were being supplied a limited number of knife options.

There were more variations of the Mark 2 than of the 1219c2. The same four manufacturers were awarded BuOrd contracts. The first generation of Mark 2s were made by Camillus and briefly included a locking nut on the pommel and scabbards lacked property marks. Later fabrication would later change to the peened and pinned type pommel and USN was embossed at the throat for scabbards; deliveries included leather scabbards. Knives delivered in 1945 were supplied with Tenite scabbards.

Within the BuAer, dissatisfaction grew regarding the five-inch hunting knife. There were many requests for the seven-inch fighting knife or Mark 2 to be substitutes. As a result, G1-43 was revised to add larger blades on the allowance list for aviation squadrons.

(Above) Aircraft carrier deck handlers move an SB2C Helldiver into position for takeoff aboard the USS Ticonderoga, November 1944. All personnel wear Mark 2 fighting knives. (National Archives)

(Right and below) Examples of generations of the BuOrd Mark 2 fighting knife and scabbard variants. (below) Pommel assembly methods.

(Above) John Schmidt AOM2/c of VT-20 aboard USS Enterprise, March 1945. Mark 2 fighting knife. Note AN-C-101 match safe secured to life preserver and shoulder holster around waist, secured to left leg. (National Archives)

(Above) Thomas Mumford ARM3/c of VC-86 aboard USS Bismarck Sea, January 1945. Mark 2 fighting knife over left hand. (National Archives)

(Below) F6F pilot Lt. jg John Purio of VF-19 aboard USS Lexington, February 1945 Mark 2 fighting knife, right hip. (National Archives)

(Below) Air Group-38 pilot in ready room of USS Essex, January 1945. Mark 2 fighting knife secured to riser of parachute harness. (National Archives)

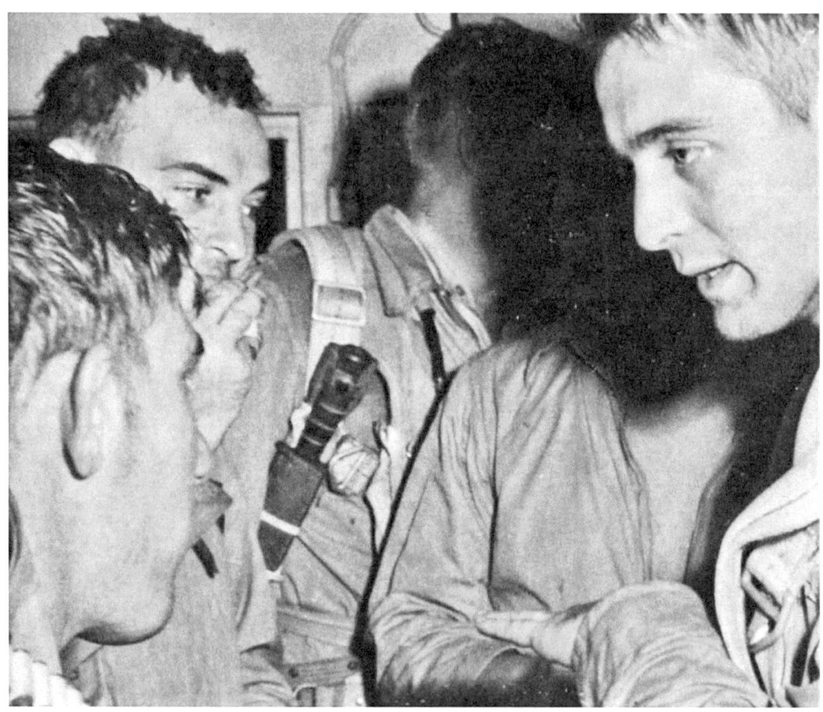

USAAF personnel did not carry hunting knives as frequently as BuAer personnel, but they were eventually listed in the allowances of the Tables of Organization & Equipment for fighter and bombardment squadrons. It is practical, however, to trace the history of their adoption. In spring 1933 a directive mandated evaluation of the requirements of an emergency kit for cross-country flights. In March of the following year a drawing for a parachute emergency kit was released titled Case, Pilots Emergency Kit, as detailed in Dwg. No. 34G3416. It was to contain rations, compass, matches, canteen, flashlight and a hunting knife. Assembly of 300 kits began, and the list of components included Knife, Hunting in leather sheath, 5" blade, Marbles #45 or equal (Procure Locally). These kits were fabricated and assembled at the Fairfield Air Depot, which purchased the appropriate knives for inclusion. The unit price for the Marbles knife was $1.10. Another listing for the emergency kit details Knives Hunting, 5", in leather sheath. While terminology changed, the term hunting knife was consistently used.

The knife was not considered for individual issue at the time; it was a component of a kit for use only in an emergency. The requirements for this type of equipment during the 1930s were low but development initiatives progressed. Coinciding with the redesigning of a suitable emergency jungle kit, bombardment groups expressed interest in an emergency kit for cross-country flights. They wanted something small that could be stored near crew positions. To satisfy the request, the Materiel Division developed a compact safety kit designated as the Case-Airplane Emergency, referenced under Dwg. No. 39G2156. It contained the same items as the parachute seat kit 34G3416 but specified a five-inch hunting knife.

In 1940 this unit was revised and redesignated as the Kit-Flyers Emergency Sustenance Type F-1, retaining the five-inch knife. It was to be issued to each crew member engaged in cross-country flying, making the knife an individually issued item but as part of a complete kit. This approach remained until the US entry into WWII.

On another front, in 1940 development commenced to assemble what was referred to as a Wilderness Kit for units operating in Alaska; design concepts progressed through 1941. Ladd Field had requested a variety of supplemental aids and tools and one item was a hunting knife. With US entry into war, details were finalized and sustenance kits were standardized as Types E-1 and E-2. The E-1 kit was the ration component that included two hunting knives, detailed by specification as Hunting Knife, approximately five inches long, complete with sheath (Marbles Mfg. Co. or equal).

At this time, operational requirements were significantly shifting as operations in the SWPA and in Asia were expanding. This greatly increasing the demand for emergency survival equipment for the vast isolated rugged terrain encountered. By the close of 1942 the Type E-5 kit was approved and quantity contracts were negotiated. It too was to contain a single hunting knife, specified as "Knife, Hunting, 4 or 5-inch blade length."

The E-5 was the last of the multi-person sustenance kits to contain a hunting knife as a standard complement. The E-12, adopted in 1943, technically included a hunting knife but it was a standard component of the E-1 that was, in turn, included in the completed assembly of the E-12.

Experiences indicated the hunting knife was better issued on an individual basis. This required significantly more knife production. They could be requisitioned as Knife, Hunting under stock number 7900-472530. Further development of sustenance kits would contain large chopping implements such as the machete and hand axe, although some would include a pocketknife.

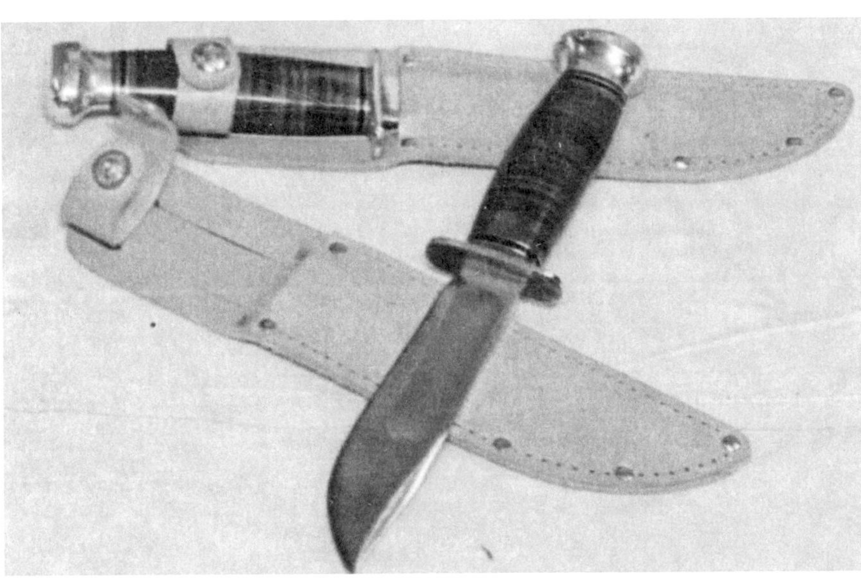

(Left) Wartime image of two Marbles Ideal #45 hunting knives in display of a USAAF sustenance kit. The BuAer established a policy for the issuance of a hunting knife whereas the USAAF issued knives upon only request. (USAF)

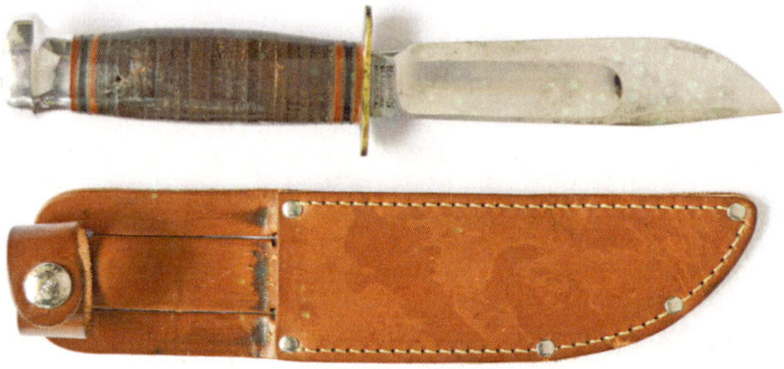

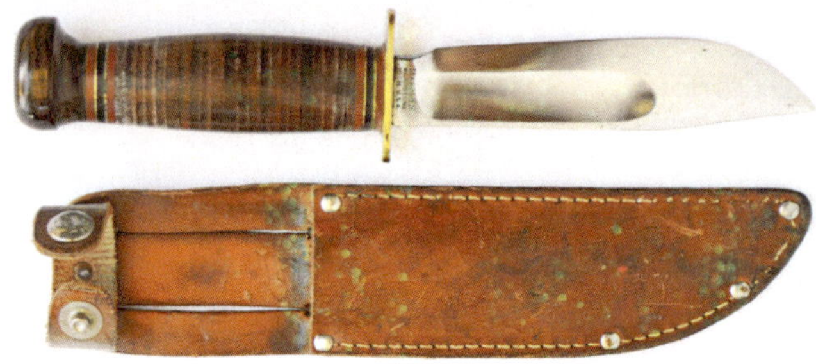

The Marbles Ideal #45 was often cited as the standard. (Above left) The original pattern included a polished cast zinc pommel. (Above) In 1943 production transitioned to a pommel made of Resinox plastic in brown or black. (Right) Resinox pommel. (Left) Modified sheath for a Marbles that includes an M-1910 belt hook.

(Below) Marbles knives worn by USAAF personnel. (Left) Commanding officer of 364th Service Group Lt. Col. H.M. Cox looks over a map while stationed at Ie Shima, Ryukyu Retto, April 1945. He wears standard khaki chino clothing with the knife secured to his trouser belt. (Right) Crew Chief Sgt. John Booth poses while carrying a Marbles as a utility tool for his duties as a ground mechanic. (National Archives)

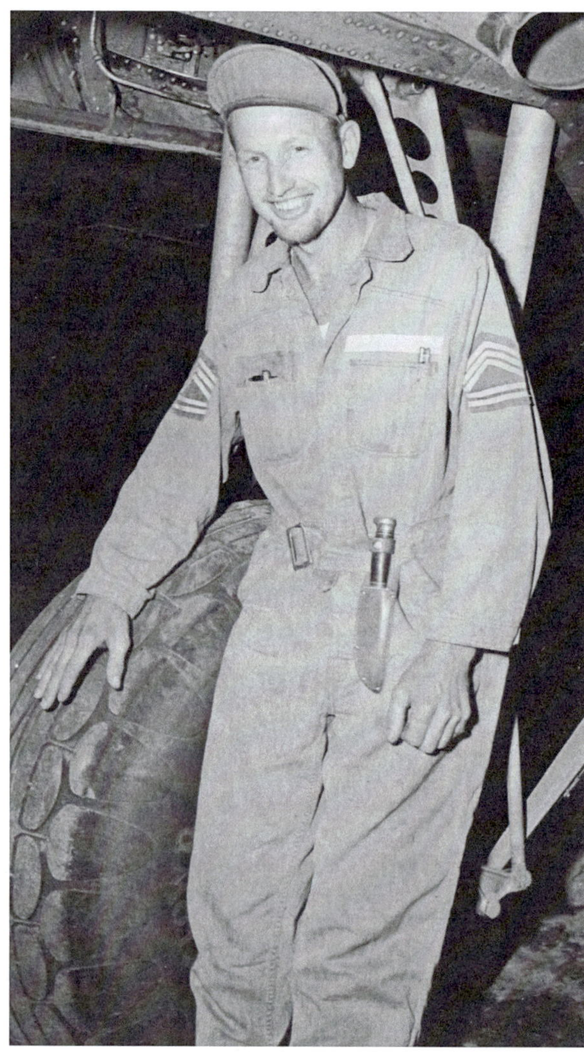

(Above) USAAF personnel carrying Marbles knives. (Left) Airborne engineer at Myitkyina Airfield, Burma, May 1944. (Right) Col. Robert Morrissey has his Marbles attached to M3 shoulder holster strap, New Guinea, May 1944. (National Archives)

USAAF emergency kits were stored aboard the aircraft or attached to the parachute. In the early stages of parachute development, a small arm was included in the kit. This was later revised for the small arm to be issued to the individual when operating in remote regions. Personnel arriving in the theater of operations, particularly in Papua/New Guinea, India-Burma and China, in the initial stages of the war, carried only a side arm. It wasn't long before a cutting tool was found to be necessary. Personal knives were also needed in the European theater.

Survival, evasion and escape aids were necessary in all theaters. Emergency kits stored aboard an airplane, in which hunting knives were a component, would not be accessible in a bailout. Crews would not stay with a distressed aircraft, meaning an individual needed to carry all necessary equipment on their person. When bailing out over dense vegetation, tree canopies and water, an individual would often need to cut himself free .

Circumstances dictated that the hunting knife no longer be carried in multi-person sustenance kits but be readily available from stock to carry on the person. The Personal Equipment Officer was responsible for distribution to all those requesting knives and other equipment for individuals.

US personnel operating in Europe carried aids boxes, maps, and currency; the hunting knife was a helpful supplement. MIS-X Detachment, ETO made notice in a 1943 E&E bulletin for briefing of intelligence officers and aircrew that USAAF personnel were in the habit of carrying large size knives on operations. They instructed that upon landing in enemy or enemy occupied territory , an evader should immediately get rid of the knife. If such a knife were found on an evader, it could be taken for a weapon and could easily involve him , if captured, in a very serious charge of carrying arms while attempting to escape in civilian clothes.

Marbles was cited by specifications, however, they only required the hunting knife to be equal to that pattern. Kinfolks, Inc. also produced knives procured by the USAAC. Their five-inch hunting knife was referenced as Model No. 925. The first generation included an aluminum guard, polished zinc cast pommel, brass spacers in the handle, and Kinfolks, Inc. stamped on the ricasso. After 1940, the ricasso stamp was changed to Kinfolks USA. With wartime WPB restrictions, the aluminum guard was revised to be of acetate but retained the cast zinc pommel. Another variant had both an acetate guard and pommel. All the pommels were secured by a locking nut. The leather sheath was the same for all.

(Left) three of the four Kinfolks models referenced as Model No. 925, all with a 5" fullered polished blade. The first generation included an aluminum guard, polished zinc cast pommel, brass spacers in the handle, and Kinfolks, Inc. stamped on the ricasso. After 1940, the ricasso stamp was changed to Kinfolks USA. A variant not pictured lacked brass spacers. (Middle) Transitional with application of acetate guard, zinc cast pommel. (Bottom) Last generation with acetate guard and pommel.

(Right and below) USAAF member undergoes jungle survival training in Florida. He wears last variant of Kinfolks that includes acetate guard and pommel. (Below) B-17 aircrew member of the 401st BS has a first generation Kinfolks hunting knife attached to the waist belt of his A-4 flight suit, England, March 1943. (Below right) B-17 waist gunner S/Sgt. William Fleming carries a Kinfolk with acetate guard and zinc cast pommel, England, July 1943. (National Archives)

Five-inch hunting knives were procured by the USAAF from Western States Cutlery, referenced as G46-5. The type purchased on government contract had the features of a five-inch fullered blade, molded Tenite guard, and pommel in a marbled pattern. The distinctive feature was the pommel was molded onto the tang. Other types manufactured by Western attached the pommel with two pins through the tang and could also include a lanyard hole. Initial contracts were executed in fiscal year 1942.

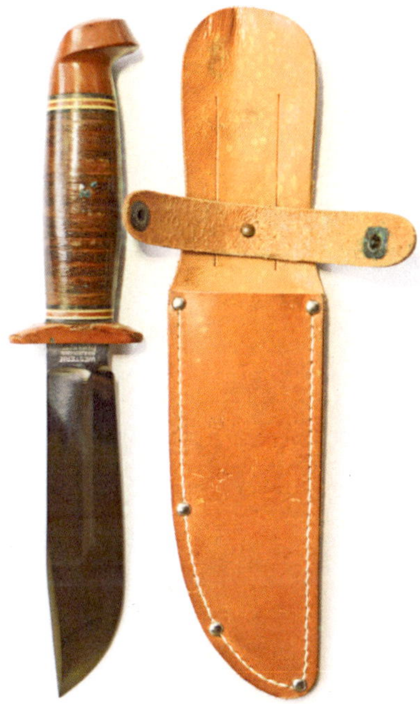

(Left and below) Western G46-5 as procured by the USAAF. One is included in a locally assembled arctic emergency kit by the ATC, Pacific Theater. Additionally includes Marbles Game Getter, c.1944.

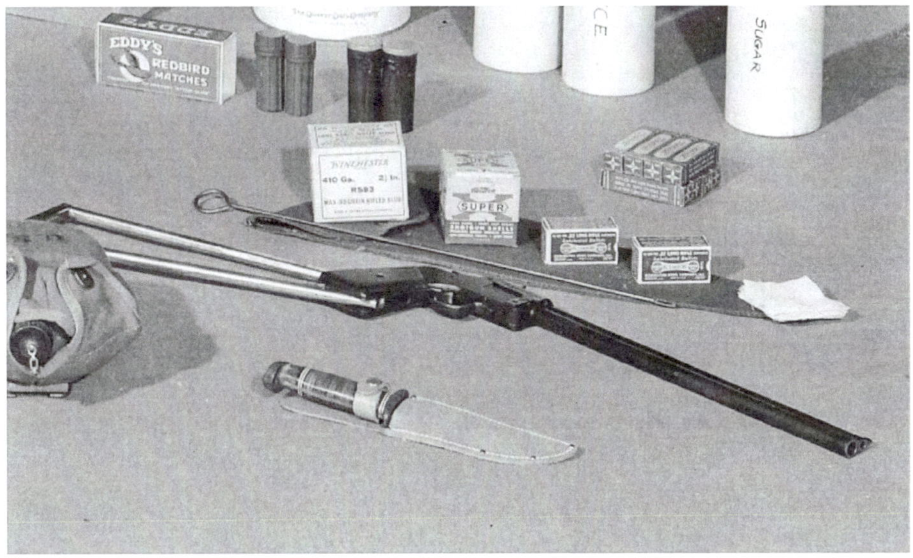

(Below) Lt. Col. Joel Pitts of the 375th Troop Carrier Group, New Guinea, January 1944. (Right) Unidentified pilot of the 90th Photographic Reconnaissance Wing, North Africa, c.1944. Each utilize Western G46-5. (National Archives)

Case & Sons also supplied knives to the USAAF, referenced under its pattern 322-5. Like Western, Case contracts were awarded in fiscal year 1942. These early procurements were of limited quantities of 5,000-6,000, perhaps just for stock purposes. However, the F-1 sustenance was still in production and distribution. Further purchases would be made for unknown quantities. Typical procurement was executed by Purchase Orders through regional Purchasing Offices located at Sub-Depots. Knives requisitioned would be based on requirements to supply units preparing for overseas embarkation to the Forward Area.

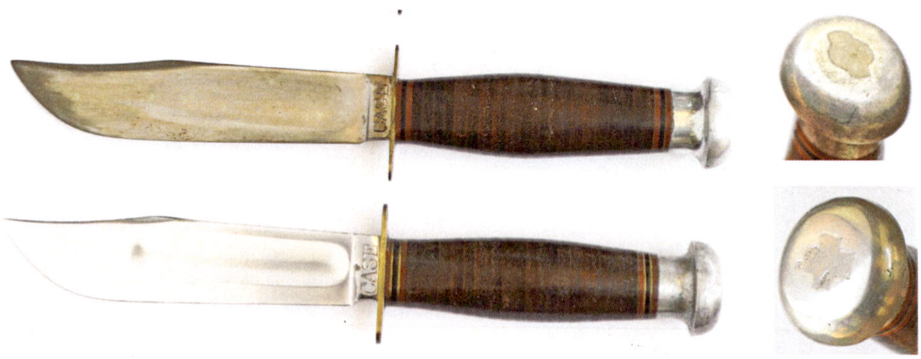

(Above) Two principle variants of the Case 322-5 and lead pommel filler. There were five variations of Case pattern 322-5. Four had a fullered blade, brass guard, and zinc cast pommel. One had brass spacers and another did not. Other variations include the shape of the lead filler in the pommel, glazed or bright finishes on the blade. Fifth variant was provided with a nickel guard.

(Right) Typical complement of a leather sheath for Case hunting knives included a central pointed tip. (Right) A type to also had been provided by Case.

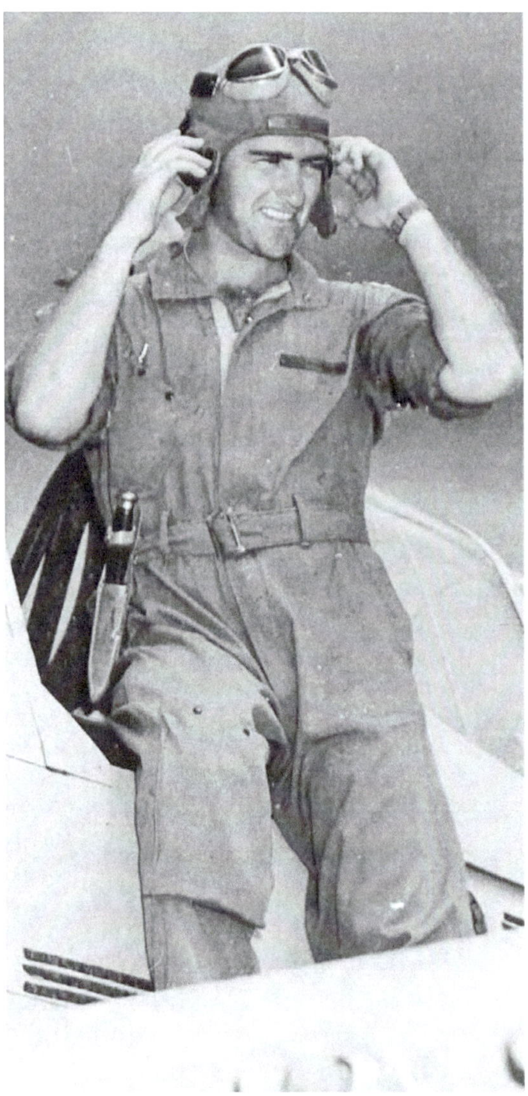

(Left) P-40 Capt. Harry Thompson of the 45th FS, first to arrive on the newly established Baker Island airstrip, Hawaii, September 1943. Wears Case type pattern 322-5. (National Archives)

(Below) USAAF personnel undergo life raft orientation in Hawaii, April 1942. Wears Case pattern 322-5. (National Archives)

The E-1 sustenance kit required two hunting knives, and 1,000 kits were placed on procurement in July 1942. It wasn't until December 1942 that a new contract was negotiated for the procurement of 16,000 more E-1 kits and 25,000 E-5s. These two contracts would require 57,000 five-inch hunting knives and, as mentioned previously, these were the only two standardized kits that included a hunting knife. It was also at this time that a new service requirement arose to issue hunting knives as personal equipment, greatly increasing the procurement schedules.

Camillus Cutlery Co. provides the mostly accurate records on the five-inch knives provided to the USAAF. These indicate Camillus provided 75,000 beginning in 1942, and they indicate that with four other contractors, a total of 332,000 five-inch hunting knives were delivered. However, records do not mention Case as another contractor; rather, they cite Cattaraugus. However, Case was known to have had received USAAF contracts or purchases. The Camillus knife was designated as 5679 L36, similar in style to the Marbles model. Camillus charged $1.19 apiece.

Three variations were provided by Camillus. Two had a fullered blade, brass guard, zinc pommel, only differing in the ricasso stamp. The first generation had a straight marking; the next had an oval stamp. In early 1943, Camillus began to fabricate a type with a steel guard and Resinox pommel, designated as L36A. The leather sheaths provided for these knives, made by Mosser Leather, were distinct, with 11 rivets.

Knives for sustenance kits were detailed as Contractor Furnished Equipment. As a consequence of a negotiated contract between the USAAF and Liberty Electric Co. for 16,000 E-1 sustenance kits, 32,000 hunting knives were required. Liberty negotiated with Camillus for the supply. The pattern of the knife ordered was new and was referenced as model 5666 L66. It had a five-inch flat ground polished blade, steel guard, and rosewood or bone handles. It was provided with the same 11-rivet leather sheath by Mosser. The order was for 35,500; half with rosewood handles and half with jigged bone. When Liberty negotiated the contract for the E-1, it also negotiated one for 12,500 knives for the E-5 sustenance kit. Because a single hunting knife was required for those kits, the remaining 3,500 model 5666 L66 knives procured from Camillus were likely packed in these units. Limited quantities of this pattern were also procured by the USMC and US Army Corps of Engineers.

(Left and below) Camillus 5679 L36 hunting knife with Mosser sheath. Strait line type first generation (top). Second generation (middle), third generation (bottom).

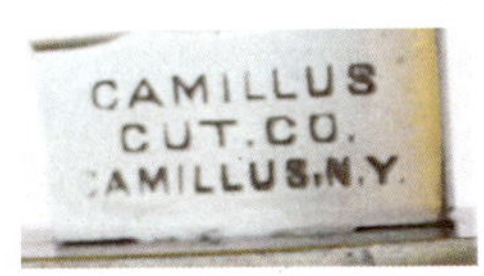

(Right) C-47 aircrew members Lt. Conrad Rowland and 2nd Lt. Joseph McWilliams of the 6th Troop Carrier Squadron wear hunting knives with leather sheaths similar to those delivered with the Camillus 5679 L36, New Guinea, c.1943. The distinctive rivet pattern can be observed near tip of leather sheath.

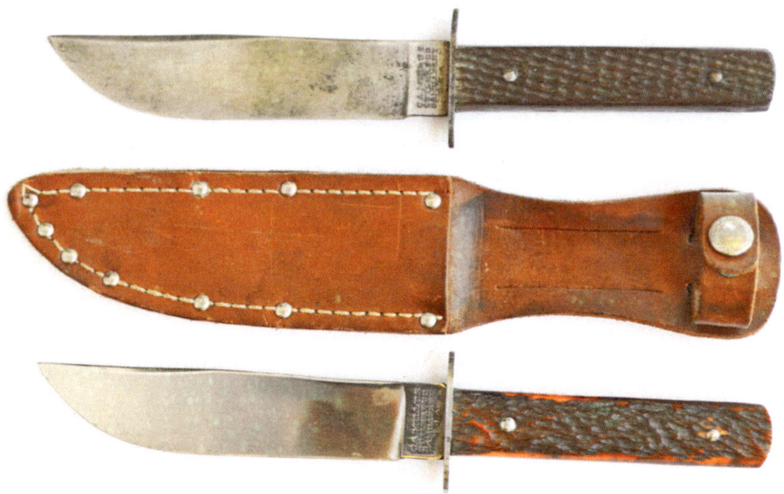

(Left and below) Camillus 5666 L66 hunting knife with Mosser sheath. (Top) Rosewood handle. (Bottom) stag handle. Ricasso stamped with strait line marking only. Steel pins used to secure handle scales.

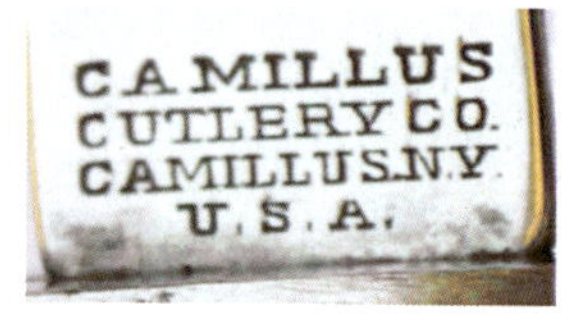

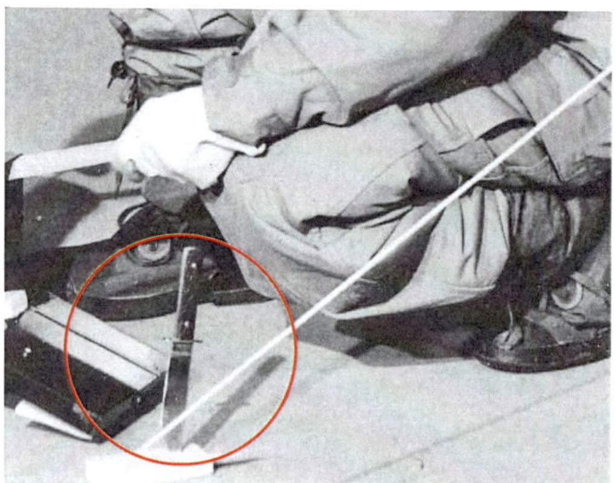

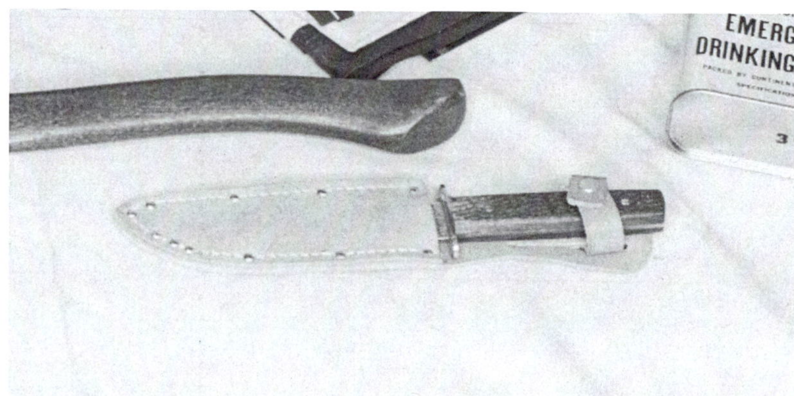

(Above) Camillus 5666 L66 hunting knife without sheath, E-12A sustenance kit display, c.1944. (Right) Same Camillus pattern displayed with components of E-5 sustenance kit, c.1945. (National Archives)

(Below) USAAF personnel undergo simulated survival training in life raft, Hawaii, August 1944. One of the items was an E-5 kit; this individual has 5666 L66 secured to flight suit. (Right) Camillus 5666 L66 hunting knife with rosewood handle. Below knife is an E-1 kit as part of assembly E-12, c.1944. (National Archives)

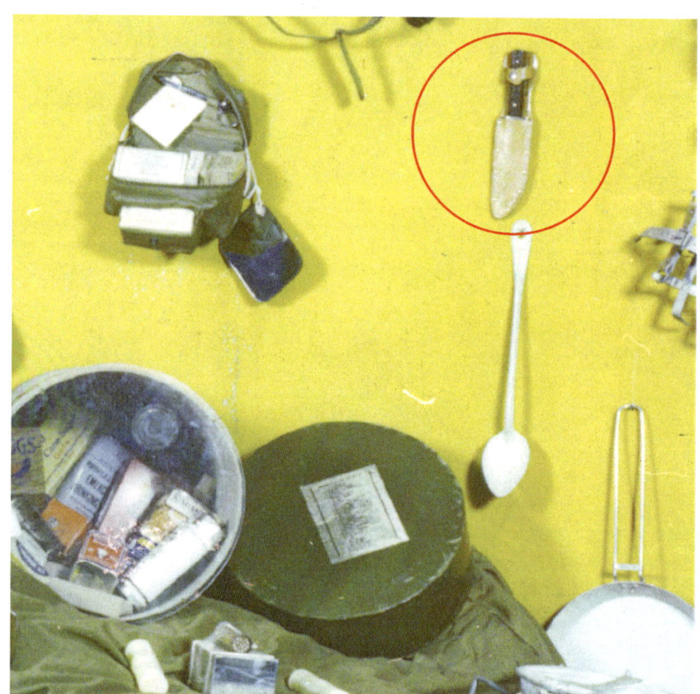

(Above) P-38 pilot Lt. Paul Gaughenbaugh, in whose shoulder holster strap is a Camillus L36.

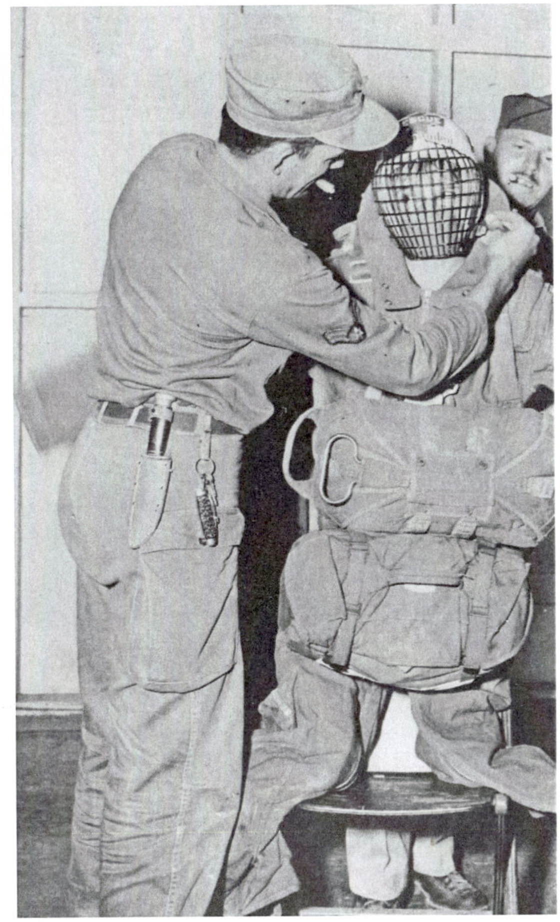

(Above) A five-inch hunting knife remained in USAAF inventory through its reformation as the United States Air Force to 1955. In the 1955 movie Strategic Air Command, starring James Stewart, he and much of the crew wear the Camillus 5679 L36 and 5666 L66. It was filmed with the cooperation and technical assistance of the USAF, so it is likely all props were provided by it. In this image, two USAF members dress a young orphaned boy in pararescue suit. The USAF member adjusting the helmet wears a Camillus L36, Charles Air Force Base, August 1954.

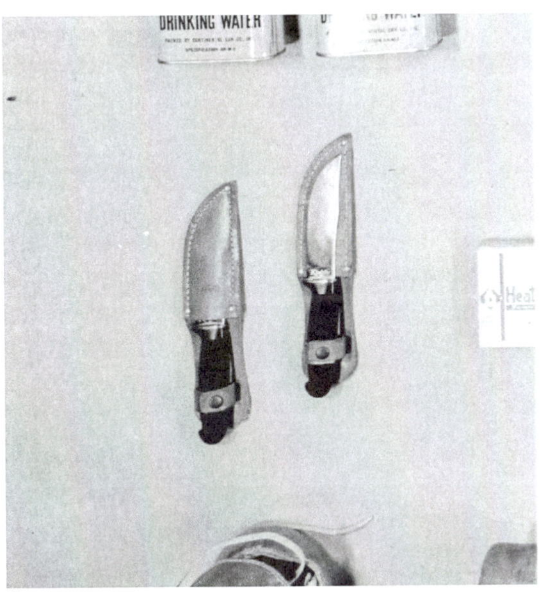

(Left and below) "(Left and below) For assembly of 1,000 E-1 kits awarded July 1942, 2,000 hunting knives were required. Early E-1s included a four-inch knife by Kinfolks. It had a flat ground polished blade and molded rubber hand. Ricasso stamp is that of the early Kinfolks, Inc. type.

(Above) B-17 aircrew member of the 359th BS, England, October 1943. Hunting knife secured to flight suit waist strap. (National Archives)

(Right) B-17 aircrew members of the 322nd BS, England, April 1943. Two aircrew wear hunting knives similar in appearance to Case. (National Archives)

War records indicate limited procurements of hunting knives for the USAAF and the US Army until 1942, when a fighting knife became required for troops in all military branches in overseas theaters. This created an initial shortage. Each branch approached the deficiency in its own way.

The USMC adopted the 1219C2, the Navy Department adopted the Mark 1 and Mark 2, the US Army adopted the M3 trench knife, and the BuAer established an authorized allowance for aviation units. Though the USAAF eventually had significant quantities of knives on procurement, sufficient quantities were not stocked until 1943. At that time, the Tables of Organization and Equipment for USAAF fighter and bombardment squadrons detailed the equipment issued to these units in preparation for overseas movement. Procurement schedules assured sufficient quantities of all the items and authorized allowances first appeared in revised T/O&Es by the summer of 1943. As a branch of the US Army, the USAAF procurements were made by the Army's technical services and entered in supply through the Army Service Forces. These materials were first handled by Air Service Command central depots, then distributed to the vast network of sub-depots. The USAAF maintained in stock the five-inch hunting knife, but those knives authorized for allowance were under authority of the Army Ordnance Department. Listed in the revised T/O&Es was Knife, Trench M3 with scabbard M6, later M8. Quantities allowed were based on the size of the unit. Issuance was to be one per individual outside the continental United States when authorized by the Theater of Operations Commander.

The M3 trench knife was adopted in January 1943 and was to be issued to any soldier not equipped with a bayonet, to meet the requirement of a fighting knife. Production through 1943 was too slow to equip all the fighting forces. Eventual production reached over 2.5 million units from nine different contractors. Initially, units preparing for embarkation from the USA were issued knives first; eventually, stock was shipped to the forward areas. Through 1944 and into 1945 the M3 knife transitioned to standard with USAAF personnel.

(Above) Example of M3 Trench Knife with M6 scabbard, prescribed for issuance per USAAF T/O&E. S/Sgt. Charles Bacon of the 40th BG wears M3 with M6 scabbard while being debriefed at an airfield in China, c.1944. (National Archives)

(Left) USAAF Maj. Lonigan draws battle plan in dirt with M3 trench knife at airfield on New Guinea, November 1944. (Courtesy Australian War Museum)

(Right) Lt. Robert DuFresne of the 79th FS, England, c. 1944. M3 trench knife with M8 scabbard secured to right lower leg. (National Archives)

(Left) USAAF 7th AF P-47 pilot stirs meal with M3, Saipan, July 1944.

At first, the M3 was provided with an M6 scabbard, then with an M8 scabbard. The M8 is a plastic scabbard provided with a heavy canvas belt loop. Production of it commenced in July 1943. By 1945 the M3 was the knife predominantly seen in USAAF aviation activities.

In a revised T/O&E dated June 1944, the allowance was changed to Bayonet, Knife M4 with scabbard M8A1. This is the earliest known date for this change; by the end of the year all T/O&Es were amended to reflect the M4. Distribution remained at one per flight crew member. The M4 was a revised M3 to a bayonet for the M1 carbine. The M8 scabbard was revised slightly by extending the belt loop and adding an M-1910 belt hook.

(Above) M3 with M8 scabbard worn by B-29 aircrew member stationed in India, c.1945.
(Above left) M3 trench knife with M8 scabbard. (National Archives)

(Below) B-29 officers on each end wear M4 bayonet with M8A1 scabbard, Tinian, April 1945.
(Below left) M4 bayonet and M8A1 scabbard. (National Archives)

(Above) B-24 copilot of the 394th BS Lt. E. Smith wears a Cattaraugus 225Q QMC knife. (National Archives)

(Above) 1st Lt. Irwin Steiner salvages parts from a wrecked B-25, Burma, c.1945. His hunting knife pommel is similar in characteristics to that of Western types. (National Archives)

(Below) B-24 aircrew of the 64th BS, New Guinea, c.1944. (Left) unidentified aircrew wears small hunting knife. (Right) Unidentified aircrew wears Marbles hunting knife. Attached to M-1916 holster are two Marbles match safes. Adjacent to holster is pocket knife and pocket watch or compass. (National Archives)

When Albrook Field in Panama commented on the design of a new pilot's emergency kit, it indicated that a machete was the most important item. Albrook's concept of a machete was based on standard QMC types; Albrook staff noted multiple times the Engineer Corps bolo, Collins No. 1005, and No. 127 styles.

Bolo type machetes typically had 15 to 17-inch blades, but with the added length of a handle, weren't suitable for carrying on a parachute. At the beginning of August 1934, the Materiel Division requested Collins Co. develop a machete type knife for an emergency jungle landing kit, which was to be a component. To be suitably carried with the pilot's parachute, the overall length could not be more than 15 inches.

Until a machete could be adopted, literature describing the contents of the new kit used the terms 15-inch bolo and, in several cases, 15-inch machete with a detachable handle. The Engineering Section requested a sample of Collins' model No. 17, leather handle, ten-inch blade machete for examination. Collins quickly submitted a sample and stated it could be furnished with a different type of handle at a lower cost. The quoted price for the leather-handled No. 17 was $20.00 per dozen. On 29 August 1934, the USAAC acknowledged receipt of the sample and, after examination, found it to be satisfactory except the handle tended to slip out of the hand due to its straight round design. Other potential suitable handle designs from the catalog were noted but inclusion of a handle guard and a means of attaching a rawhide wrist loop were emphasized.

On 7 September, Collins mailed a sample of the No. 17, redesignated as the No. 18 machete. It had an overall length of 15 inches and a horn handle. The horn material was chosen because it was impractical to bore a hole in stacked leather. Collins also supplied what it referred to as a plain No. 17 sheath. The price quoted for the new knife remained at $20.00 per dozen. The sheaths were offered separately at $6.00 per dozen, and the No. 10 sheath was also quoted at $11.00 per dozen.

A week later, the Engineering Division notified Collins that this new design was satisfactory with modifications. The horn handle provided by Collins was the No. 216; the USAAC requested the handle pommel be ground off, reducing the length about a half-inch for better stowing in the container. The USAAC also requested the handle guard not be over 3-3/8 inches as provided on the previous model. These changes were detailed along with the complete assembly of the machete in USAAC Dwg. No. 35B1647. With the approved drawing, Collins agreed to its manufacture and that it be ordered as #18 Machete, short pattern, brass guard, #216 horn handle with end rounded. The pattern was adopted and 300 jungle landing kits were authorized for assembly in August 1935.

Most of the other kit components had been purchased months prior, so with the machetes ordered, the 300 kits only needed to be assembled. In December 1935, 180 additional machetes were shipped to Panama and 102 were shipped to the Philippines. If any more were required, it was directed that they be Locally Procured. The USAAC purchased the 35B1647 at a cost of $1.48 each.

The Collins No. 17 dates to at least the American Civil War. It was often designated as Bowie No.17 in catalogs. It had a round corrugated leather washer handle and a ten-inch blade. No. 17s and No. 18s were provided in various lengths of blades from 12 to 16 inches. Both models sometimes had ornate handle designs made of bronze. Each model was provided with two fullers parallel to the spine. Purportedly, the No. 17 designation had been dropped in 1926. However, literature through the 1930s still reflected its availability. In correspondence between the USAAC and Collins, the No. 17 was specified by both entities. Collins eventually clarified that the model was to be referenced as the No. 18. It is not clear what a plain No. 17 sheath was; possibly Collins meant to reference the No. 13. Another common sheath provided by Collins was the No. 10, identified as having a frog for the belt loop and a nickel-plated tip. Both the No. 13 and No. 10 had similar tooled leather features on the face of sheath.

After negotiating machete features, the USAAC adopted the design referenced under Assembly Dwg. No. 35B1647. The new No. 18 had a nine-inch blade, red brass 4-3/8 inch guard, and green horn handle with rounded end and drilled lanyard hole. No. 216 handle was secured with five brass pins. It included what is referred to as a Sword Stamp on the ricasso. Collins & Co. was arched over the straight-line Hartford/No. 18. The Sword Stamp was used on military purchases dating back to the Civil War. Due to limited requirements, only 600 of these machetes were purchased for jungle landing kit assemblies.

Initial stocks of the 35B1647 were depleted rapidly; Wright Field alone purchased projected requirements. Technical Order 13-5-8 dated 30 May 1936 directed that additional machetes be purchased locally, if required. Its nomenclature by technical order was Machete, Collins Co. No. 18 or equal, short pattern, brass guard, with No. 216 handle. Rounded end had been removed from nomenclature. Collins had by then marketed this new configuration commercially.

(Left and above) Initial Collins type machete presented to the USAAC for evaluation; No.17 with round leather handle and No.10 leather sheath. (Above) Domestic Spanish logo for export to Spanish speaking countries.

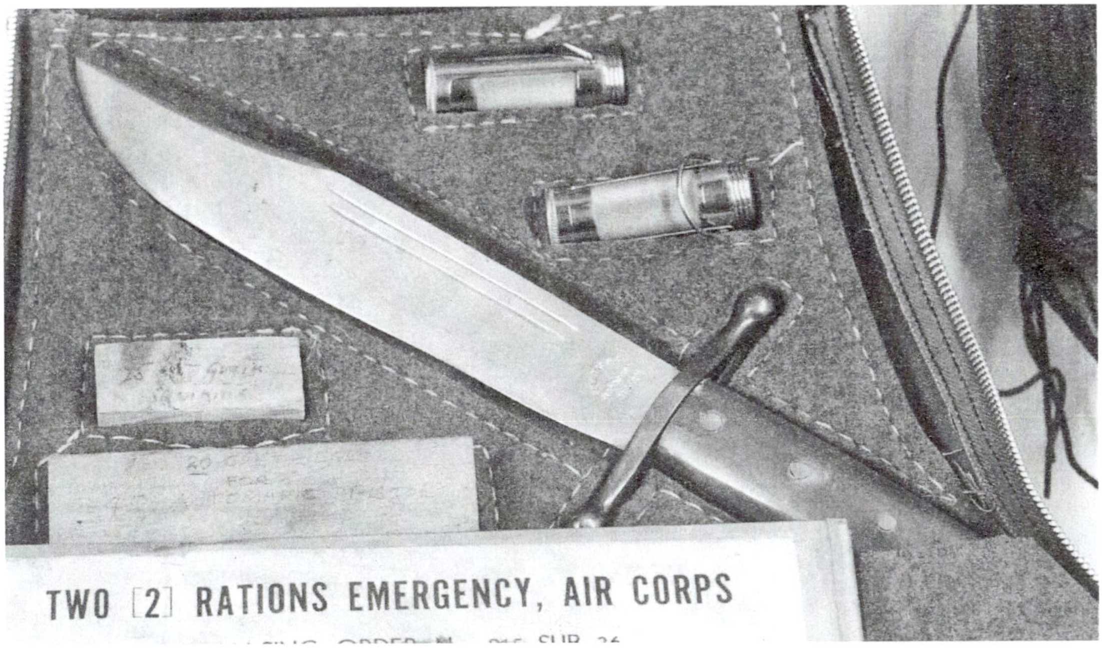

(Left) Experimental parachute back-pad kit, c.1936. Kit includes 35B1647 No.18 machete with sword stamp. Blade is same as provided on the No.17. Note narrow width of blade.

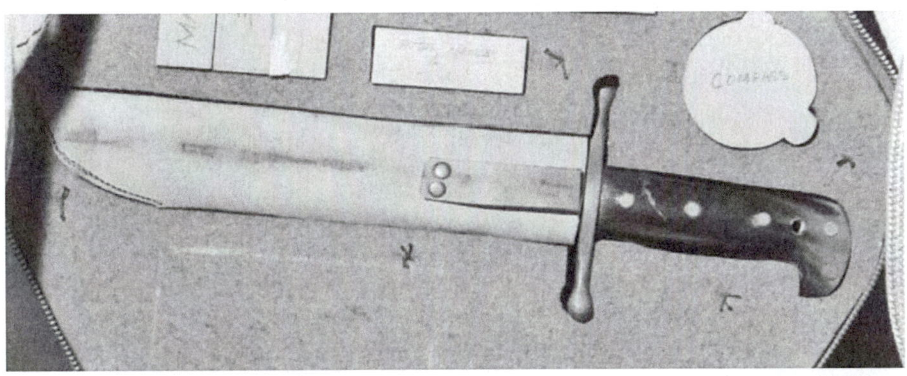

(Right) 35B1647 included in both standard and experimental 38J5217 parachute emergency kit. Each includes modified No.10 leather sheath. These early production kits were cut for a blade width of two inches,standard for No.17 and No.18 production. Cut increased to 2-3/8 inches in the later procurements to accommodate commercial No.18.

In 1937, Wright Field was developing a new superseding jungle back-pad kit referenced under Dwg. No. 38J5217, which was in production by the close of 1939. The 35B1647 machete became a standard component. Assembly of these new jungle kits occurred at the Middletown Air Depot. Any additional purchases of the machete were shipped from Collins to this location for assembly. Jungle landing kits still in stock were being dissembled and the machete provided for the new parachute kits.

At the time of the adoption of the back-type jungle kit, it had already been concluded that the Collins bowie was not suitable for jungle use as a machete. In response, proposals for a new design were being contemplated. The new direction would greatly affect any further procurements of the No. 18 with the intention of making it obsolete. In 1941, when the Middletown Air Dept. requested more bowie knives, it was denied, and directed instead to provide a five-inch hunting knife as an interim. The premise was that a new jungle type kit and new type of machete were in initial development to supersede the 38J5217.

However, when the United Sates entered WWII, the new jungle kit was still in its experimental stage. To satisfy immediate requirements, an emergency procurement of the former pattern of jungle kit was authorized, which included the retired bowie type machete until the new machete became available.

It was at this time the bowie type machete was designated as Machete, nine-Inch with Leather Sheath, assigned stock number 7900-486870, and entered in supply for requisition and replacements. Prior to WWII, administrative infrastructure was centralized; with the requirements of war, decentralization became necessary. Assembly of jungle kits was diversified to regional depots at embarkation points where the machetes would be requisitioned for assembly of parachute emergency kits.

By the spring of 1942 the Type B-2 jungle kit was adopted as standard; this kit included the new ten-inch folding machete per Dwg. No. 42D7462. However, production of this machete was delayed. By August 1942, quantities of B-2 kits were ready for shipment but still lacked the appropriate machete. It was then directed that all jungle kits, B-2 and 38J5217, be equipped with the nine-inch machete until the 42D7462 became available. At this time 28,000 nine-inch machetes were in stock from the emergency procurement of December 1941 to supplement.

Between the time the USAAC began procurement of the Collins No. 18 machete and US entry in WWII, the ricasso marking changed. Early specimens included the Sword Stamp. Manufacturing by Collins transitioned into Collins & Co. arched over the logo and straight lined Legitimus/Made in USA/No. 18. The logo is an arm holding a hammer cradled in a crown. For the emergency procurement, Collins provided the No. 18 with a green horn No. 216 handle, yellow brass guard, brass pins and domestic stamp. As supplies of horn and other critical materials dwindled, Collins transitioned to fabricating the machetes with phenolic plastic handles and steel pins. Both were delivered with the No. 13 sheath.

(Below) Collins marketed the No.18, as designed by the USAAC, for commercial sales. The sheath provided was that of the No.13 tooled leather type, which remained the standard. Hand made blades evolved to wider width. This example is of the type provided on local purchases from USAAC installations. Standard green horn No.216 handle and sword stamp.

(Left and above) Collins No.18 with domestic branding crown and hammer. This branding was typical on machetes provided for emergency procurement in 1941. This example has USAAF serviceable tag.

(Below) In jungle environments, especially in the early campaigns in the Solomons and New Guinea, survivors found coconut milk to be a valuable resource. The machete provided in back-pad kits was well suited for coconut splitting. USAAF personnel enjoy a quick meal after harvesting the meat from coconuts, Hawaii, c.1943. The individual in center is using a green horn handle Collins No. 18 machete; others use the standard five-inch hunting knife. (National Archives)

(Right) Transitional Collins No.18 with phenolic black plastic handle and steel pins. No.13 sheath was provided on all machetes.

(Above) Collins No.18 machete with green horn handle and No.13 sheath in display with experimental B-2 jungle emergency kit, c. 1941. (National Archives)

(Right) Capt. William Wells of 6th Troop Carrier Squadron wears a Collins No. 18 machete over left hip, New Guinea, April 1943. (National Archives)

(Below) B-24 aircrew, New Guinea, c.1943. Handle of Collins No. 18 machete over left hip, second from left. Individual middle, has hunting knife with M-1910 belt hook modification. Far right, individual wears money type belt with survival aids. Most aircrew wear locally procured shoulder holsters. (National Archives)

(Above) USAAF personnel undergoing jungle survival training; two members are using nine-inch machetes. Stamped side of the blade faces the camera, c.1944. Each exhibits the domestic hammer and crown marking. (National Archives)

(Below) Two aircrew of 6th Troop Carrier Squadron carry Collins No. 18 machetes, April 1943. (Right) B-17 aircrew member has a Collins No. 18 nestled behind his M-1916 hip holster, New Guinea, January 1943. (National Archives)

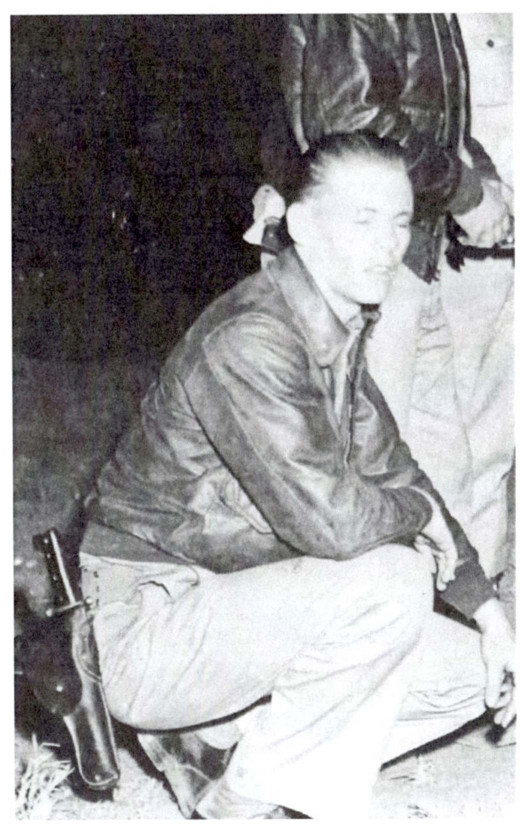

(Above left) USAAF aircrew member at a Caribbean airfield, c.1942. He holds a Collins No. 18 machete.

(Above) Sgt. John Pavesick types morning reports on Ie Shima, near Okinawa, May 1945. Hanging from the tree is a web pistol belt with Collins No. 18 machete. 38J5217 jungle kits were never removed from service, or Pavesick received his machete from stock. (National Archives)

(Left) Unidentified USAAF personnel on New Guinea, c.1943. Two individuals wear Collins No.18 machetes; one with added retaining strap over handle.

Prior to US entry into WWII, the nine-inch machete was no longer required for procurement, leaving a limited stock on hand. With the emergency procurement in December of 1941 for the 38J5217 jungle kit, these machetes were placed back into demand. Manufacturing was expedited and other contractors were invited to produce them. Case & Sons negotiated for the manufacture of the 42D7462 and, to supplement the USAAF, the company manufactured a similar machete it titled simply Machete No. 1. The appearance resembled the Collins No. 18; it was provided with a plain leather scabbard. Case's design closely resembles Dwg. 35B1647 and it is likely it used that drawing as its pattern. Two variations exist: either brass or steel rivets were used in the Resinox handle.

Nine-inch machetes were placed into stock and requisitioned when assembly was required. Requisitioning may not necessarily reflect they were for the assembly of back-pad kits but rather for individual distribution. Due to their necessity in jungle environments, PEOs or those organizations responsible for survival and evasion indoctrination could request or procure them via local purchase orders.

Like Case, Kinfolks fabricated a pattern based on characteristics of Dwg. 35B1647. Two variations exist, one of which was void of manufacturer's mark. The handle was Resinox with poured lead rivets; the machete was provided with a plain leather scabbard.

(Left and above) Type manufactured by Case & Sons. Same features as the No.18 with 4-3/8" brass guard and 9-1/2" blade having two parallel fullers. Production figures unknown.

(Left) USAAF B-24 aircrew member of the 65th BS on Dobodura airstrip, Papua, New Guinea, March 1944. Wears 9" machete similar to the case handle pattern or Australian made. (National Archives)

(Below) P-38 pilot 2nd Lt. Kenneth Sparks of the 39th FS with Case type machete. When this photograph was taken he was an Ace credited with destroying nine enemy aircraft, March 1943. (National Archives)

(Left and below) Type manufactured by Kinfolks; company pattern designation and production figures unknown.

(Right) B-24 aircrew member of the 328th BS wears machete with handle pattern similar to Kinfolks type, Libya, c. 1943. (National Archives)

(Below) Member of the 376th BG also wears machete with handle pattern similar to Kinfolks type, Libya, c.1943. (National Archives)

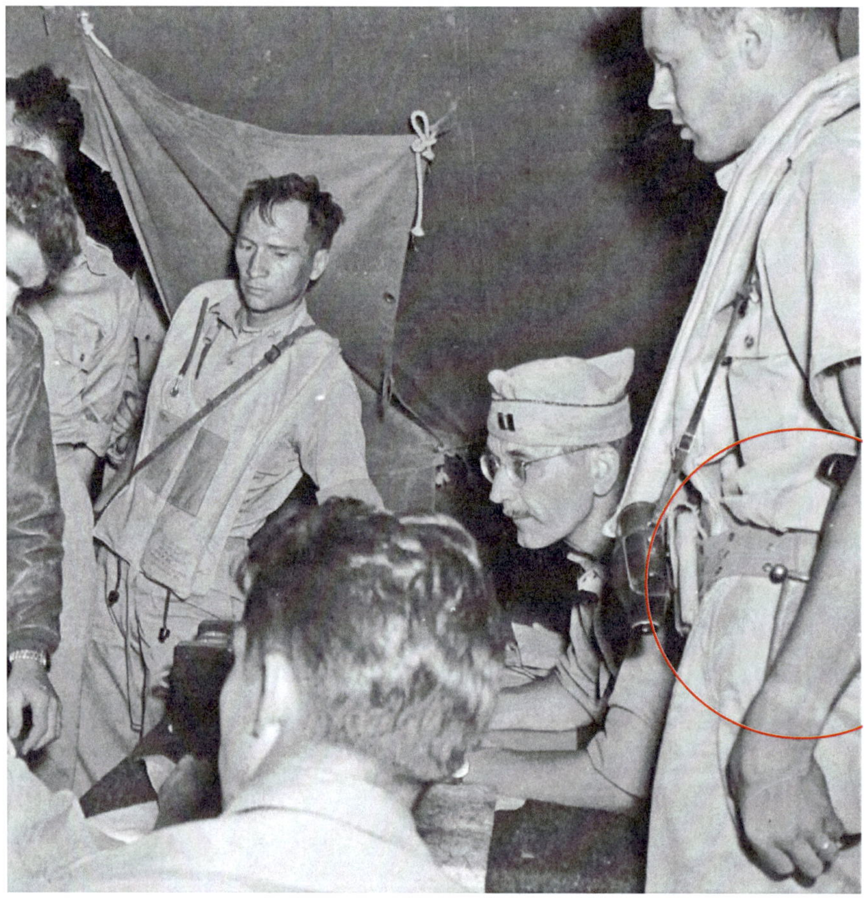

Significant assistance was reciprocated by the Australian Government, especially the Royal Australian Air Force, which worked in coordination with the USAAF throughout the SWPA. The Australians had the advantage of being familiar with the terrain, so prepared and indoctrinated their pilots and aircrew for survival, evasion and survival. Elements of the USAAF 5th AF began arriving in January 1942, soon followed by survivors of the overrun Philippines. Coordinated efforts between the USAAF, US Navy, RAAF, RNZAF and RAF halted the Japanese advance, depriving Japan of the complete overrun of Papua/New Guinea.

Companies in Australia and New Zealand manufactured knives and machetes for their respective military forces and eventually sold some to Americans. Three known companies made a nine-inch machete: Gregory Steel Products, Marsden Products, and Mars Machine Tools. All the blades were a bowie type but varied in other characteristics, including round stock brass or flat steel guards and handles made from wood, plastic, rubber, or cast alloy. Some had identifiable stampings on the blade and guard, others did not.

The RAAF seemed to operate like the BuAer in that it preferred providing supplement gear as personal equipment. The concept was to have all the equipment easily accessible in the event of any emergency. A tool such as a hunting knife or small machete was necessary in a variety of situations.

(Above) Some Australian machetes resembled the Collins, but many differed. (Right) Machete manufactured by Gregory Steel Products. Mars produced a similar type (left). These are just some samples produced in Australia. They were used by American forces, handle characteristics can closely resemble the Case pattern; making it difficult to distinguish in images.

(Left) RAAF Warrant Officer R.A. Blades utilizes 9" machete paired with two pouches of emergency provisions, c.1943.

(Below) Marsden machete worn by 5th AF aircrewman Bud Yokley. Inscribed on the scabbard is his service timeline from induction through the stations in SWPA and home.

(Above) Kittyhawk fighter pilots of the RAAF No. 76 Squadron carry a complement of nine-inch machetes and emergency jungle kits attached to their pistol or trouser belts, c.1944. (Courtesy Australian War Museum)

(Right) Flight Sgt. P.M. Sheahan of the RAAF No.74 Squadron casts a vote while stationed at Milne Bay, Papua, c. 1943. He wears an Australian 9" machete on his right hip. (Courtesy Australian War Museum)

(Below) Beaufort bomber aircrew members of No. 6 Squadron RAAF being transported from the flight line at Milne Bay, Papua, after a night attack over Rabaul, November 1943. One member carriers a Gregory 9" machete. (Courtesy Australian War Museum)

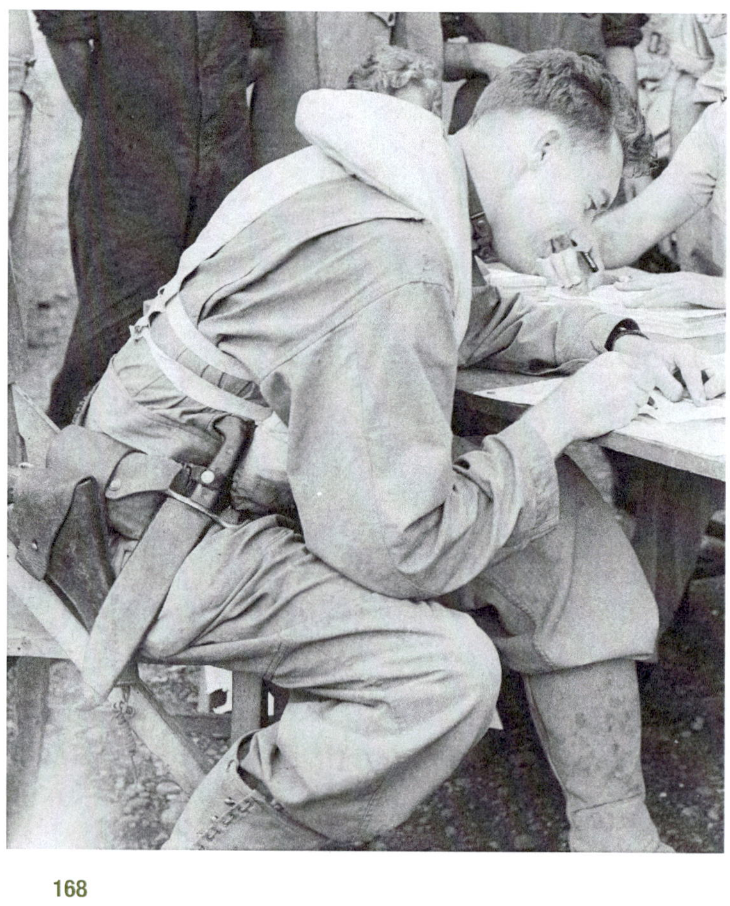

In the India-Burma Theater, the Gurkha Kukri knife proved useful as a jungle knife for airmen. In correspondence to the 10th AF, MIS-X and E-Group indicates their availability for issue and private purchase. It was stated that they were available for about 16 rupees from the Mercantile Association, Dehradun, India. This is one location; a Mercantile Association was a conglomerate of business owners with many locations.

(Above) RAF personnel examine and negotiate purchase of a kukri in Calcutta, India. (Courtesy Australian War Museum)

(Above) B-24 RAF aircrew members of No. 355 Squadron wear kukris as survival tool if forced down, India, November 1944. Crew member on left wears a Beadon Suit. (Courtesy Australian War Museum)

(Below) Lt. Robert McClung, USAAF, of the 51st FG with kukri, India, April 1943. (National Archives)

(Below) USAAF 768th BS pilot undergoing interrogation after bomb strike, India, November 1944. Kukri secured to pistol belt. (National Archives)

The Collins nine-inch No.18 machete was adopted by a decision of the Engineering Section of the Materiel Division since it met dimensional requirements. Little thought was given to its functionality. However, as early as July 1937 the command at Albrook Field had found it to be highly unsatisfactory for use in Panama; the same perspective was held by the Philippine Command. It was too light, too small, and unsafe to carry. Albrook recommended development of a larger bolo style machete; to conserve space it would have a removable blade stored in the handle.

Working with this concept, the Materiel Division drafted a drawing of a machete that met the parameters referenced under Dwg. No. 38D5613. The machete in this drawing had a bolo blade and wood handle that would attach to the blade by two thumb screws. The concept was to have a blade of 13-15 inches with a short tang. When attached at the end, the handle extended the length approximately five inches. When discussing the design with Collins, they noted the arrangement would be highly unsatisfactory for long or continued use and requested a change to the blade design. Progress on a new design extended into the spring of 1941 and the Materiel Division reached out to other prospective manufacturers for design concepts.

One of those was The Christy Co. The Materiel Division requested a machete design analogous to Christy's knife design, with a retractable sliding tubular framed handle. It did not materialize. During this time the blade shape was revised to be like a paring knife. This design was advanced by The Union Cutlery Co., which submitted a pattern of a foldable type. In lieu of a detachable handle, it was secured to the blade by a pivoting pin, and a locking spring would hold the blade in the open position. Union assigned it Part No. 3131-10. The sample included either cocobolo or rosewood wood handles and a leather sheath.

In a new design, the machete's blade shape was changed to be similar to a ten-inch linoleum knife blade with the drooped pointed tip rounded off. A requirement was also issued for a means to carry the machete. Union Cutlery designed a leather sheath, and the machete would be secured by a strap at two points with a sheet metal insert to protect the blade edge. In December 1941, Union Cutlery's design was approved, and drawings and specifications began being prepared. The Materiel Command assigned it Dwg. No. 42D7462, approved on 10 January 1942. It was named Machete, Jungle Emergency Back-Pad Kit.

Few changes were made from the specifications proposed by Union Cutlery; Dwg. 42D7462 merely formalized its design. The leather sheath was eliminated but the stamped sheet metal blade guard was retained. The handles were changed to be made of Resinox in lieu of wood. Material was added to the tip; the added weight improved chopping and digging characteristics. The primary factor behind the design changes was to increase safety and decrease possible injury. Accounts of the use of the nine-inch machete indicated there had been incidents of the machete puncturing through the back-pad kit. Although no serious injuries occurred, this design helped eliminate that possibility.

After finalizing sources of materials, Dwg. 42D7462 was released for procurement 30 March 1942. By April, companies were in negotiation for contracts. In August, it was reported that initial deliveries were not expected for another 30 to 60 days. Because of these delays, the former nine-inch machete was used to supplement those B-2 jungle kits being delivered. With the approved drawing for a new superseding jungle machete, the felt insert for the revised B-2 jungle kit was cut to accommodate the machete in its folded position.

The construction of the new ten-inch folding machete included a polished steel blade with a black sheet metal guard, black plastic handles and an integral spring steel liner. Though Union Cutlery Co. designed the new folding machete, it appears they received no contracts. This was not unheard of; there are many records of a company painstakingly engineering an item but receiving no awards for quantity production. Three companies manufactured the 42D7462 machete: Case, Camillus and Cattaraugus. The bulk of the production was executed by Camillus with a quantity of 114,000 from awards in 1942. Case and Cattaraugus produced an additional 180,000 for a total of 294,000. Production in 1942 was slow due to a delay in the delivery of the spring steel liners and locking mechanism; the first deliveries by Case were recorded in October with a quantity of fifty dozen or 600 machetes. Once the machete entered the supply system its name was revised to Knife, Machete, Folding and it was assigned stock number 7900-486880.

Technically, the need for the 42D7462 machete was for 349,000 units, matching the quantity of B-2 kits on order. However, with the standardization of the B-4 kit and the proposed cancellation of contracts for the B-2, the machetes from the disassembled kits were to supply the new superseding kit. Production of the machete was pending the outcome of that dilemma, resulting in quantities of the B-2 being delivered empty. It is possible more were ordered by assembly depots.

By the close of 1943 there was a major supply deficiency: an excess of B-2 kits without the new machete and only a limited procurement of the B-4, none of which had been delivered. To meet the machete requirements, an additional 100,000 were placed on procurement in June 1944, split between Camillus and Cattaraugus. Machetes were supplied to Companion Bags as Contractor Furnished Equipment for the order of 25,000 B-4 kits, making the requirement for the jungle kits only. The need was for 37,000, resulting in the cancelation of the remaining contracts.

Confusion followed the adoption of the B-4 emergency kit, resulting in an interdivisional debate on the status of individual emergency kits. The B-4 had replaced the B-1 and B-2 kits, but procurement had been withheld because of outstanding orders for them. A new emergency vest kit was then proposed, and the B-4 was again placed on pending status.

As part of the development of the proposed vest kit, the machete was going to be replaced by a new multipurpose folding tool better suited for integration in a vest. By the close of 1943 the vest kit was on its way to becoming standard, superseding all the previous parachute kits. This roughly coincided with the completion of B-2 deliveries, creating a need for the procurement of 100,000 machetes in June of 1944. Though the B-4 was superseded by the Type C-1 vest it was found optimal to revise it to better complement the C-1.

This new cushioned seat kit was designated Type B-5. Several configurations were assembled and submitted for evaluation and field testing. For one variant, a ten-inch folding machete was proposed. Nine samples of the C-1 vest and B-5 seat cushions underwent field testing at Eglin Field from 25 May 1944 to 2 June 1944. The result was not favorable for the seat cushion assembly, although development of it continued. Ultimately the B-5 was not standardized. For the nine samples of the B-5 field tested, the machete was packed with the poncho hood and wool socks. Following the field test, it was recommended that the machete be removed from the seat kit since a knife performing the same function was already included in the vest.

(Left) Design presented by the Union Cutlery Co. with the leather sheath. Interestingly, this image was used in the Air Sea Rescue Equipment Catalog printed in 1945 and erroneously used as a visual reference description hunting knife.

(Right) Case and Cattaraugus ten-inch folding machetes 42D7462. Open, closed and sheet metal guard.

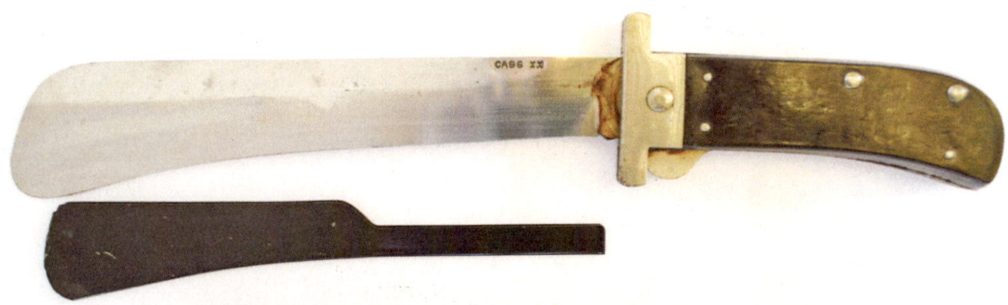

(Below) Friction fit cutouts were provided in parachute emergency kits B-2, B-3 and B-4 for machete in folded position. Additional cutouts provided for pocket compass and Scout knife.

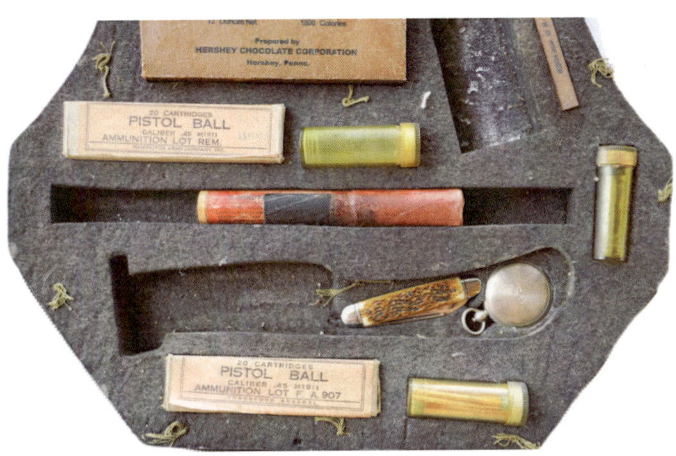

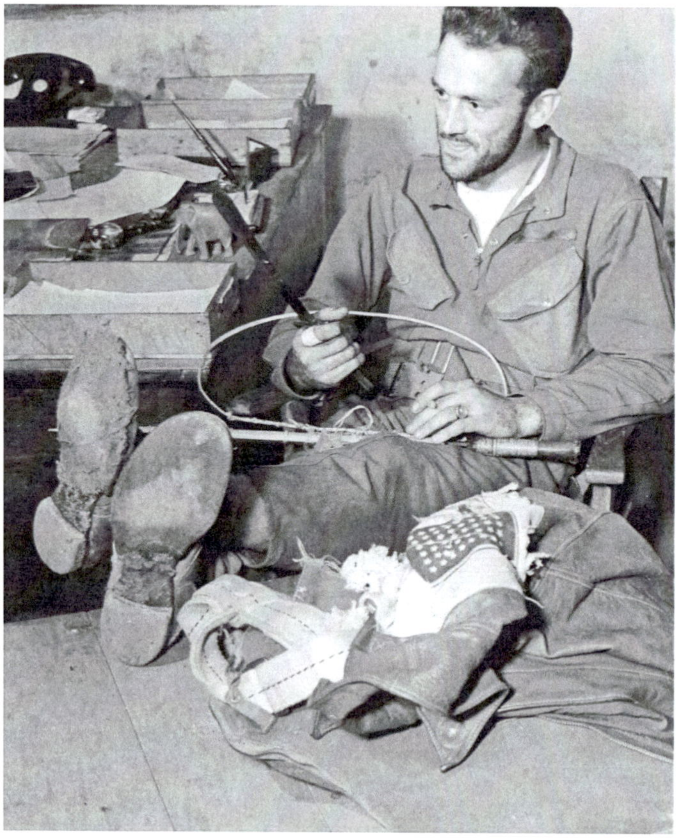

(Above) 1st Lt. Harold Robbins of the 89th FS relaxes in the A-2 office, Upper Assam, India, April 1944. He is being interrogated after his return from nine day walk-out after bailing out from his plane. In right hand is ten-inch machete from jungle kit on table. (National Archives)

(Above) USAAF person in jungle training uses 42D7246 along with D-2 gloves and mosquito headnet from jungle kit. (National Archives)

(Below) One of the nine samples of B-5 kits used for field testing to include the 42D7246 folding machete. Packed with hood and socks. (USAF)

(Below) Unidentified B-24 crew member of the 43rd BG, Dobodura Airstrip, Papua, c.1944. This crewman carries the folding machete attached to M-1923 cartridge belt. Probable pockets carry survival aids. (National Archives)

The field test conducted at Eglin Field was not favorable to the new two-bladed large knife either. The final report of the C-1 vest was released 8 August 1944. It included similar statements to those included in the June report and elaborated on suggested changes and performance. The final recommendation was that the large knife with saw should be replaced by a machete, and for it to include a sharpening stone, complemented by the pocket knife. It also recommended the handle be modified to an oval shape and a more pronounced ball for the end of the handle added to make it safer to use when chopping with a sweaty hand.

Subsequently, development of emergency kits and machetes again shifted. The intent was to simplify the volume of multi-person sustenance kits into one airplane kit that would complement the C-1 vest. An additional seat kit was not found suitable, and the B-5 was abandoned. But development of a new universal sustenance kit that could contain essential items not contained in the C-1 vest continued. The first step was to revise the design to make it more ergonomic. With the assistance of the Cruver Mfg. Co., initial steps were taken. The problem with their hooked pommel design was that it dug into the palm of the hand making it uncomfortable to use.

By this time, a series of tests was required before new equipment was adopted. Lack of testing had been a deficiency in the early years of the war. Testing of the concept machetes with the new Cruver handle were set at a 1000 stroke goal in hopes that any flaws would become apparent. Six models were submitted for testing, all using handle designs provided by Cruver. The bulk of issues occurred in assembly such as engineering of the guard and internal parts becoming loose. Testing continued into July 1945.

As part of the redesign process the machete was required to have a sheath with a wet sharpening stone. Those machetes submitted for field testing included a sheath but it was not scrutinized as much as the machete itself. The directive to develop a basic airplane kit included the folding machete with a sheath. The kit was to supplement three men and provide items not carried in the C-1 vest. However, only one machete was included and carried using the sheath provided. The contents lists it as Machete with Sheath (Cruver Mfg. Co. type, including sharpening stone). This kit was standardized as the Type E-18 sustenance kit in August 1946. Shortly thereafter the ten-inch folding machete manufactured by Imperial appeared, that included a sheath with a belt loop only and a sharpening stone.

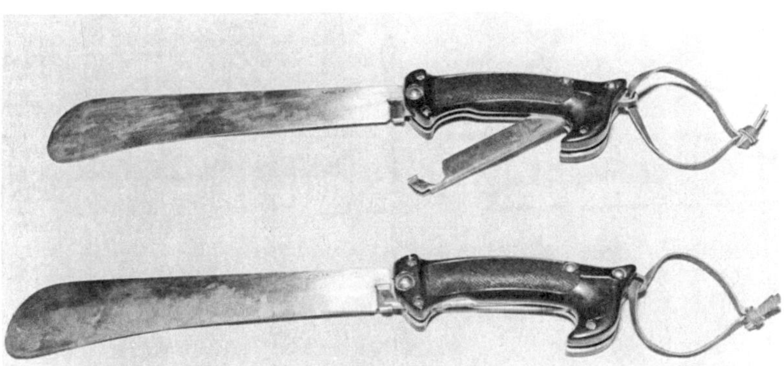

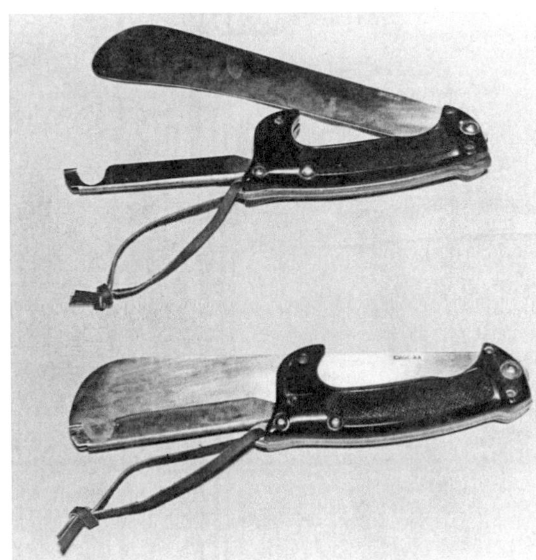

(Above and right) First concept of revised model of 10" folding machete with rounded, slip-proof hook handle with thong and integral blade guard, November 1944. (USAF)

(Below) Second generation Cruver machete with rounding at all edges, February 1945. Hooked pommel removed. (USAF)

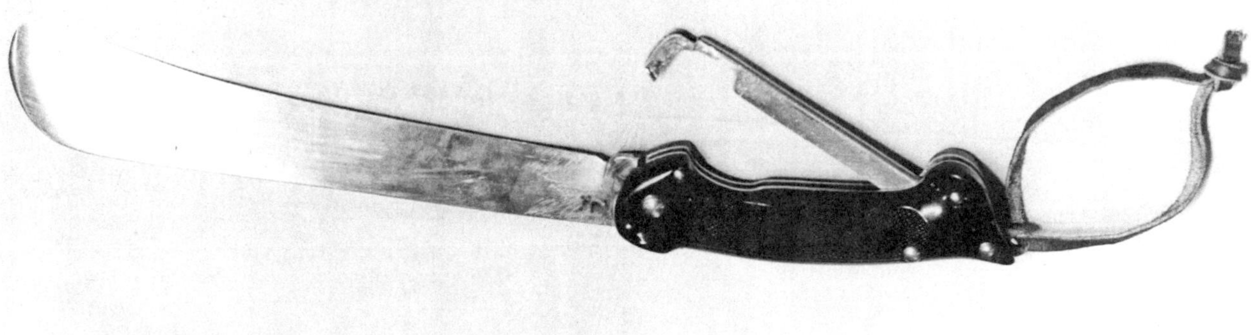

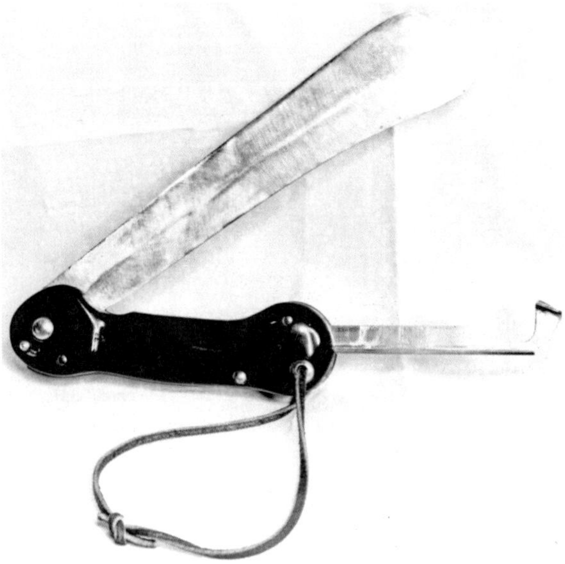

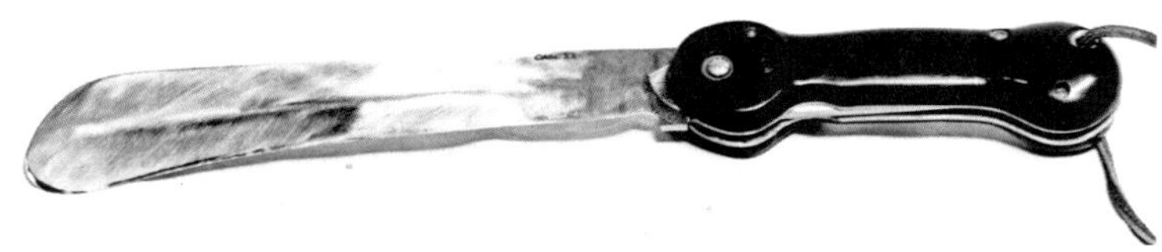

(Left and above) This is a further improved version of the 10" folding machete tested with differences from the previous examples, March 1945. Has close resemblance to the post war produced type by Imperial. (USAF)

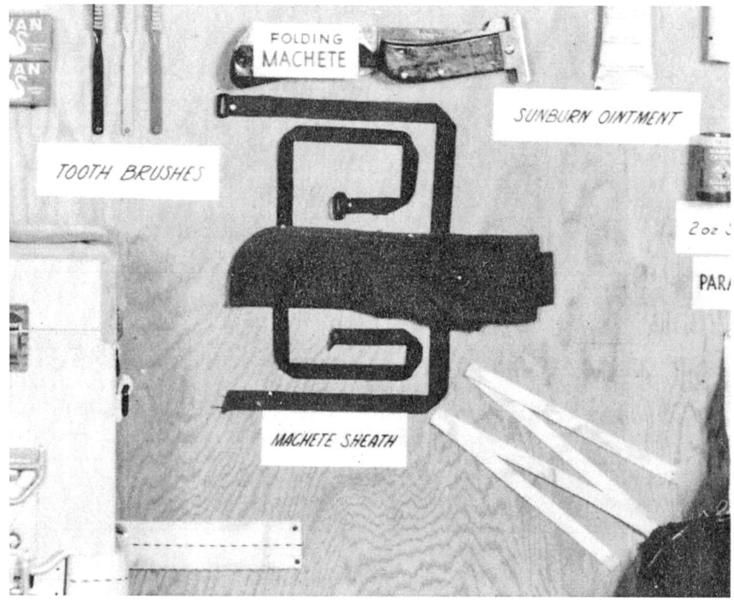

(Right) Concept assembly of Basic Airplane Kit, including 42D7246 machete, March 1945. As part of the concept, a sheath was included, worn on belt and secured to leg. (National Archives)

(Below) The British would go on to produce a ten-inch folding machete with a lanyard tab on the end of the handle. Four companies produced these, and all used Broad Arrow property marks. These machetes were intended for inclusion in the Beadon Flying Suit. The lightweight cotton suit was designed as a jungle survival kit, with specialized pockets on the coveralls; it was used in conjunction with a backpack. The suit pockets carried most of the survival items on the display board. The backpack carried supplemental items. It was used in Southeast Asia (Far East) and available to flying units in the SWPA. The Beadon suit was a recommended item for use by American units; records indicate their requisition by commands of the 10th AF. The complement includes a Chindit pattern leg machete. It is unclear if the suit was modified with a pocket or if the machete was stored in the backpack when the folding machete became available. (National Archives)

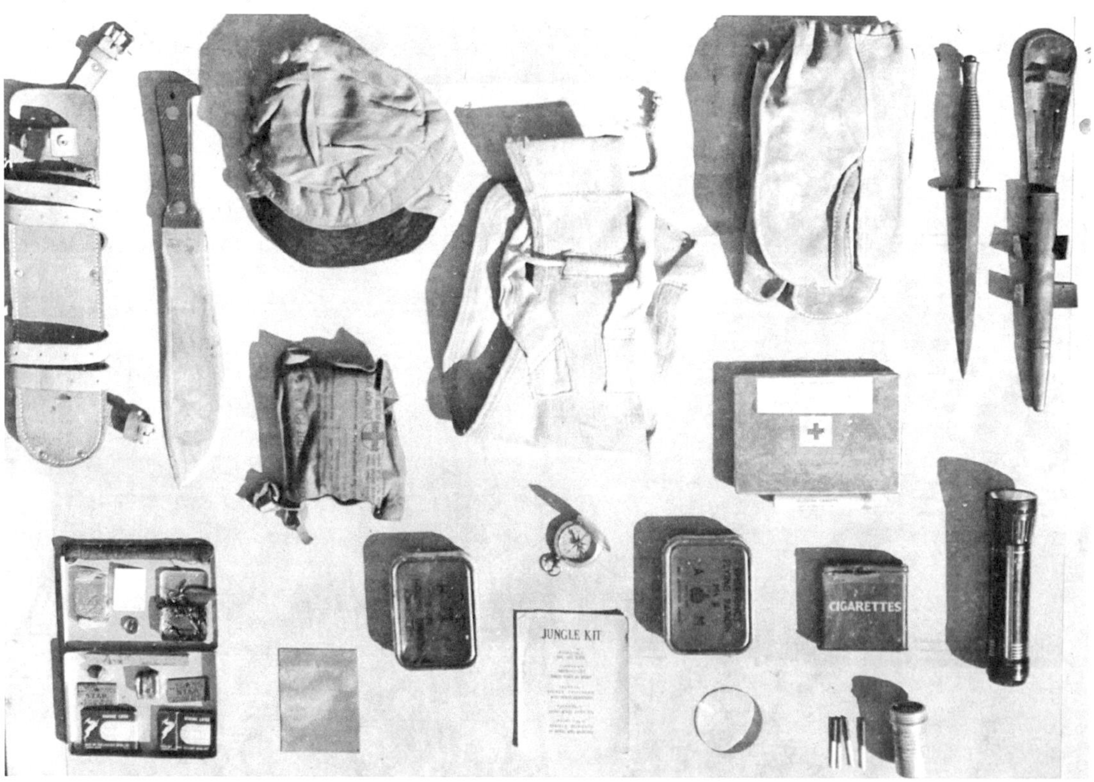

(Left) USAAF member demonstrating the Beadon suit, India, November 1944. (National Archives)

(Above) RAF aircrew wear Beadon suits, India, c.1944. Kukri and Chindit machete being utilized. (National Archives)

Emergency kits for the USAAC had to be compact, limiting the types of tools that could be included. The USAAC first settled on five-inch hunting knives, then nine-inch machetes, and finally ten-inch folding machetes. Following an initiative to improve emergency supplies for multiple people, a series of sustenance kits was developed. The first of these was the Type E-1 ration unit, which included two hunting knives. It was to be used in conjunction with the Type E-2 kit, which contained such tools as a rifle, cooking stove, fishing kit and a machete. The initial procurement of 1,000 E-2 kits in July 1942 included Collins & Co. model No. 127s, provided as Contractor Furnished Equipment.

The No. 127 was a commercial machete viewed as unable to fulfill a USAAF requirement. However, with other designs of sustenance kits requiring such an item, the Materiel Command issued a directive to design a machete for emergency kits. The new knife copied a 16-inch knife used for many years in the South Seas, which in turn imitated a native Philippine bolo. The new machete was intended for general use in clearing brush and chopping wood but could be used as a weapon and carried on the belt. Tests had shown this machete to be superior in shape and balance to other types tested.

The machete's blade was shaped like a bolo, 16 inches long with a parkerized finished. The handle was molded on the tang by an injection process of Lumarith celluloid by Cruver Mfg. Co. This durable handle withstood extreme temperatures and environments. Each knife was provided with a heavy leather scabbard and a belt loop.

At the time, most aircrew wore standard cotton or wool trousers with a waist belt. The pistol or cartridge belt was not regularly used at the time by the USAAF, and the loop was not designed to accommodate the larger size. In November 1942, Materiel Command stated the sources of supply were Union Cutlery Co. and Collins & Co. Only Collins manufactured this machete, assigned model No. 1253. This utility machete is void of any identification.

The bolo was designated as Knife and Sheath Assembly, 16-Inch, Utility referenced under USAAF Dwg. No. 43E10628, it was intended to replace the commercial one supplied in the E-2. In October of 1942 it was included in the revised specifications for the E-2 and any future kits requiring such an item, which included Types E-8, E-10, E-12 (E-2 was a component), and E-14. Each of those specified Knife and Sheath in accordance with Drawing 43E10628, listed as Contractor Furnished Equipment.

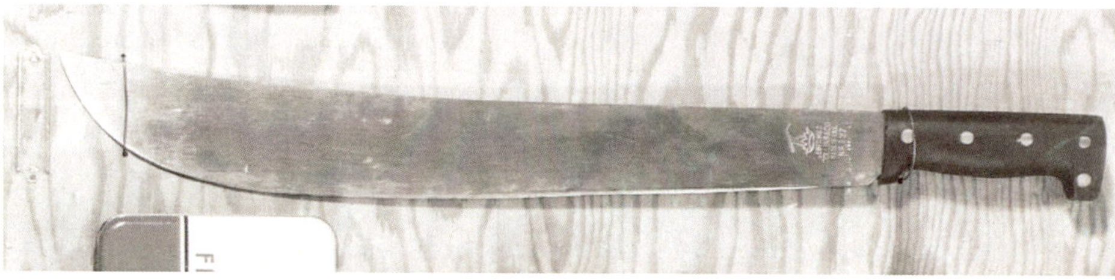

(Left) The No. 127 included in the E-2 had a 17" polished steel blade and a No. 3 black plastic handle. A scabbard is not on the E-2 kit display board but Collins No. 13 leather scabbards were likely supplied. These machetes were also equipped with green horn and stacked leather No. 3 handles. The black plastic handle was associated with US Government purchases.

(Below) Prototype Collins No. 1253, 43E10628. Scabbard fabricated from heavy duck fabric, leather reinforcement at tip and throat. (USAF)

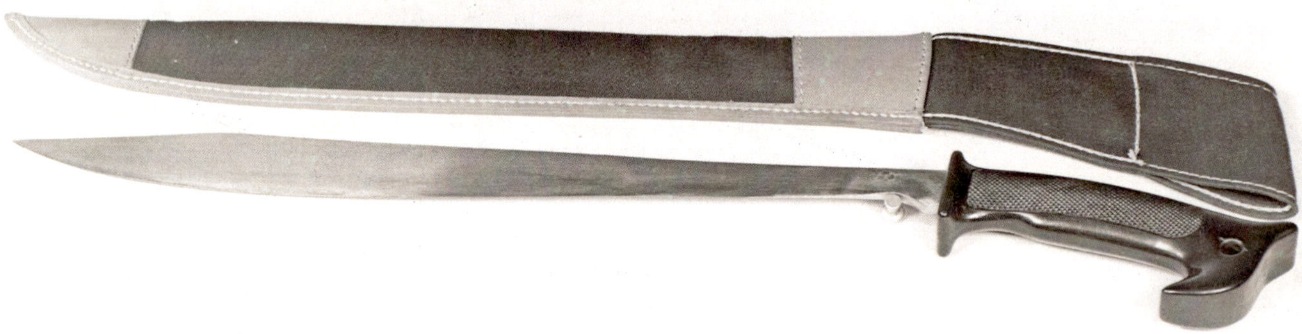

(Left and below) Two patterns of 43E10628 were produced, but the principal pattern tapered to a point. The other had a beveled square tip that shortened the blade to 14-1/2 inches. No information explains the purpose of the square tip. Two examples can be observed here in an E-8 sustenance kit.

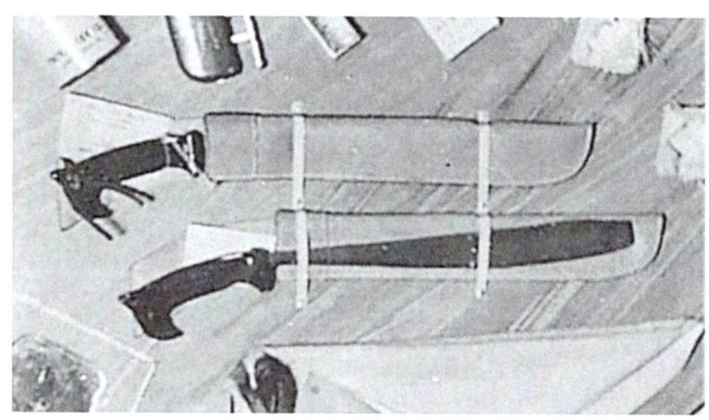

Experiments conducted by the Materiel Command on the new general-purpose machete were not completely thorough. In January 1943, Ladd Field in Alaska obtained samples of the 43E10628 and put it through rigorous field tests, concluding it to be unsatisfactory. The machete was then sent to AAFSAT for review in the spring of 1943. The AAF Equipment Board report released 15 September 1943 concluded that the knife was difficult to use for long periods. The plastic grip injured the hand, and the blade was springy and therefore unable to take or hold an edge. The conclusion was that the standard QMC M-1942 machete was more satisfactory for general utility purposes and recommended discontinuation of the 43E10628.

Since the 43E110628 failed to chop wood, a hand axe was added to sustenance kits as a supplement, and the E-2 was revised in June 1943 for this item. The M-1942 machete then replaced it. The M-1942 manufactured by Collins was referenced as Model No. 37. It had an unpolished, 18-inch straight back blade and a No. 6 phenolic resin handle. In June 1942, the QMC issued specifications for the M-1942. Its development and adoption paralleled the USAAF bolo.

The M-1942 replaced the 43E10628 in all sustenance kits. The specifications would subsequently be revised generically describing Machete, 18-inch blade, with plastic scabbard. The machetes purchased by contractors and the USAAF differed from those delivered to the QMC, which had standard Collins commercial markings with No. 37 stamped on the blade and came with a No. 6 handle. First leather, later duck canvas scabbards were provided with those for the QMC, to assure durability and to eliminate deterioration.

The USAAF designed a plastic scabbard and designated it as Sheath, Machete, Carrying, which was detailed under Dwg. No. 44D9059. This scabbard (manufacturer unknown) is a one-piece molded plastic with 3/8-inch holes on the front and back. AAF caduceus was Integrated into the mold, and the scabbard had molded lugs to accommodate an M-1910 type belt hook. The belt hook reflects the more common usage of pistol belts by operational personnel.

The M-1942 and 43E10628 were listed in supply catalogs, each labeled Knife, Machete. Neither carried a stock number, consistent with them being Contractor Furnished Equipment (CFE).

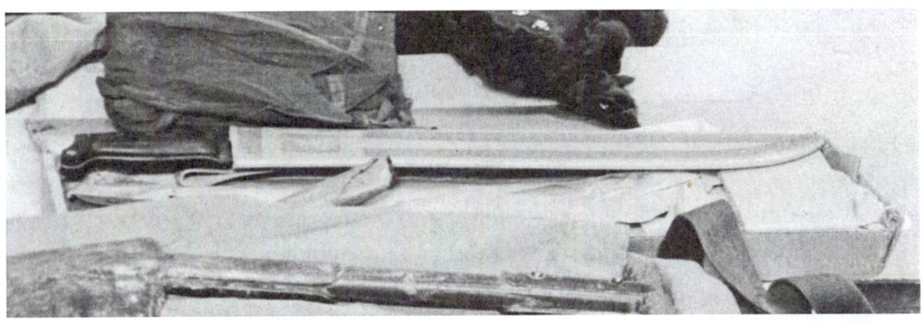

(Right) Prior to the availability of the 44D9059 scabbard, 18-inch machetes were delivered with the Collins No. 13 scabbard. This example is in a display for an E-10 sustenance kit.

(Above and right) Features of USAAF procured M-1942 (No. 37) and scabbard 44D9059. Machete lacks property markings, includes Collins model number and domestic black label. Distinct No. 6 handle.

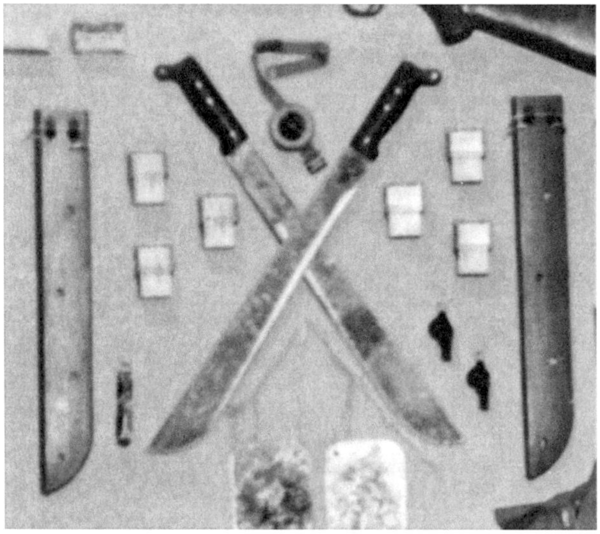

(Above) Machete M-1942 with scabbard 44D9059 included in E-14 sustenance kits; two each were to be included. Supplemented by hand axe. Original specification cited inclusion of 43E10628. Revised under Amendment-1, 11 April 1944. Substituted with description Machete, 18-inch blade with sheath, Collins Company, Collinsville, Connecticut, or equal.

The US Navy BuAer struggled significantly to adapt to the wartime environment since it had no standardized personal survival equipment prior to the war. However, in 1937 BuAer personnel in Panama improvised a large knife for jungle terrain. This concept was revisited in 1942. The machete adapted for this kit had the characteristics of a Collins machete with a No. 6 handle; likely an exported pattern had been sourced locally and modified. Length was around 15 inches with the blade around ten inches. For quick access, it was secured to the exterior of the pack. Should the parachutist get caught in the dense jungle canopy, the machete was immediately available.

Naval air operations were executed early in 1942 to help the defenses around New Guinea and protect the volatile supply line to Australia. Small carrier task groups harassed Japanese installations in the Gilbert and Marshall Island groups and attempted to loosen the Japanese foothold in the Solomon Islands, which led to the first carrier-to-carrier air battle in May 1942 in the Battle of the Coral Sea. This engagement ended in a draw but thwarted the Japanese invasion of Papua/New Guinea.

It became apparent that, if aircraft went down, the crew would face forests of trees and bamboo, swamps, rice fields, sand banks, beaches and mud. A heavy cutting tool was deemed necessary in these environments and the need was continually pressed as air operations and occupation increased. The hunting knife was the initial solution but many strived for a machete as well. Some Naval and Marine units were able to obtain USAAF jungle kits, but the principal focus was on obtaining a machete. Commands worked to obtain suitable tools and on submitting recommendations for the BuAer to adopt a machete for issuance. With the occupation of Guadalcanal and the expansion of bases in the Solomons, nine-inch machetes were procured and issued.

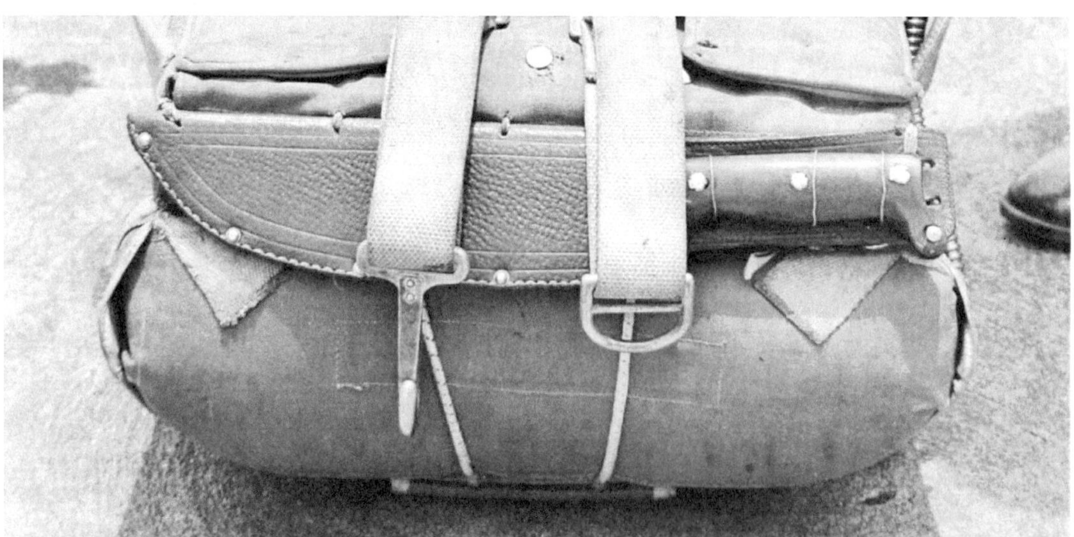

(Left) Locally modified machete attached to emergency parachute pack, Panama Canal Zone, c.1937.

(Above) 9" machete used by Lt. jg Steve Hyarady, TBF pilot of VT-12, aboard USS Saratoga, November 1943. Handle pattern is like that produced by Case. He also carries a canteen of water. (National Archives)

(Above) Pilot Lt. jg Bowen of VF-13 aboard the USS Franklin, October 1944. On his pistol belt is a Collins No. 18 machete. The machetes were purchased by the squadron while stationed at their port of embarkation in California. (National Archives)

(Left and below) This Collins No. 18 machete once belonged to Lt. Cmdr. F.T. Green, ACIO with Air Group 10. He participated in the Battle of Santa Cruz in October of 1942 while serving aboard the USS Enterprise. He became the intelligence officer for Air Group 80 in February of 1944. The tooled No. 13 scabbard has the addition of a handle strap; on the back are his name and association with Fighting Squadron 10 (VF-10). (below) VF-10 pilot aboard the USS Enterprise, February 1943. Over left hip is a tooled leather No. 13 Collins scabbard. (National Archives)

(Above) Further examples of 9" Collins machetes worn by Marine aviators. (Right) SBD Dauntless gunner at operations shelter on Henderson Field, Guadalcanal, April 1943. Behind his personal side arm is a Collins No. 18 with the distinct No. 216 handle. (Left) Ace and Medal of Honor awardee Maj. John L. Smith of VMF-223 with Collins No. 18 behind his M-1916 hip holster. (National Archives)

(Below) Marine squadrons operating from Guadalcanal show machetes carried on the person, illustrating their vital importance. USAAF personnel stationed in New Guinea also used machetes. It is likely the machetes were requisitioned from the USAAF or purchased from Australian sources. (Right) VMSB pilot has a machete secured to cross strap of his shoulder holster centered on the back, Guadalcanal, April 1943. (Left) VMF-124 pilot resting near the airfield holding a Reising submachine gun and in his right hand a 9" machete. (National Archives)

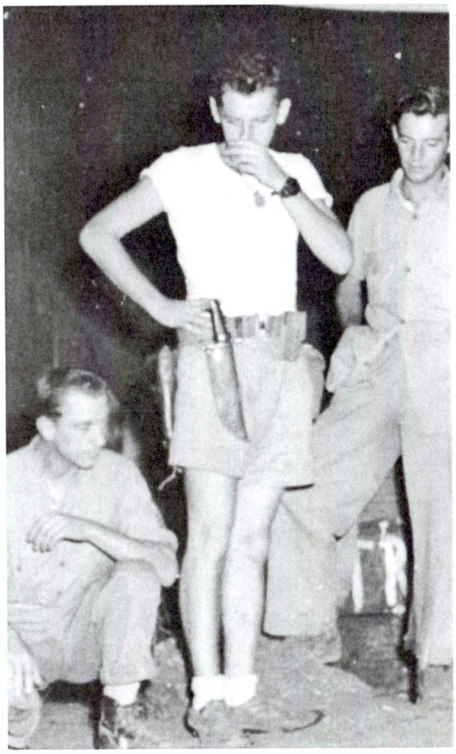

(Above) Marine SBD gunner wears blade manufactured in Australia while passing the time near operations tent on Guadalcanal, April 1943. Taper of the pommel was a distinct feature of Marsden and Mars machetes. Some scabbards had a handle strap and a rivet near the throat; here, camera flash has highlighted the rivet. (National Archives)

(Above) Unidentified land-based Marine VMTB aircrew, c.1943. Individual on right utilizes 9" machete among notable other accoutrements. Individual standing third from right utilizes Robeson 5" hunting knife.

(Above) In 1942, the USMC Quartermaster procured some quantities of the Collins No. 191 machete in a 21-inch blade length and with a No. 9 handle. By the close of that year the USMC had decided to adopt the 18" M-1942 pattern No. 37. Here, example of a No. 191 as procured by the USMC but cut down and with a custom-made scabbard. The material is waterproof nonspecular sea blue aircraft fabric. Other features indicate it was made in a parachute loft. On the back is stencil of the name HISE. The machete can be conveniently carried on a belt with the loop provided. Hise was likely an aviator. (Robin Bartel Collection)

When Collins, Case and Kinfolks nine-inch machetes made it into circulation, quantities of the new jungle back-pad kit and limited quantities of the ten-inch folding machete were in service. The nine-inch machete remained in the supply system and were temporarily supplied with the Type B-2 kit until further quantities of the 42D7462 were delivered. These companies also sold these machetes on the resale market, making them available to individuals of any military branch.

Another machete available for both issuance and purchase was a pattern provided by Western, designated as the Bushman No. Bx54. The blade was identical to the other manufacturers' blades but had a flat stock steel guard. The handle was a molded resin with steel pins; a variation had a brass guard and brass pins in the handle. The scabbard provided with these machetes had a handle strap and rivets. There was no evidence these were procured by the USAAF.

(Left) Western 9" machete. They provided leather sheath with rivets and handle retaining strap.

(Below left) Pilots of VF-23 carrying Western 9" machetes on USS Langley flight deck, Spring 1945. This was a squadron-level procurement. (National Archives)

(Below right) Lt. Cmdr. Donald White of VT-9 wears Western Bushman on right hip along with Norfolk shoulder holster. (National Archives)

The 1917 model bolo knife was adopted in WWI as a utility tool for digging and clearing away fields of fire by machine gun crews and infantry units. It and the 1910 model bolo remained in Army Ordnance inventory into WWII. It was repurposed as a survival tool. Some Naval aviation units found it suitable as a compact alternative to a personal machete.

Both models of bolo knives were maintained at Navy Department supply depots and were requisitioned with scabbards from NSD Oakland using stock number 74-B-735 for the 1910 model and 74-B-737 for the 1917 model. Stock numbers were also assigned to the bolo only (74-B-734) and scabbard only (74-S-67). Class-74 designated the item for Landing Forces. Emergency back-pad kits were not always available and the original kit was not provided with a machete.

The bolo was a supplemental item necessary for the jungle environment. The V-44 machete was provided in the standard M-592 back-pad kit but was only provided for pilots; it was distributed to aircrew if supplies were available. Additionally, the M-592 was not available in quantity until 1944. Due to cramped aircrew positions, such as in the ball turret of a TBF/TBM Avenger, a back-pad kit could not often be worn. Squadron level initiatives provided what was obtainable.

(Above) Images from VF-17 while stationed on Bougainville Island in the Solomons, March 1944. (Left) Ens. G.M Keller Jr. (Right) Other pilots with 1917 Model bolo applied as survival emergency tool. (National Archives)

(Below) Bolo hanging in USN officers' quarters of VD-1, Guadalcanal, December 1943. (National Archives)

(Below) Pilot Ens. Ralph Cole Jr. of VT-14 wears bolo, USS Wasp, August 1944. (National Archives)

(Below) Unidentified SBD pilot of VC-24 or VB-98 wears USMC hosptial corpsman bolo as personal machete.

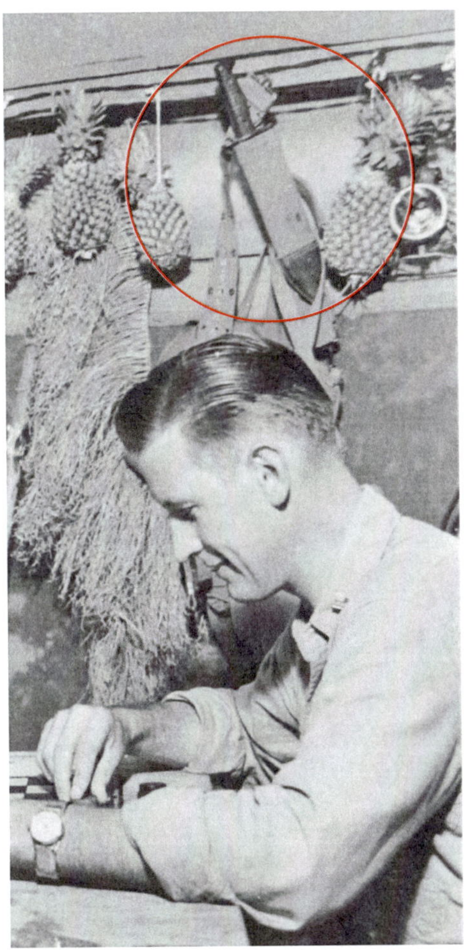

Initial focus was on design of an emergency parachute pack, NAF-69233 and the subsequent Original type did not include the provision of a machete type tool. It was recommended that a machete be included in the design of a standard jungle emergency kit. The nine-inch machete was preferred, since that what was available for evaluation. BuAer and Marine aviation units obtained USAAF materials with an aim to using the machete in the jungles of the Solomon Islands and Papua/New Guinea. As illustrated, other types were applied for the same purpose.

Integrated into the design of the standard M-592 back-pad kit was a machete pocket. Through kits' design process, BuAer then adopted a pattern based on USAAF Dwg. 42D7462, referenced under BuAer Dwg. No. 8662. It eliminated the folding feature but retained the basic profile and overall dimensions. W.R. Case & Sons Cutlery Co. was the sole approved manufacturer and designated the pattern number V-44. These machetes were provided as Contractor Furnished Equipment at $2.71 each. They were not maintained by the ASO until the introduction of the Model-A pararaft case, January 1945. When it became available, service branches disassembled the M-592 and AN-R-2, placing specific components in the new seat kit. The machete was placed in stock and available for requisition under stock number R74-K-200.

The V-44 machete had a ten-inch blued blade with a sheet metal edge guard with walnut grips pinned to the tang. Each machete was wrapped in wax- or asphalt-coated Kraft paper prior to being put in the M-592 back-pad kit. For the Navy, the blade was treated for corrosion resistance and a lanyard was provided to reduce loss.

(Above) V-44 as supplied with M-592 back-pad kit. Provided as CFE.

(Right and below) Instructors and cadets at NAS Pensacola demonstrate some of the basic principles of its use. One cadet uses the spine to process mollusks while another is chopping a limb; the machete was equally effective in clearing brush and digging. (National Archives)

The requirement for an edged survival tool was not proprietary to the US Forces. Allied Forces faced the same circumstances and adopted small machetes. One example, from the RAAF, was a leg machete and was often referenced to as such. Earliest reference to it appears in the booklet Instructions to Air Crews, published in late 1942, which describes its basic features and intended functions. Clearly, the RAAF viewed it as a necessary item, issuing it to flying personnel.

The common features of the machete include an 8-1/2-inch square-tipped steel blade with a two-piece alloy or wood handle, having a heavy leather sheath that was strapped to the user's leg. Contrary to common belief, the booklet Instructions to Air Crews makes note that the tip had been squared off for the purpose of digging; being blunt allows the machete to serve as a shovel. A drooping point at the tip was for penetrating coconuts.

It is not defined when RAAF first adopted the leg machete or when it began to be issued to US Forces, but records indicate it became a universal staple into 1944. Evidence suggests that Scott Trojan (Scott and Sons Pty Ltd) began production of the type with wood handle starting in January 1942 under Australian Army contract.

In the summer of 1943, ComAirPac recommended that every plane be furnished with a machete type knife having at least a ten-inch blade heavy enough to be used as an axe. They also state that machete type knives must be obtained from commercial sources. This falls in line with what has been illustrated.

Organization in the SWPA was in part decentralized until September 1943 with the official formation of the Military Intelligence Service, Section-X, SWPA. This was created as a centralized intelligence body established to aid the return to Allied lines of Allied personnel, who may become isolated in inaccessible or enemy territory. Its primary concern was for combat aircrews. As a liaison office, they would brief all incoming intelligence officers into that theater of operations. In addition, members from that office would conduct lectures and training courses on survival and rescue. This would include distribution of survival and evasion aids. MIS-X, SWPA procured, developed and distributed a significant amount of evasion and survival materials; to be covered in another volume.

The leg machete was one item procured by MIS-X for distribution, assigned nomenclature was Knife, Jungle, with Sheath. It was through MIS-X that its use was promoted to all organizations coming into the region. In a monthly report from November 1944, MIS-X states that approximately 4,000 jungle knives had been received from the manufacturer during the past two months. All of those were being delivered to operational units as rapidly as transportation facilities would permit. However, these leg machetes were being distributed much earlier than September-November.

The Navy liaison attached to the office mentions in March 1944 report that a larger knife, with a much stouter blade and handle, such as those now being issued to USAAF and RAAF pilots, be recommended for Navy flight crews. It had been found that such a machete type knife was most easily carried and most rapidly available if strapped on the leg. At the time of the report, Commander, Naval Forces, Northern Solomons acknowledged receipt of 400 RAAF jungle knives and sheaths from MIS-X, as per their request.

It was through MIS-X records that some measure of production quantities can be calculated. As of 1 January 1945, 7,087 had been delivered to the Far East Air Forces; encompassing USAAF units. In May of 1945, the total count increased to 11,782 for the FEAF. Navy units had received 1,487 as of 1 January, and a total of 1,916 of May 1945. Marine aviation had accounted for 1,100 delivered as of 1 January; this number remained static through 1945. From this total, MIS-X alone had procured at least 14,798 leg machetes. This was only one accounting record, not accounting for unit and command level requisitions or purchases outside of MIS-X.

Accounts suggest the basis of issue was for like that of the E-3J kit; two per bombardment aircraft or one for pilots in multi-seat aircraft. It also falls in line with the suggestion by ComAirPac for at least one per aircraft. Officers would have first priority, but it is known that they were worn by enlisted personnel.

Extracted from an Emergency Rescue Report from 22 June 1945, B-24 right waist gunner, Sgt. Kahl of the 372nd BS, was the last to bail-out over the Celebes, Netherlands East Indies. He carried on his person matches, some medical supplies, canteen, E-17 kit and two knives. He also had a .45 pistol and Australian leg machete. One account of an enlisted man with this tool.

Use of the machete was included in Navy accounts from VC-84, VC-85, VC-86, and VPB-149. VC-84 and VC-86 operated from two different aircraft carriers and sailed from California to support operations in the Marshall Islands, then on to support roles in the Philippines. VPB-149 similarly traveled from the west coast of the US to Pearl Harbor, on to Midway Island and finally to the Philippines. Advanced MIS-X offices were established in the NEI and Philippines where this type of equipment could be obtained. Additionally, MIS-X liaison officers were brought aboard carriers prior to air strikes conducted in the SWPA. These squadrons, as well as VC-85, also used the E-3J kit assembled in Brisbane, Australia by MIS-X. Units stationed in the SWPA, to include Navy and Marine, would have more direct access for local requisition and purchasing.

A variation can be observed in the construction details of the handle and the blade of the leg machete. Most examples are void of any identifiable markings. Contractors were East Bros.., Sydney, Australia, JM Avon and Co., Gregsteel and Scott Trojan. Gregsteel was a New Zealand-based manufacturer, and with Scott Trojan, produced a type with wood handle.

Scabbards also varied slightly but conformed to the same pattern. Manufacturers included Goldseal, Sydney, and Bonney & Clark Ltd. Makers, as well as an unknown company that made an unmarked version. All three have differing rivet patterns.

(Above) RAAF pilot with leg machete; J.A. Farquharson of No.75 Squadron on Tarakan Island near Borneo, June 1945. (Courtesy Australian War Museum)

(Above) RAAF No.457 Squadron pilot, Labuan Airstrip, North Borneo, c.1945. (Courtesy Australian War Museum)

(Left) Machete manufactured by East Bros., per blade stamp, with unmarked scabbard.

(Left) Branded sheaths: Bonney & Clark Ltd. And Goldseal.

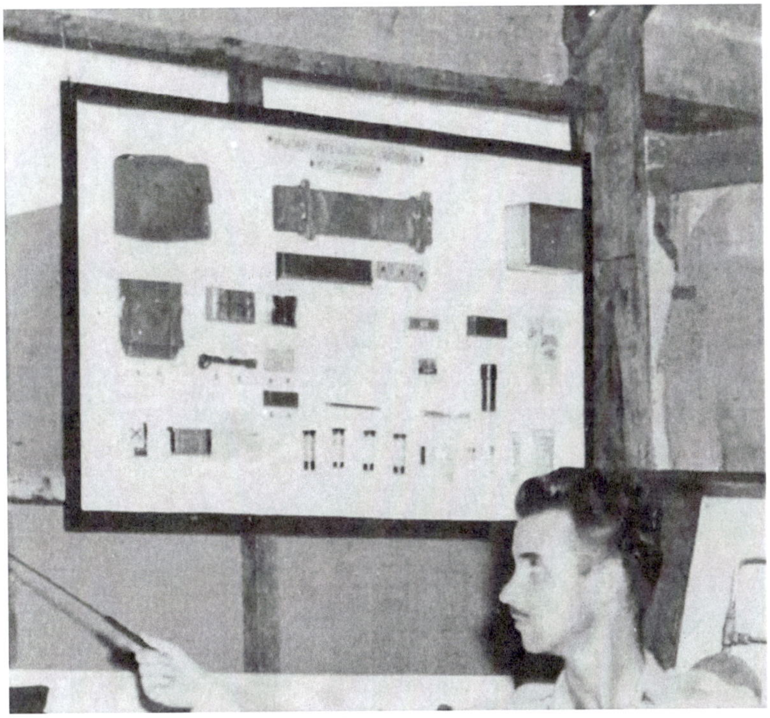

(Left) MIS-X display board exhibits jungle knife (leg machete), E-3J, and booklet Survival Hints, c.1944.

(Right) TBM pilot Lt. Martin Ginsburg of VC-86 on board the USS Bismarck Sea, January 1945. He exhibits both the jungle knife and E-3J. (National Archives)

(Below) B-24 pilot Capt. Robert Wallace and copilot 2nd Lt. Harlod Myers of the 13th AF wear leg machetes, Philippines, December 1944. (National Archives)

(Below) JM Avon machete with Goldseal leather sheath. USAAF stamped into handle. Circumstances of providing property marking unknown. MIS-X was an independent entity and not associated with the USAAF.

(Above) TBM enlisted personnel VC-90, USS Steamer Bay, c. 1945. Each wear leg machetes. Individual on left is Parke Bloyer ARM2/c. He additionally wears 6" Western hunting knife.

(Above) Ens. Harold Chapman of VC-84 on board the USS Makin Island, January 1945.

(Far right) Jungle knife attributed to Curtis Stonelake ARM1/c. Stonelake served as a radioman on PBY Catalinas for VP/VPB-33 from its commission to decommission serving 33 months from the Solomons to the Philippines. Bonney & Clark Ltd scabbard.

(Right and below) Machete used by Henry Champagney AOM2/c of VPB-149. On 12 March 1945, while on a search mission, his PV-1 Ventura aircraft ditched for lack of fuel off the coast of Cebu Island, Philippines. He was quickly picked up by natives and escorted to the headquarters of guerilla forces. For two days, he evaded the Japanese until picked up by a PBY operated by the USAAF 3rd Emergency Rescue Squadron.

An Australian leg machete appeared in multiple frames from a private film recorder by 2nd Lt. James T. Sykes, USMC. Sykes flew Corsairs with VMF-212 from Green Island, then with VMF-223 from Bougainville. Commentary, added later after it was digitalized, Sykes in the film notes only that machetes were purchased by the squadron.

In two circumstances in which leg machetes were used by USMC aviators, their units were stationed in the Solomon Islands. After the seizure of Guadalcanal and after some control was obtained, squadrons operated in the forward combat area for 4-6 weeks, then were given a week's leave in Sydney or Auckland. Personnel were then eased back into combat with 2-4 weeks of training, absorbing replacements at Espiritu Santo or Efate. Leg machetes could be obtained during Australian rotations.

(Right) This leg machete has been attributed to 1st Lt. Richard M. Robinson USMC, corsair pilot VMF-218. His squadron was initially stationed on Green Island in July 1944, attacking targets around Rabaul, Papua/New Guinea. They moved to Leyte Island, Philippines, in November 1944 and flew close air support for ground forces, air cover for convoys, and as bomber escorts throughout the Philippine Campaign. On 18 December 1944, while supporting Army ground forces on Mindoro Island, Robinson was shot down by friendly 20mm anti-aircraft fire and was picked up by a PT boat after parachuting into the ocean. (Paul Falkner Collection)

(Below) Pilot of VC-84 Ens. J.D.W. Hubbeling. Machete is secured to his inner right lower leg. (National Archives)

An item specifically designated an emergency tool was the fireman's axe; there were crash, emergency, and fire types. Some were for ground crews and others installed in aircraft. An axe was part of the standard complement for airfield crash and fire trucks. However, installation was not universal in heavy, very heavy and transport aircraft. When they were installed, they could be equipped with a standard fireman's pick headed axe, but these were found to be unwieldy in the cramped confines of the fuselage. A small hand axe was referenced as Axe-Emergency Hand was a useful alternative; the type specified was an Official Boy Scout type made by Fayette R. Plumb Inc. as model No. 680. This axe became the standard in aircraft, typically installed near the pilot's position; later, an axe was stored near the rear most exit.

In 1942 a new axe was developed, referenced under USAAF Dwg. No. 42D8331 and designated as Axe, Fireman's Small Hand (Emergency) to overcome the deficiencies of the traditional hand axe. It was 16 inches long and had a vulcanized nonslip handle that would insulate a user up to a 20,000 volt shock. The pick end delivered the initial thrust, providing an opening for the blade edge. The blade edge, in turn, permitted penetration of the aircraft skin in a manner that allowed for easy withdrawal. This design was superior to a hand type that would get stuck in the aircraft skin.

Crash axes were intended for equipping rescue crews to break into the aircraft to retrieve the occupants. It was only later, in 1944, that this new axe, 42D8331, was installed in aircraft. At that time, aircraft were also prescribed to be marked with color coded bands on the exterior of the fuselage for entry points. Technical Order 01-1-212 of 31 May 1945 prescribes that broken orange-yellow colored bands be applied to entry points at areas near all crew positions. Each was to be marked with one inch letters, CUT HERE FOR EMERGENCY RESCUE.

(Left and below) Though intended for emergency entry into aircraft, axes provided inside could prove useful in survival situations. In this simulated survival scenario, an aircrew member is using the crash axe to remove sod from the tundra, creating a message to potential rescuers. The two known manufacturers of the 42D8331 were American Fork & Hoe Co. and D.L. Auld Co. Crash axes were provided as Government Furnished Equipment to aircraft manufacturers and could be requisitioned from stock under number 4500-043000.

(Right) Axe, 42D8331 secured in bracket behind the pilot's seat in a C-46, c.1945.

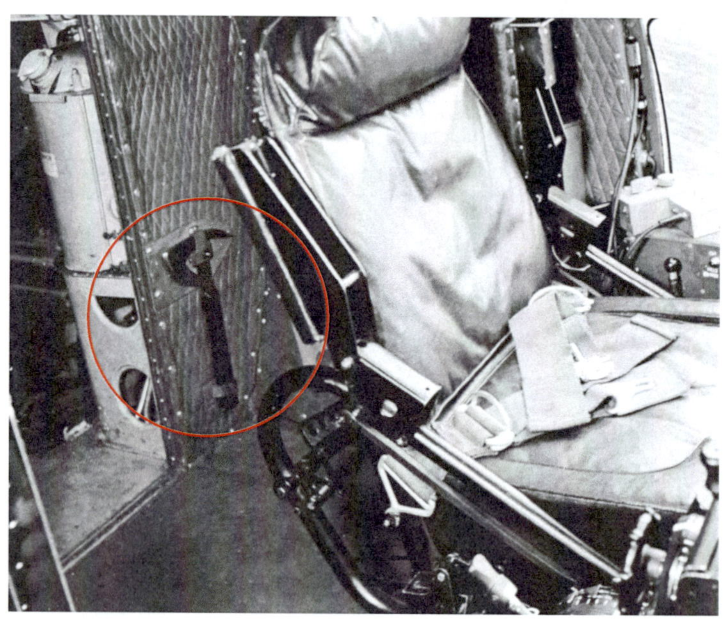

(Below) Axe, 42D8331 manufactured by D.L. Auld. Variant with grey rubber handle. The head pattern would change in the post war years.

The smallest World War II knife was most often referenced as a Boy Scout or Scout type knife. In some cases this name was accurate, but it was more commonly known as a pocketknife or jackknife by the US Navy. It quickly gained importance as an item of survival gear valuable for cutting lines, rigging sails, opening rations, cleaning fish, and dozens of other uses aboard a life raft or boat. On the beach or land a knife was indispensable. The number of companies that provided pocketknives on US Government contract and purchase orders is unknown, but there is a record of principal manufacturers.

Purchases may be categorized by the services of the US Army and US Navy, each made up of departments that procured pocketknives for many different purposes and in many different designs. For the US Army: Quartermaster Corps, Engineer Corps, Signal Corps, Ordnance Dept.. and USAAF. For the US Navy: Bureau of Aeronautics, Bureau of Medicine & Surgery and Bureau of Supplies & Accounts.

A great volume of pocketknives and jackknives made their way into aeronautical use. These companies provided the majority of pocketknives for the USAAF and US Navy: Camillus Cutlery Co., H. Boker Co., Inc., Imperial Knife Co., Pal Blade Co., Ulster Knife Co. and the Utica Cutlery Co.

Other companies, including Union Cutlery, Case, Cattaraugus, Robeson Cutlery and Schrade Cutlery, had various patterns of knives in circulation prior to the war that were obtained on small purchase orders. They produced the TL-29 for the Signal Corps and other types for the QMC.

The folding knife, a jackknife, was first used in the early 19th century by the US Navy in the age of wooden ships adorned with ropes. They were manufactured and imported from England. By the Civil War, the US Government began procuring them in some quantity from US-based companies. This single-bladed jackknife remained consistent until WWI.

It was a rope knife, typically with a 4-½ inch handle made of bone or hardwood and having either a clevis or a drilled hole for a lanyard. A single, stout spine, 3-1/2-inch square tip or sheepsfoot blade was either etched or stamped US NAVY and designed to withstand hammering from wooden mallets.

By the time of WWI, the rope knife had evolved into a general utility tool used by all branches of service, with modifications. The handle length was decreased to 3-1/2 inches and a clevis becoming the standard. A second manicure blade was added, and for easier opening a notch was added to the profile of the handle.

This era ushered in other patterns of knives for the military; for example, the Signal Corps designed its own. Two patterns emerged that took on some of the same characteristics: they had two blades of about 3-1/2 inches. One version included a spear point main blade with an accompanying pen blade. The other design is now better recognized as the electrician's knife TL-29. It included a screwdriver that could be locked open by a back-spring lock instead of the pen blade.

Representative pocketknife patterns in circulation at the beginning of the 20th century in the US military. (Top to bottom) US Navy Rope knife (blade etched US NAVY), WWI utility knife, WWI Signal Corps (pre-TL-29).

From WWI to the 1930s, commercial pocketknife patterns evolved. The next significant military contribution would be the standardized four-bladed Engineer Corps pattern, established in 1937 although the pattern had been in use prior. This design influenced the manufacture of folding knives for the war effort.

The Navy Department revised and adopted a new general utility jackknife in accordance with Specification 41-J-4, issued c.1940. This new knife had many of the features of the two-blade utility knife, Signal Corps type, and other commercial patterns. Replacing the rope knife for standard stock and ships stores was a 3-½-inch two-bladed jackknife that included a spear-tipped main blade and pen blade with a handle that was made of jigged bone. It had a clevis and easy opening feature. The 41-J-4 was requisitioned with stock number 41-J-50 but also purchased from ships' stores at a price of $0.20. It became the new standard of utility knife aboard ships and as an emergency and utility tool. Its procurement was not proprietary to the US Navy; it was also purchased by the QMC with and without a clevis for resale to service members. The rope knife was discontinued but remained in service and on procurement as a USCG approved sailors' jackknife for the Maritime Commission or Merchant Marine.

Five companies (Pal, Imperial, Utica, Camillus, and Kingston) manufactured the 41-J-4 jackknife, producing more than 1,700,000 until it was superseded by the 41-J-5 in 1944. Initial production included stag handles and brass linings. The latter was removed in Amendment-B, dated 1 September 1942, which required all-steel construction. Imperial and Kingston manufactured versions with cellulose acetate-molded scales.

(Right) Examples of the 41-J-4, illustrating the jigged bone and acetate scales. Easy opening feature is the crescent moon cut into handle.

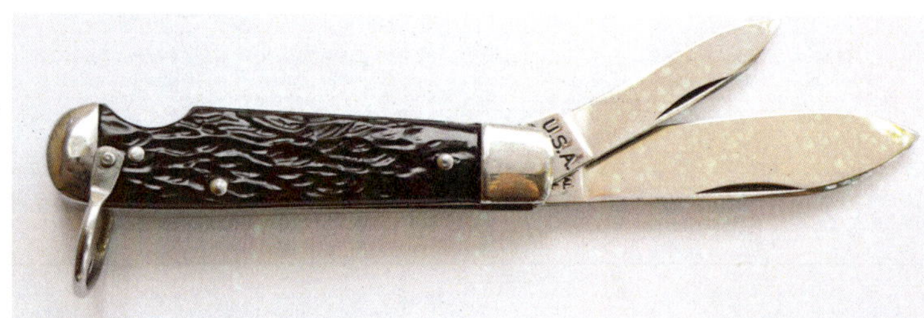

(Below) Steward in pantry of USS Yorktown, June 1944. 41-J-4 secured to trouser belt loop. (National Archives)

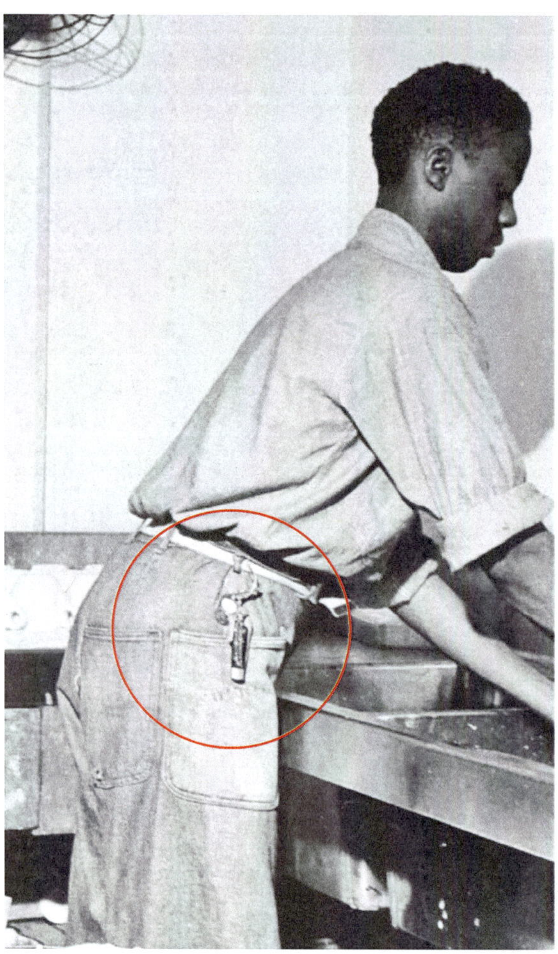

(Below right) Wire checker at his post near the flight deck on board the USS Monterey, June 1944. 41-J-4 secured to the trouser belt loop. (National Archives)

When the 41-J-4 was issued, the BuAer adopted a jackknife for inclusion in life raft repair kits under Amendment-N of Specification M-3, 27 November 1939. The repair kit description states it was to include a Scout type knife with heavy cutting blade, can opener, and screwdriver. This was revised in Amendement-1 25 February 1940 to state that knife was to be equivalent to official Boy Scout Knife as shown in the issue of the Local Council Exchange of the Boy Scouts of America.

The Council Exchange lists three models: Remington model No. 1496, Ulster model No. 1502, and Universal model No. 1585. Universal was a pattern manufactured by Landers, Frary & Clark. The official BSA knife was the standard until the issuance of NavAer Specification M-575. Quality standards for these knives were high, meeting military requirements. Life raft contractors were required to purchase the knives as CFE. BuM&S Specification 41-J-001 for a four-bladed knife specifically has in the title Jackknife "Boy Scout" type. They also adopted the BSA pattern.

(Right) Illustration in 1939 Local Council Exchange of the Boy Scouts of America showing official BSA three- and four-bladed knives manufactured by Remington Arms Co. Caption states they are regulation model with large stag handle, made of high carbon crucible steel, and carefully tempered to ensure a keen cutting edge. The four-bladed type was the principal model, comprised of a heavy cutting blade, screwdriver, bottle opener, can opener, leather punch blade, and a clevis.

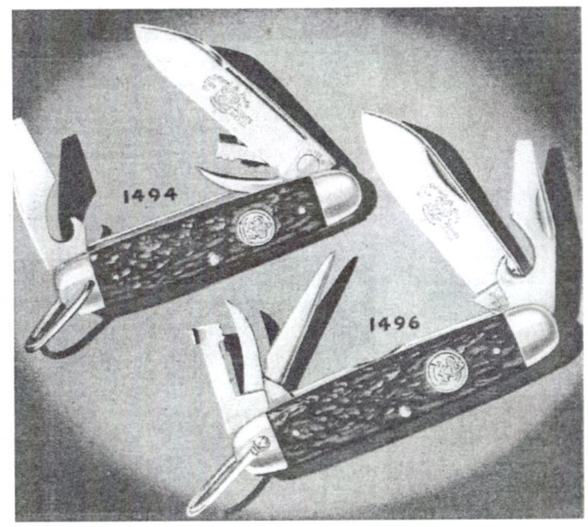

(Left and below) Official BSA pocketknives. (Top) Remington, (bottom) Ulster. Etched on the main blade of both knives is the BSA crest.

(Below) Standard US Navy life raft accessories, including jackknife. (Below left) Jackknife included in accessories of US Navy M-3 series life raft has characteristics of Ulster model that includes the BSA crest. (National Archives)

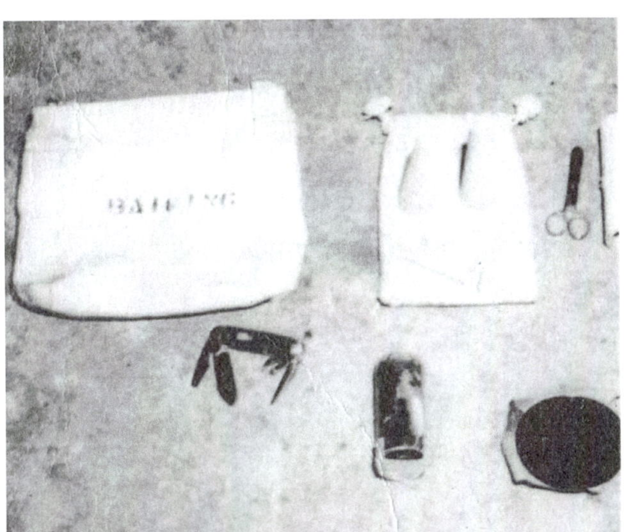

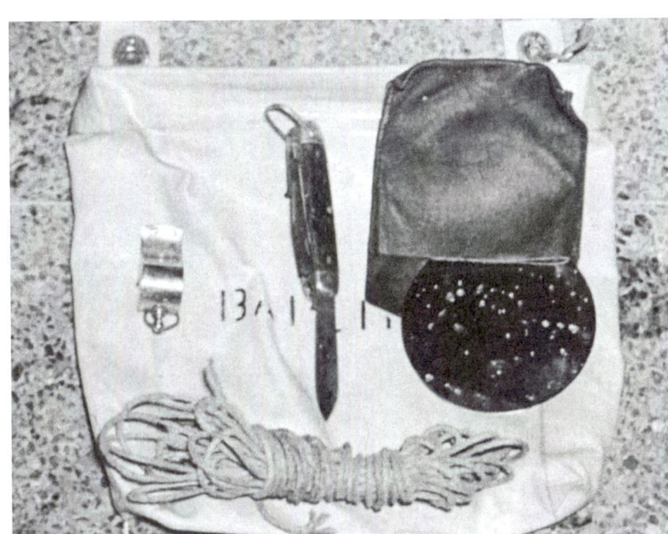

Official BSA knife stock was quickly absorbed by the war effort. Contractors were advised that the Boy Scout knife was a standard and that all types procured must be equal to the quality of materials and finish. In January 1943, Camillus submitted to the BuAer a knife it was manufacturing under US Army Specification 17-170, known as: Knife, Pocket, Engineer, 4-blade with clevis, for approval to include in life rafts. It was accepted and life raft contractors were so notified. Knives delivered to the BuAer had a blank shield instead of being stamped USA. Also, due to War Production Board regulations, all brass and nickel-plated parts were to be replaced by polished steel. This knife was purchased directly from life raft contractors for the BuAer.

Camillus began production of the 17-170 in 1940 with the distinctive feature of a stamped USA shield. Until about October 1942, production included brass linings and spacers, after which the knives were all steel. The 17-170 remained in production until the end of hostilities. By the close of 1944, restriction on the use of brass had been lifted and brass was integrated back into pocketknife manufacturing.

(Below) These variations are from Left to right: early production with 4-line Camillus tang stamp, late production (1944-45) with 3-line Camillus tang stamp and Identically fabricated knife but with a stamped USN shield.

(Below) Variations of Camillus 17-170. USA shield with four and three-line tang stamps. The other, without shield, was a general-purpose knife for resale.

The BuAer lost trust in contractors to provide an appropriate jackknife after reports of inferior deliveries. With pending development of emergency kits that would also include a jackknife, specifications were drafted for a type that could be procured by the ASO and drawn on from purchase orders by contractors. On 10 March 1943 the BuAer standardized a jackknife, referenced under NavAer Specification M-575, titled Life Raft Knife. At its issuance, the only standard equipment requiring it was the M-3 series life raft, hence the nomenclature in the specification. The BuAer entered negotiations with Imperial submitting their specifications.

The knife resembled the four-bladed BSA type but with a clip pointed main blade, and a flat steel clevis stamped USN. It also resembled the US Army Mountain knife. Imperial recommended the switch to a spear type main blade like it was providing for BuM&S and USMC. Additional recommendations included switching to a round stock clevis and placing the property mark on the tang of the knife. BuM&S contracts for the 41-J-001 specified the property mark MD-USN be stamped in that location.

Accepting those recommendations, the BuAer revised M-575 under Amendment-1 10 June 1943 for the main blade to be a spear type and the clevis to be round stock. The property mark was not applied. After other fine adjustments under that amendment, the M-575 was ultimately no different from the 17-170, 41-J-001, 41-J-4 knives or those manufactured for the USMC. All BuAer specifications requiring a jackknife detailed the inclusion of the M-575 as the standard.

Following initial procurements, new emergency kits M-592 and M-594 were finalized for adoption, both of which required the M-575. On 16 September 1943 Amendment-A became effective revising the knife's name to Knife; Emergency, Pocket since it was then no longer proprietary to life raft equipment. It was specified that the can opener and combination screwdriver and cap lifter be made of spring steel in lieu of flat stock steel.

Jackknives maintained by the ASO were referenced under stock number R41-K-365. Other Scout-style jackknives were made by Boker and Pal. Ulster continued to produce its official Scout knife but eliminated the shield. These four companies retained US Navy contracts as with the QMC for resale. Camillus and Imperial would have the most significant negotiations with the Navy Department and BuAer. Boy Scout type knives like these were purchased by life raft contractors and the ASO for stock purposes conforming to specification M-575.

The M-575 became a standard component for the M-3 series life raft, M-592 back-pad kit, M-594 Shipwreck kit, and the ADR-1. Life raft contractors purchased knives from many approved sources. Contractors for the M-592 and M-594 negotiated jackknife supplies with Imperial Knife Co. In 1942, Imperial and the Ulster Knife Co. entered a joint venture as the Imperial Knife Associated Companies to produce knives for the US military. It was known on paper as the Kingston Cutlery Co. Kingston's knives had Made in USA stamped on the tang. The knives delivered in accordance with M-575 were manufactured of all polished steel parts but had variations in can openers and jigged bone or acetate scales; some had brass liners. One variation has KINGSTON etched on the main blade.

(Left) US Navy approved suppliers for life rafts and emergency kits. Left to right, Imperial, Boker, PAL, Ulster. All steel construction and stag scales.

(Left) The knives by Imperial, Ulster, and Kingston were both engineer and general-purpose types procured by the QMC. After Imperial added a new safety can opener, Ulster and Kingston followed suit. Knives manufactured with this can opener had all-steel construction and variants with brass liners.

(Right) Jackknives could be conveniently carried by an individual such as this VD-5 aircrew member, January 1945. (National Archives)

(Below) Imperial was the prime contractor for the BuAer supplied through Kingston. The examples here were supplied as detailed under Specification M-575. Types can be found with stag and acetate scales. Tangs stamped Made in USA. Common feature includes stud on can opener, other variant provided without stud.

Two types of Kingston M-575 jackknives included in back-pad kits. (Above) Example from an original back-pad kit with acetate scales. (Above right) Jigged bone variant from 1943 contractor Belber Trunk & Bag, model M-592. Knives for emergency kits were required to be supplied with a three-foot lanyard, which is missing from the M-592 knife, a factory oversight. Jackknives were sealed in an asphalt foil-lined wrapper to prevent corrosion while in storage.

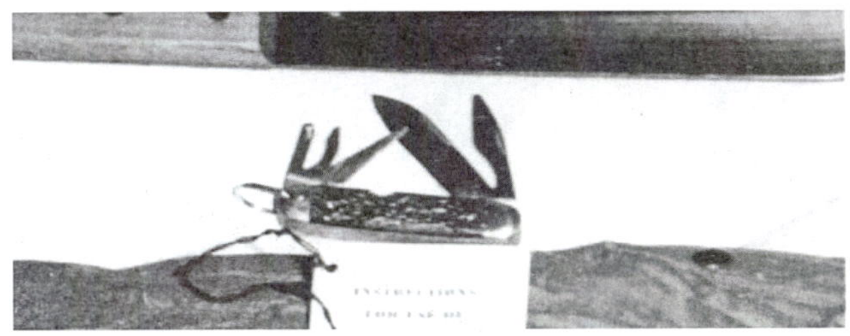

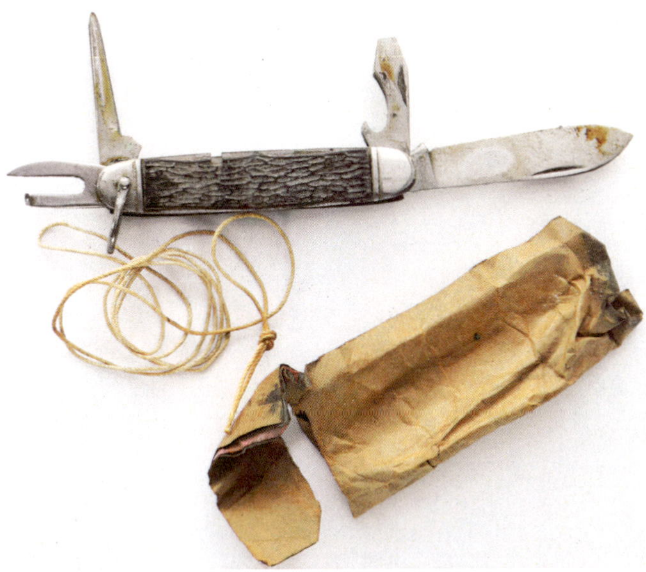

(Left) Camillus engineer type pocketknife with acetate scales, provided in a 1944 Belber M-592 back-pad kit. Approved as M-575. (Below) Display of 1944 Belber M-592. Closer examination revealss a bent pull on the can opener, a typical feature on Camillus engineer pocketknives.

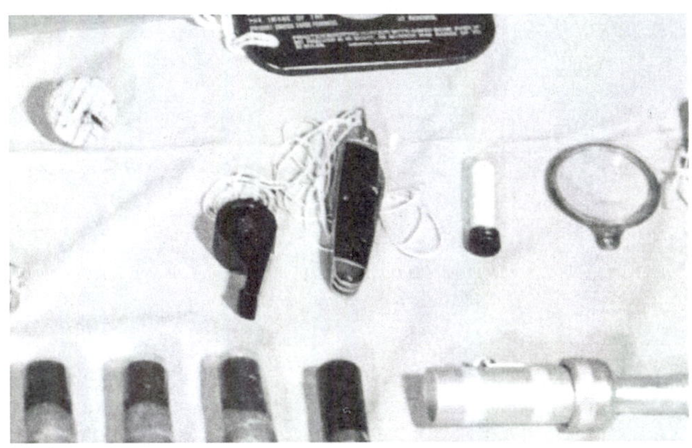

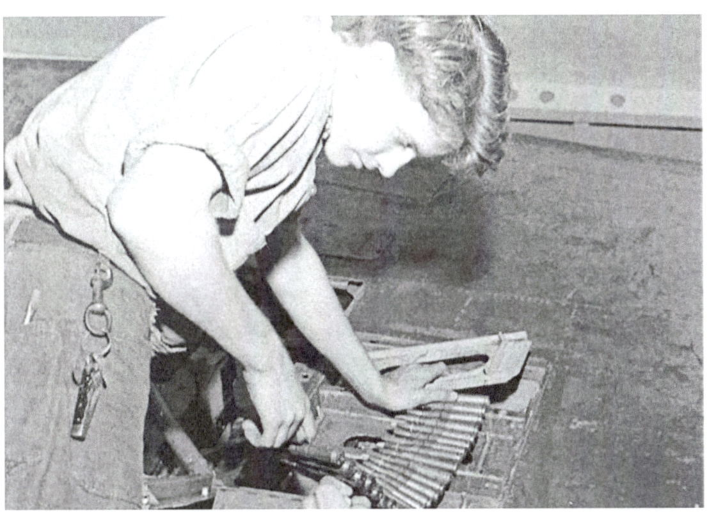

(Below) 41-J-4 manufactured by Imperial with acetate scales included in another 1944 contract M-592 assembled by Belber. (Left) Deck crew on USS Hornet with 41-J-4.

Another pattern of knife was procured in accordance with Specification 41-J-4 that had three blades and was referenced as 41-J-4, 3-bladed. It was primarily purchased by the ASO as a utility tool like the electrician's knife TL-29 for repair and maintenance facilities. These were being procured prior to the issuance of M-575 when the only other available specification was 41-J-4. It is listed as Jackknife, Boy Scout, three blade type and included one heavy cutting blade, one screwdriver, and one bottle/can opener. It shared the same stock number, R41-K-365.

(Right and below) Imperial was the primary contractor for the three-bladed knives, both on its own and through Kingston. The company primarily delivered its standard Scout knife. Its commercial knife included plastic scales jigged in a unique pattern. These three-bladed knives can be seen in the back-pad kit assembled by ComAirPac and in a display of life raft accessories. Its construction was not from quality materials and was an example of the inferior types that easily corroded or broke, prompting the BuAer to draft M-575. Type with shield stamped Scout is same as pictured in ASO supply catalog for Class-41 material.

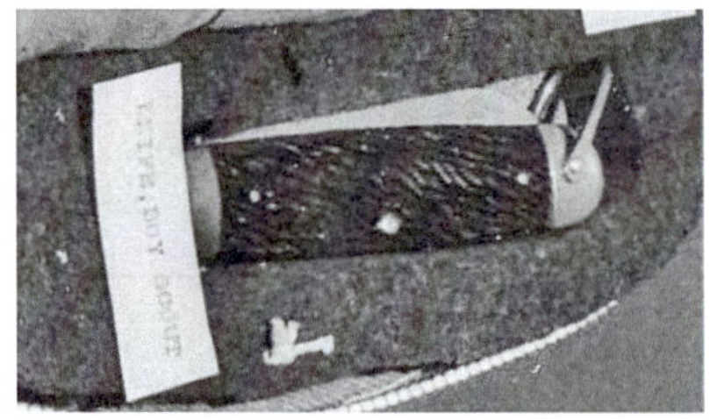

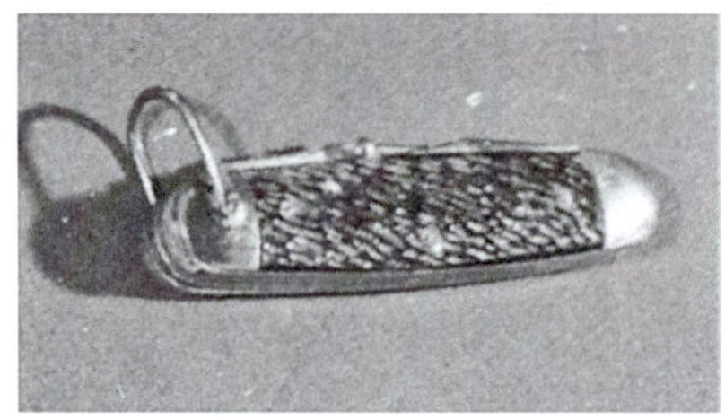

(Below) Manufacture transitioned to construction details prescribed in 41-J-4. The four-bladed knife became the preferred standard, while three-bladed patterns were viewed as working utility tools. Procurements made in 1944 from Imperial and Camillus included versions that included a stamped USN shield and without. Those without stamped shield were for ships' stores as a resale item. From Left to right: Imperial with long pull, Imperial with short pull, Kingston with stamped Made in USA stamped tang.

The 41-J-4 was the standard Navy Department jackknife, available from stock at Naval Supply Depots and for purchase. The jackknife had become a standard issue utility tool for sailors under the same premise as the fixed-bladed type. After its use increased, it was found to be deficient in that it had limited use.

To meet a wider range of circumstances requiring a knife, a new specification was drafted effective 14 April 1944 as Specification 41-J-5 and titled Jackknives, Utility. It superseded 41-J-4 and upon its issuance all Navy contracts for the 41-J-4 were cancelled. The new knife had a large blade, a small blade, a can opener, and a combination screwdriver/bottlecap opener. It was required to have a shackle or clevis fitted to one end and an imitation stag handle. The small blade was to be affixed to the shackle end with the large blade to the opposite side. The specification describes the use of brass liners and nickel-steel parts. Pre- and interwar manufacturing practices included brass and nickel-steel components, but, to comply with WPB regulations, manufacturing transitioned to all polished steel.

The 41-J-5 immediately entered mass production by five contractors: Camillus, Imperial, Boker, Pal, and Ulster. Camillus and Imperial were awarded the bulk of production. From 1944 through 1945 Camillus produced just over 4,000,000 and Imperial reached 2,500,000. The other companies cumulatively produced close over one million. The production of this knife was equal to the volume of the Engineer type 17-170 procured by the QMC, of which Camillus manufactured three million. Figures for the other companies are unknown. Ships' stores resold the knife at $0.65. Four-bladed knives were negotiated at an average cost of $0.52-$0.57 each.

The 41-J-5 resembled the 17-170 but the leather punch was replaced by a pen blade affixed to the shackle end. Blade placement was indicated in the specification, but can opener and screwdriver placement were not. Their positions were on different ends, depending on the manufacturer. Camillus integrated its new can opener, long shackle and a new long screwdriver. Imperial applied its new safety can opener in the shape of an eagle claw or beak.

Through 1944, the US Army and Navy extended requirements for pocketknives creating a shortage. Contractors of the USAAF C-1 vest were authorized to requisition knives on Navy contract from Pal and Imperial. The BuAer also reported a shortage and a delay in delivery of M-575s. To compensate, the Chief of the BuAer authorized the substitution of the 41-J-5 and 41-J-4 on all contracts for life rafts and the M-592 until the M-575A became available. Those maintained by the ASO were referenced under the same stock number, R41-K-365. By 1945, the 41-J-5 became the unofficial standard, although NavAer specifications continued to require the M-575. The 41-J-5 became the Navy Department's standard for all Bureaus until superseded by the MIL-J-818 in 1948.

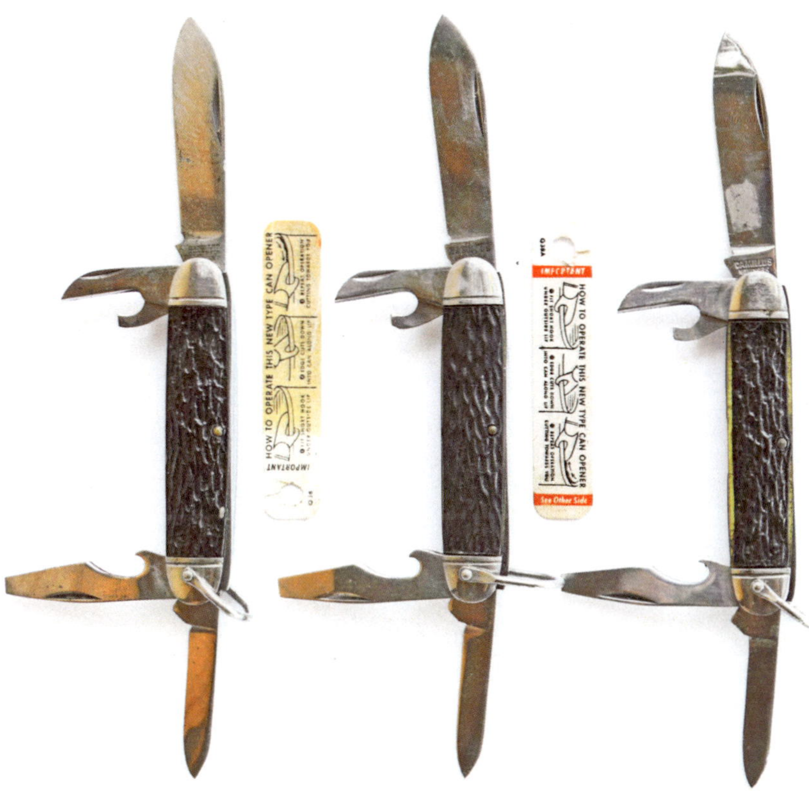

(Left and below) Examples manufactured by Camillus. Each knife provided with instructions on new type can opener. Left to right: 4-line with short clevis, 3-line with short clevis, 3-line with long clevis. (Below) comparative between long and short clevis.

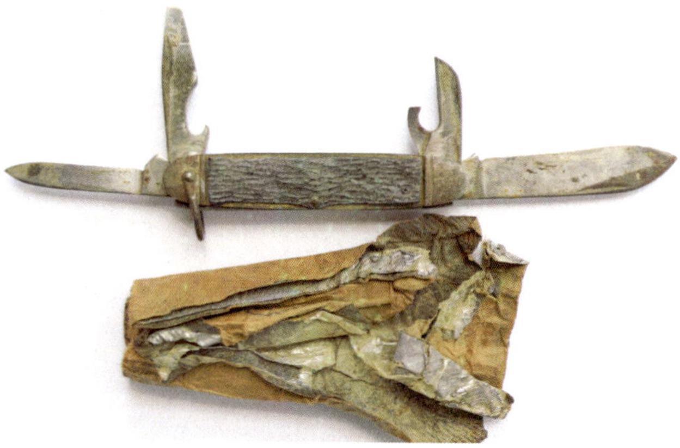

(Above) Camillus 41-J-5 provided in a 1944 contract M-592 assembled by New England Bedding.

(Above right) Imperial 41-J-5 provided in a 1944 contract M-592 assembled by New England Bedding. Note steel pins to secure scales.

(Right) 41-J-5 manufactured by Imperial with either brass or steel pins in the scales.

(Below) Ulster purportedly produced the 41-J-5 but its production was likely delivered through Kingston. Its jackknife was manufactured in accordance with specification but used jigged bone scales. Pal produced multiple variations, all in accordance with Navy specification. One Pal variation included jigged bone scales and imitated the Remington can opener design. Pal Blade Co. continued to produce pocketknives with Remington parts after acquisition. Pal also manufactured a version, but in accord with 17-170 it had a leather punch blade. Left to right: Pal with Remington can opener (all steel construction), Pal manufactured in accordance to 41-J-5 with acetate scales and a type by Boker.

The BuAer first considered the jackknife a utility or repair tool provided in life rafts. As the war progressed, the knife also took on the role of an emergency tool. Having proved its worth, it was integrated into back-pad emergency kits, due to its compact size, although a machete was later added.

As emergency equipment became personal items, jackknives became commonly worn on the person. They were suitable alternatives to the fixed-blade hunting knife. After the introduction of the PK-1, jackknives were listed as optional Pilots Personal Equipment. The small jackknife was easily secured to the individual, ready for any emergency aboard an aircraft or in a ditching or parachute landing. Equipment deemed immediately necessary was kept in flight suit pockets or secured to the life preserver or other means.

(Right) Complement worn by VF-20 pilot Lt. jg. E.A. Phillips comprises a whistle, reflector, and jackknife, USS Lexington, January 1945. These items were available from the back-pad kits, and in this case probably from the Original kit. (National Archives)

(Below right) Lt. jg J.B. Chandle of VF-82 wears an assortment of tools on his life preserver, including a Mark 2 knife, standby compass, whistle, life preserver light, and jackknife secured to oral inflator tubes. (National Archives)

(Below) Unidentified Corsair pilot of VMF-312. Adjacent to hunting knife and M-1916 hip holster is jackknife. Pistol in cloth wrap. (National Archives)

(Above) Ens. William Friday of VF-3 gives one of the better illustrations of Pilots Personal Equipment, USS Yorktown, February 1945. When the back-pad kit was considered for elimination, more items were carried on the individual or in the pararaft, which was eventually standardized as the PK-1. The life preserver was issued with two sea markers and a whistle. The sea marker packets were on the backside of the vest to prevent them from snagging. A side arm with shoulder holster and a hunting knife were standard issue; Friday wears the Mark 2. The jackknife was secured to his life preserver (optional). A locally fabricated emergency kit was secured to vest front. The shape of one pocket indicates it likely contained tablet ration M-539B. The large pocket likely contained smoke signals and a signaling mirror or other rescue items. Sewn to his flight jacket is the CBI shoulder sleeve insignia. The Yorktown supported operations in the Philippines, flying sorties over the area known as French Indochina. The CBI patch was distributed by ACI officers. (National Archives)

(Above right) VB-20 SB2C Helldiver gunner William Martin ARM2/c with 41-J-4 secured to his life preserver, March 1945. (National Archives)

(Right) VB-20 gunner Carlton Walker ARM2/c aboard the USS Lexington, February 1945. (National Archives)

(Left) Ens. J.C. Dixon of VF-17 at Bougainville Island airfield, March 1944. Jackknife with lanyard secured to life preserver. He additionally wears locally fabricated emergency vest. (National Archives)

(Below) Unidentified aircrew member of a US Navy PB4Y-2 Privateer, c.1945. Secured to trouser belt is a jackknife similar in appearance to 41-J-5.

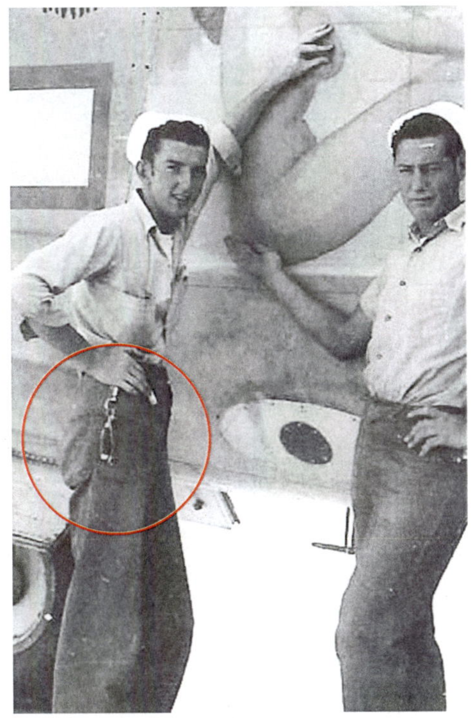

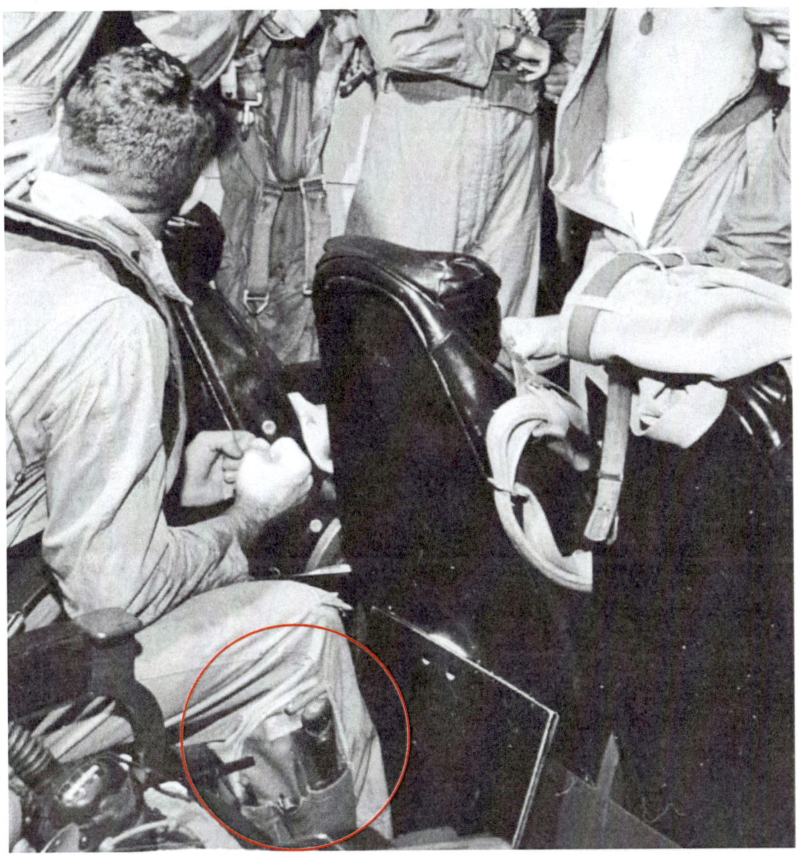

(Left) VF-80 pilot in the ready room of the USS Ticonderoga, November 1944. Jackknife is in the applied lower leg pocket next to the hunting knife. (National Archives)

(Below) Comedic cartoon illustration exaggerating the equipment a pilot or aircrew member carried. Attached to belt is large hunting knife.

(Above) Unidentified gunner of VMSB-231 with simplified personal complement of 23P6 life preserver, one sea marker packet and a jackknife, Marshall Islands, c.1944. (National Archives)

Unlike the US Navy, the USAAF never drafted specifications for pocketknives. It preferred hunting knives, and only after adoption of parachute emergency kits were pocketknives considered a standard item. The kits developed in the 1930s first included a five-inch hunting knife and a nine-inch machete. It wasn't until the standardization of the Type B-1 parachute kit in 1941 that a folding knife was adopted. That knife was a switchblade manufactured by the Shapleigh Hardware Co., referenced as Hunting Knife.

The Scout knife was first adopted for use in the Type B-2 jungle kit and remained standard and used with the Type B-4, eventually replacing the switch type in the B-1. It would further be integrated into the Type C-1 vest. Specifications for the B-2 and B-4 simply cite: Knife, Pocket, Three Blade, with their relative drawings citing a Boy Scout knife.

Adopted was a three-bladed Scout knife commercially referred to as a Cattleman, which consisted of three cutting blades. Those supplied to the USAAC had a spear point main blade, sheepsfoot blade, and a pen blade on the opposite side. Camillus and Imperial are recognized as prime contractors for the USAAF three-bladed Scout knife. Other companies, such as Case, Pal, Utica, Ulster, and Cattaraugus, manufactured that pattern but sold them to the QMC as a resale knife. There were also potential US Navy purchases. Camillus began limited manufacture of the Scout type knife in 1940 and 1941; it is not clear if these were under USAAC contract. The most significant purchases from these two companies were in 1942, with 775,767 from Camillus and 540,000 from Imperial.

The knives were requisitioned from supply under stock number 7900-473900 and simply titled Knife, Boy Scout. Their procurement declined in 1943 as ample numbers were in stock and those kits requiring the knife were en route to becoming Limited Standard and obsolete. All kits requiring a Scout knife transitioned to the four-bladed engineer design like that adopted for the C-1 vest.

Camillus produced the most extensive variety of knives. Its original assembly included jigged bone scales or handles with a blank shield, brass linings, and nickel steel parts. The deletion of the shield occurred by the end of 1941, when brass linings were also eliminated. The bulk was manufactured with all-steel parts and bone handles, which later transitioned to imitation stag. One variation had rosewood scales but with a three-line tang stamp indicating a late manufacture.

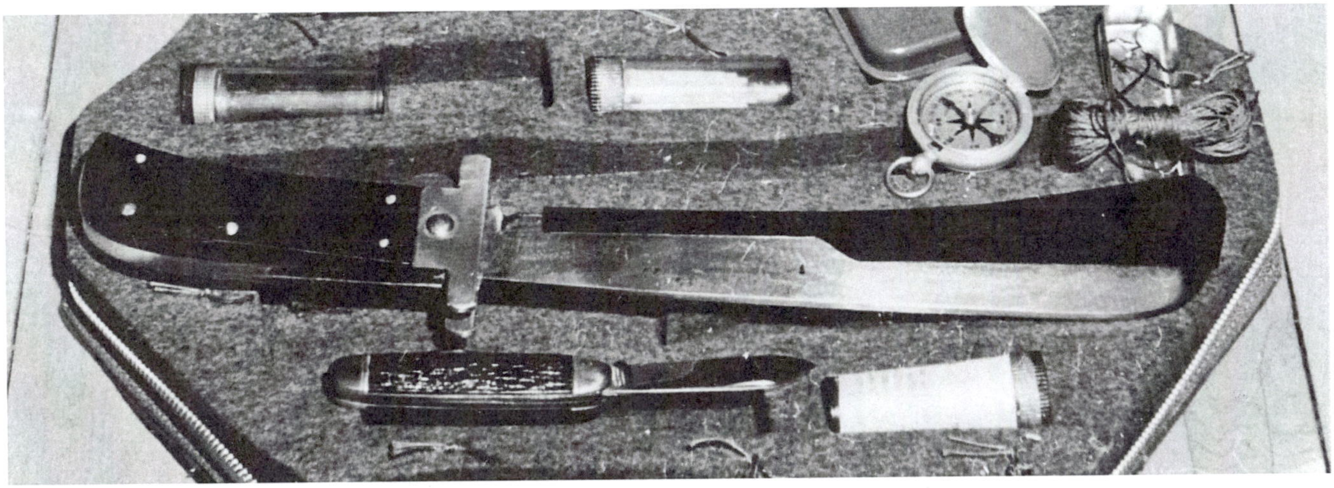

(Above) Felt inserts in the B-2 and B-4 were cut to accommodate the Scout knife, which was stored with the compass below the machete. Here, a Camillus Scout knife with B-2.

(Left) 4-line Camillus Cattleman as originally provided on USAAC contract.

(Below) Variations of Scout knives provided on USAAF contract by Camillus. Left to right: Long and short pulls on the main blade with stag scale, acetate scale type and rosewood scales.

(Above) Scout knife manufactured by Kingston (left) and Scout knife manufactured by Ulster (right).

(Above) Imperial Scout knives. Left has jigged bone scales and brass pins, right is of all steel construction and acetate scales. The Imperial line adhered to construction changes, first having bone scales and brass linings; later, all-steel and imitation stag handles.

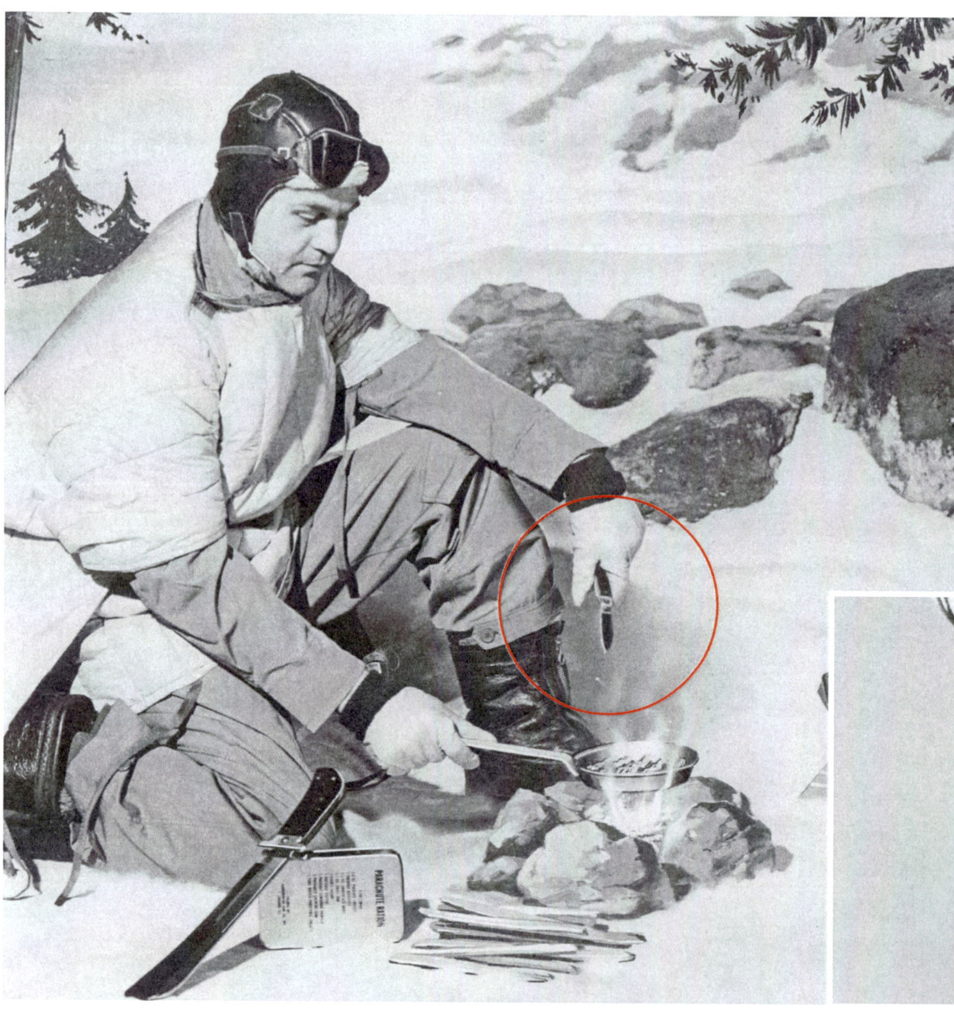

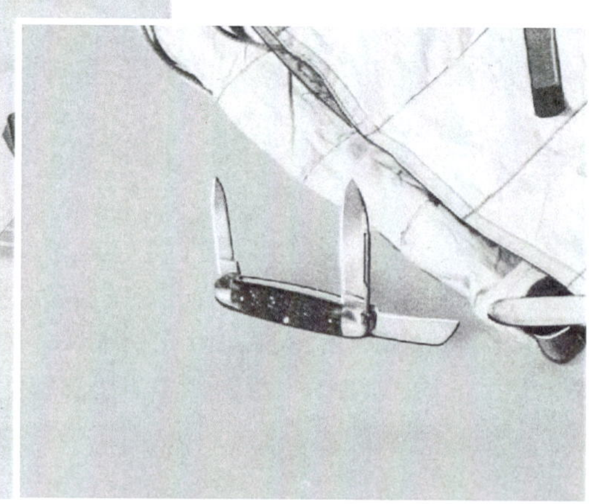

(Left and inset) Volunteer simulates cooking meal with frying pan from B-4 kit. For preparation and utensil, a Scout knife is used, left hand. Folding machete at his right. He dons the quilted poncho and D-2 gloves. (Below) 3-bladed Scout in display of accessories for B-4 kit. (USAF)

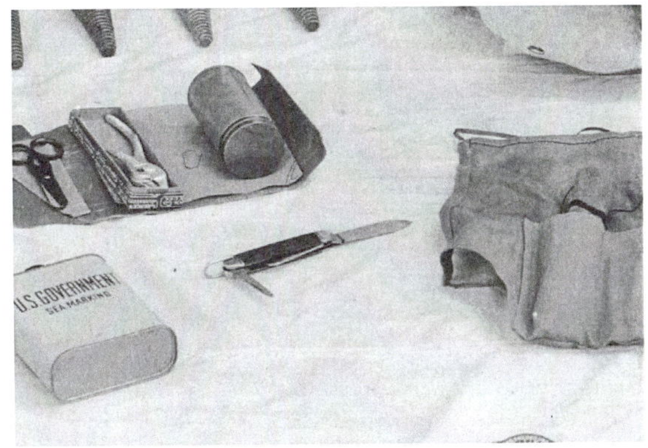

(Above) Following life raft revisions, the Scout knife was officially adopted in December 1942. To prevent loss, a clevis was added for attachment of a lanyard; all other details remained. Camillus and Imperial retained contracts for these revised knives. Example of Camillus life raft knife with all-steel construction, acetate scales and a clevis. Detail to note is placement and attachment of clevis. (Right) Wartime image exhibits the same knife, displayed with the accessories of a USAAF multi-person life raft.

(Far left and left) Life raft knives manufactured by Imperial (left) and Kingston (right), each made from polished steel and acetate scales. Colored image is from a training aids poster for Type A-3 life raft.

(Below) USAAF B-25 Copilot Lt. Thompson was rescued after a forced ditch somewhere in the Solomon Islands; around his neck is the life raft knife, April 1944. Like the Navy, the USAAF required contractors to supply the knife as CFE. This pattern of knife was superseded in 1944 by floating life raft knife, Dwg. No. 45B18786.

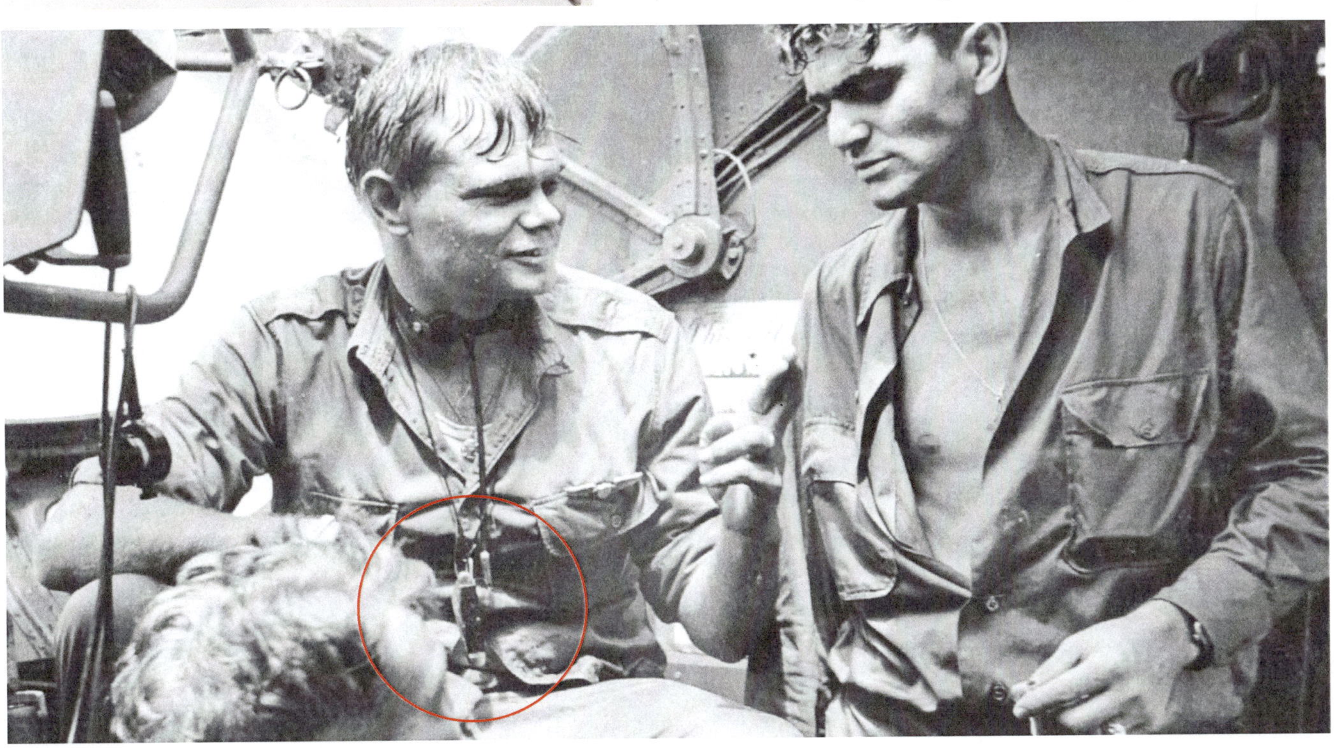

USAAF standards required a Scout knife be included in parachute emergency kits and life rafts. The latter was to have a shackle or clevis. Specifications for the multi-person sustenance kits would progress to include a pocketknife but of the engineer type. Types E-10, E-11, E-14, and E-15 would require a pocketknife. The specifications for the E-10 and E-11 cited the same entry: Knife, Boy Scout model (4-blade). The specifications for the E-14 and E-15 cite: Knife, Pocket, Boy Scout Type, with large blade, can-opener, and screw driver and interestingly specifically mentions Boker as a manufacturer. By description, the knife was to be the Engineer 17-170.

(Below) These wartime images are of pocketknives in E-14, and E-15 multi-person sustenance kits. (Left) E-14 with Engineer shield marked USA. (right) E-15, knife in accordance with the 17-170 Engineer knife.

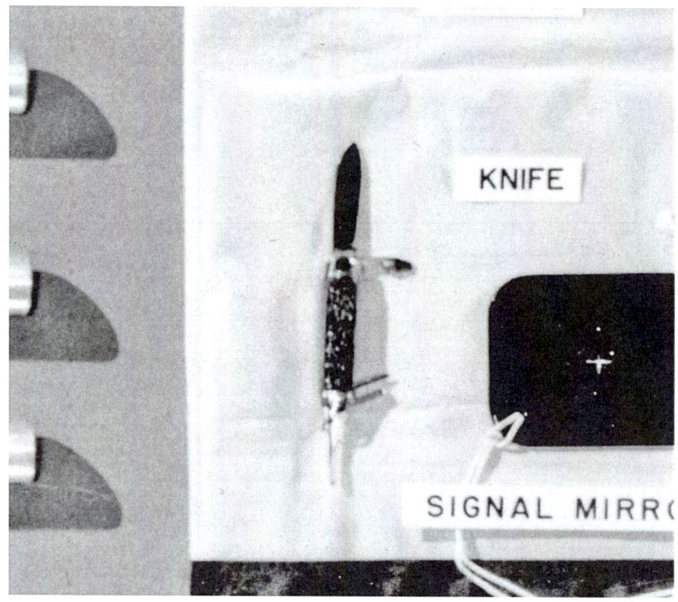

(Below) These wartime images are of pocketknives in E-11 multi-person sustenance kits. They are both commercial Scout type knives with pearl handles, associated with construction method by Imperial.

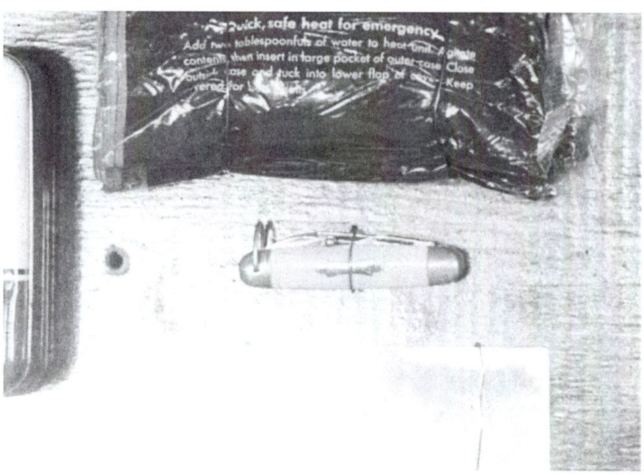

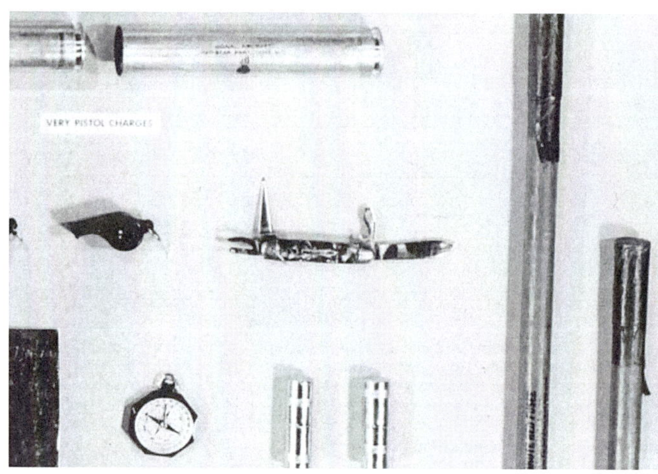

(Left) Camillus produced a 4-bladed Engineer type knife of all-steel construction, acetate scales but with a shield stamped AAF. The examples have a 3-line tang stamp placing its manufacturer by late 1944. It is not clear for what purpose these were produced, but potentially for inclusion in E-14 or E-15 sustenance kits, assuming the shield stamp signifies Army Air Force.

Operational experience had shown the knife to be a valuable life raft tool, but the existing type was prone to corrosion if mishandled, or loss if the raft were to capsize. When evaluating the RAF clothing and life raft knife in 1943, US personnel found the RAF knife features to be desirable, as they eliminated accidental punctures and had the ability to float. The AML copied the design for an emergency knife on board aircraft. Another knife of an almost identical design was incorporated in life rafts.

In January 1944, directive TMI 04-15-1 instructed the British type of knife be installed on AAF life raft equipment. It prescribed that it be installed on the bladder of all types of multi-person life rafts.

The curved emergency knife came first, and the life raft version copied its design, referenced under Assembly Dwg. No. 45B18786 as Knife, Floating. They shared the same handle and scabbard designs but differed in blade shape and attachment method.

The life raft version had rubberized fabric backs on the leather scabbard for cementing it to the bladder of the raft and the blade had significantly less curvature. Amendment-A of the life rafts Type A-3 and E-2 dictated that the knife be applied at the factory. Active service personnel requisitioned the knife under stock number 7900-470970. Western States Cutlery is the only known manufacturer; initial contracts were awarded in April 1944 prior to the release of the formal drawing.

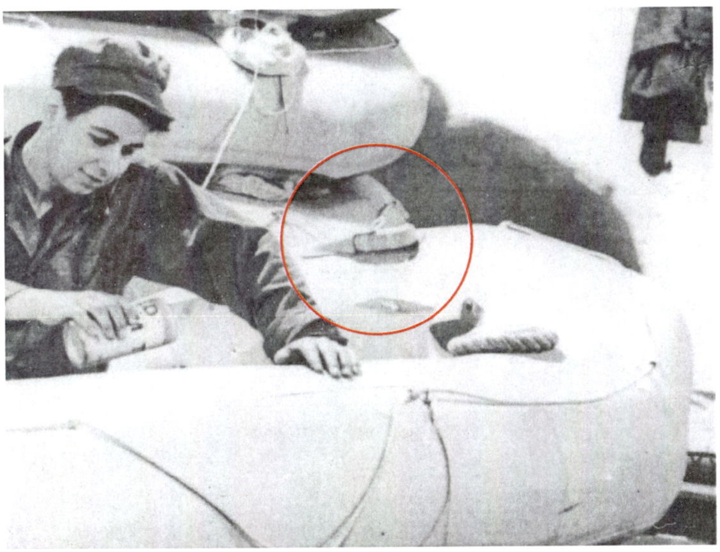

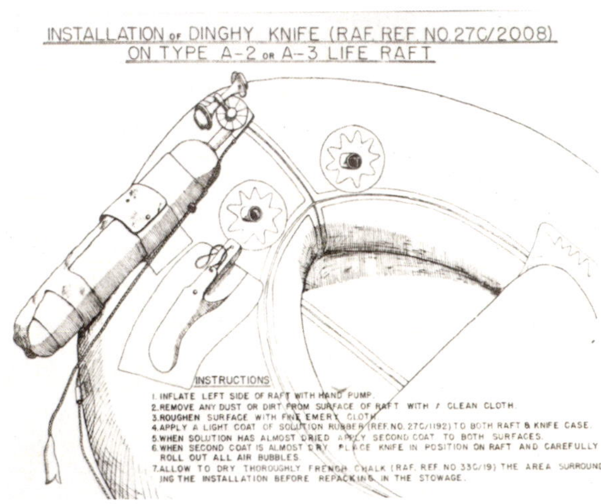

(Above and left) Drawing attached to directive TMI 04-15-1. (Left) British life raft knife attached as prescribed by drawing to USAAF life raft.

(Left and below) British version (top) is exhibited above the USAAF type (bottom). (below) Materiel Command image illustrating placement and details of new life raft knife 45B18786.

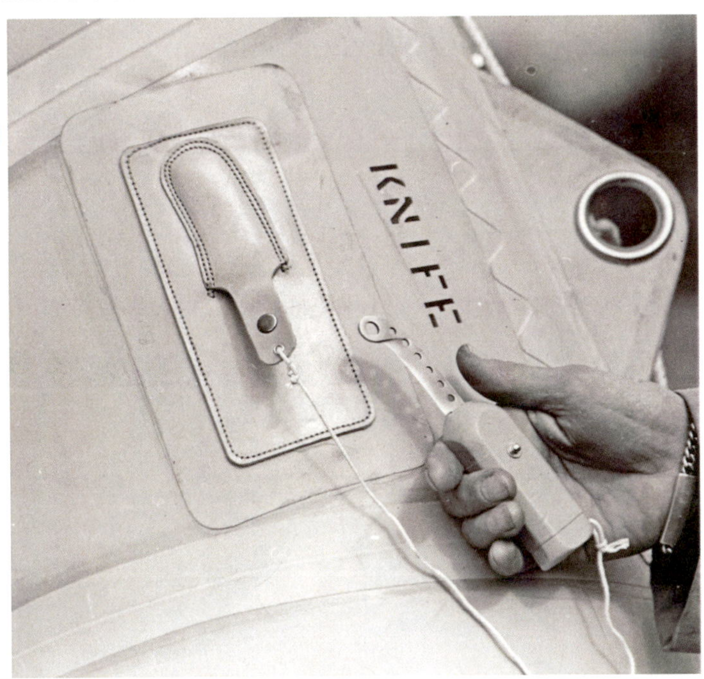

Standard emergency parachute kits for the USAAF and BuAer progressed to include a small machete and a jackknife, supplemented by a hunting knife worn on the person. By the close of 1943, both services developed new emergency equipment, which presented new requirements.

In April 1943, the USAAF initiated development of a new vest kit. The intention was to create a universal individual sustenance kit in the form of a vest with pockets for the components. The standard parachute kits included a three-bladed Scout knife and a ten-inch folding machete; the machete created the challenge of integration in the proposed sustenance vest.

The initial recommendation was to create a pocket on the back to accommodate the machete, but this was quickly dismissed as adding too much bulk. George Peterson, vest designer, emphasized that each item was to be compact. His solution was to design an oversized collapsible knife with three blades, designed to fit into a pocket with the dimensions of 6 x 1-¼ inches. His knife was a copy of the three-bladed one in parachute kits. He then consulted with Colonial Knife Co. and by the middle of June 1943, Colonial had samples for evaluation. After further study and consultation with Colonial, the knife was redesigned to have a master cutting blade of 4-1/8 inches, dagger blade of 4-3/16 inches, and a saw blade of 4-3/8 inches. It was detailed in a drawing submitted on 15 July 1943. The small Scout knife was retained and contained in one of the vest accessory pockets, but, with the elimination of the machete, a replacement was required. The new large folding knife assumed that role. The robust master cutting blade could be used for splitting wood by hammering across the spine of the blade, the saw was capable of cutting wood and metal, and the dagger blade (with its fine point) was of general use.

The vest kit was in development for the remainder of 1943, during which the dagger was eliminated. The new two-bladed configuration placed the main cutting blade and the saw on the same side with a locking spring on the outside of the frame, which locked the cutting blade in the open position. Colonial continued to revise knife details, by which time the company had unofficially titled it the "Jiant" Jack-Knife. By January 1944, Colonial engineered a design suitable for mass production. Samples were submitted and approved on 14 January 1944, referenced under USAAF Assembly Dwg. No. 44D5159 and titled Knife Assembly-Vest Type Emergency Kit, All Purpose.

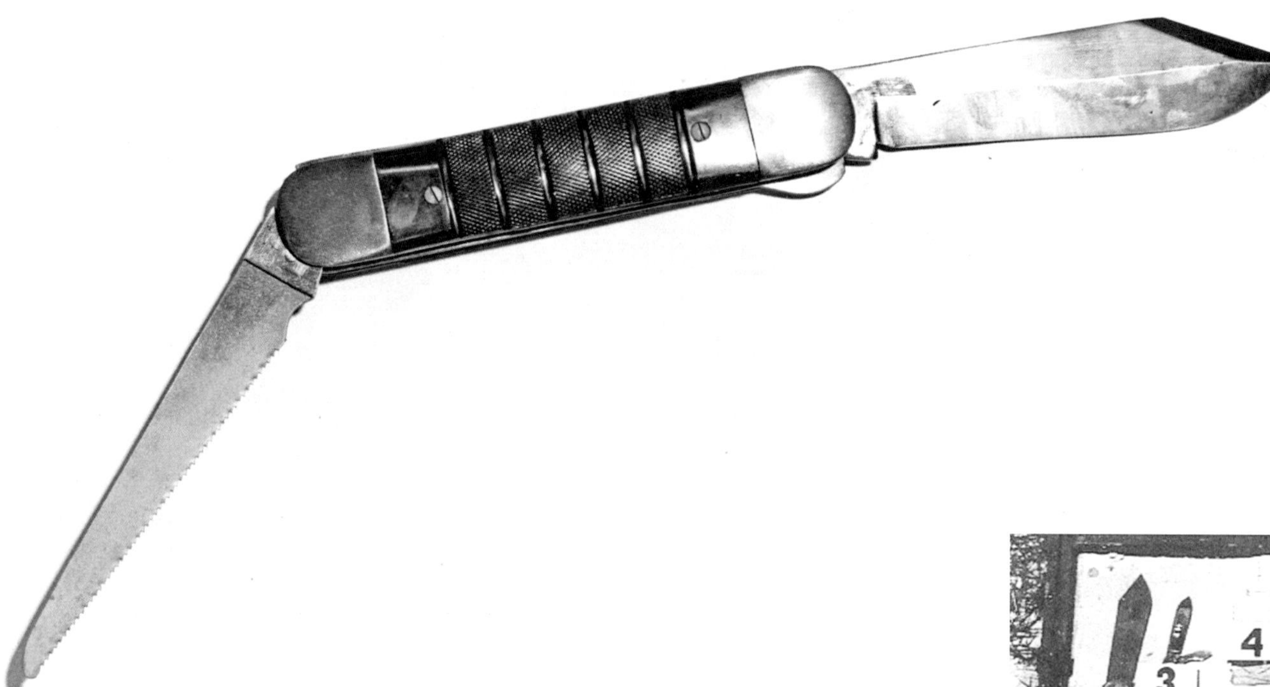

(Right) 3-bladed version in display of components for proposed vest kit (labeled as no.2). Note: saw and dagger blades housed opposite main cutting blade. Knife was of all-steel construction, including the handle, to reduce use of critical materials.

(Above) Revised version as adopted under Dwg. No. 44D5159. Finalized blades were at opposite ends with the locking spring in the center inside. Scales were changed to be made from a phenolic plastic referred to as Tenite-II. Steel parts had a rust preventive finish applied as detailed in Specification 57-O-2c, Type II, Class A, Grade I.

Colonial was awarded all the initial contracts for the jackknife with quantities equal to the purchases of the newly designated Type C-1 emergency sustenance vest. All negotiations and purchases were conducted by the vest contractors and provided as Contractor Furnished Equipment.

By fall 1944, with increased procurements of the vest, resident inspectors noted problems with alkali blooms on all parts of the knives. To prevent further corrosion issues, requirements for rust preventive were increased to Type III, Grade A, Type I as detailed in revised Assembly Drawing No. 45D9060 dated 24 November 1944, which made no engineering changes to the knife. Drawings 44D5159 and 45D9060 required salt spray corrosion tests, however, when the revised drawing was issued Peterson verbally authorized Colonial to treat the surfaces with a nonreflective black oxide finish, thereby eliminating the need for salt spray tests. Alternatively, conversations were taking place regarding sealing the knives in a waterproof wrapper. In April 1945, contractors were required to seal the knives in 30 lb. Kraft-asphalt lead foil packets, the same material as used by the BuAer.

Colonial was the prime contractor until January 1945 when United Machine Tool Co. was approved for manufacture. United took over production and supply to C-1 vest contracts. The company had on contract 154,000 knives, equal to the amount of outstanding C-1 vests on procurement. Colonial's production was allocated to the BuAer.

By October 1944, alternate sources for the manufacture of the jackknife were being explored. Lang & Lang submitted a sample that was approved. Colonial then submitted a formal complaint to Wright Field accusing Lang & Lang of intellectual property theft by altering Colonial's knife and drawing and submitting them as its own. The resolution of this dispute is unknown, as is whether Lang & Lang produced the knife in quantity. The USAAF listed it as Knife, Folding, Hunting and assigned no stock number, instead listing it as CFE. Each knife was purchased at $2.23.

Colonial submitted samples of the Jiant Jack-

(Left) Folded all-purpose jackknife as procured by the USAAF, late production type in asphalt wrapper.

(Below) Jiant jackknives manufactured by Colonial and United Machine. USAAF procurements lacked clevis. Unmarked versions have been observed, manufacturer unknown. Unmarked types share construction details of Colonial.

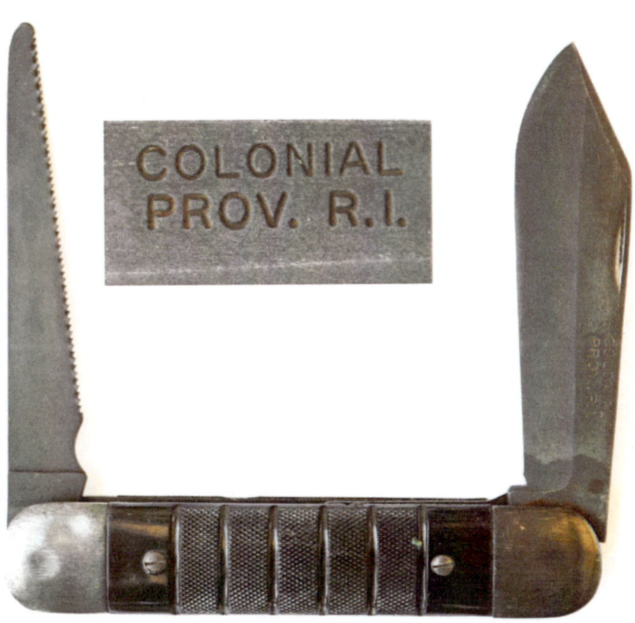

Knife to the BuAer while in development with the USAAF. The company kept the BuAer updated on all progress in hopes they would find the same knife or another of use in emergency equipment. The BuAer expressed some interest in the knife and made limited purchases for field testing purposes. However, the BuAer encountered the same issue as the USAAF when it initiated streamlining of the M-592 and AN-R-2 pararaft into a compact universal seat kit.

The design process continued during 1944 while the BuAer evaluated alternatives to standard equipment, including this knife. By the summer 1944, a working prototype of the combination seat kit was prepared for testing. The concept was to eliminate the back-pad kit and those items not included in the seat kit became the Pilots' Personal Equipment. Partly influenced by the development of the large jackknife, the V-44 machete and M-575 jackknife were deleted. The machete was too bulky to efficiently integrate into the new assembly and too cumbersome to be worn as personal equipment. The Colonial all-purpose jackknife compensated for the machete and was small enough to make the M-575 redundant.

Upon standardization of the new PK-1 pararaft in November 1944, a bid was submitted for the manufacture of 50,000 all-purpose jackknives, equal to the amount of PK-1s projected for procurement. After the award of the PK-1 contracts in January 1945, procurement of the jackknife was authorized. However, at the time the knife design was again being revised. As detailed in BuAer Dwg. No. 9253, the jackknife included a clevis and was provided with a sheath. It was titled as Knife, Emergency, Aviator's.

After these changes were negotiated with Colonial, contract NXSA-94964 was released for the manufacture of 100,000 jackknives on 28 February 1945. As detailed in Technical Order 24-45, the PK-1 became the new standard for all crew positions in carrier based aircraft, complemented with Pilots' Personal Equipment. These personal items included two sea markers, two distress signals, a life preserver, .38 cal. pistol, shoulder holster, wrist compass, life preserver light, headnet, fishing kit, whistle, first aid kit, and a Jiant jackknife. The knife replaced the old standard of a hunting knife. Of the procurement of 100,000 knives, 50,000 were intended to be issued with the new PK-1 and, upon the new directives to eliminate the issuance of the back-pad kit that contained the machete, the remaining were placed in stock and requisitioned as needed via stock number R41-K-465. Both the V-44 and M-575 remained in inventory as items of optional equipment.

(Left) The jackknife procured by the BuAer retained the design details but added a clevis. A sheath with a belt loop was also provided. Colonial was the only contractor for the BuAer. (Left) Pilot Ens. Douglas Millard of VF-23 clowns around with other pilots on board the USS Langley, c.1945. One of his accruements is a Jiant jackknife.

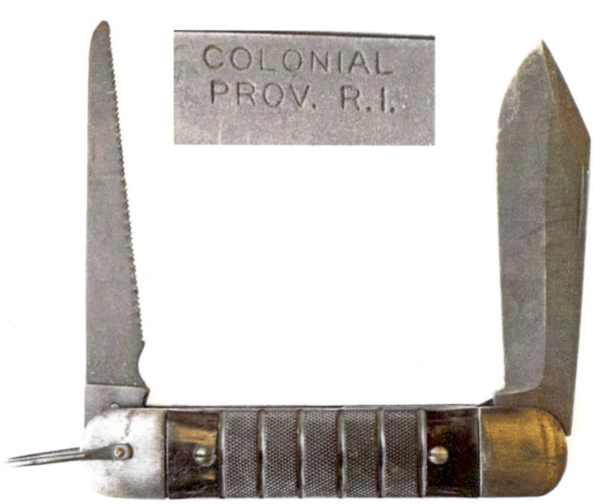

COMPASSES

Compasses are vital for navigation on land and at sea – and therefore were integral to emergency rescue equipment during wartime. A compass was first adopted as standard by the USAAC for use in its new pilot's emergency kit in 1934. However, during this era, compasses were not included in such items as life rafts, being proprietary to USAAC land units. A variety of compass types, meeting a range of specifications from casual commercial conditions to technical governmental requirements, were utilized by both aeronautical services during WWII (e.g., Pilot Emergency Kits detailed a Pocket Compass, and the Jungle Seat Kit included the US Army Engineer Prismatic).

Prior to the United States entering WWII, the USAAF had been preparing specifications for several parachute emergency kits. While the prismatic compass was preferred, due to the volume needed, manufacturing logistics, and cost, the pocket watch type of compass – more readily available and less expensive – was formally adopted.

Upon the onset of war, Keuffel & Esser discontinued manufacturing pocket compasses in favor of technical surveying and engineering instruments for the US Government. Similarly, Waltham Watch Co. stopped production of compasses and switched to wristwatches. In spring 1942 the newly renamed Longines-Wittnauer Watch Co.- previously Wittnauer Watch Co. until 1936- began production of the pocket compass for the USAAF under Assembly Dwg. No. 1941. Its model was almost a direct copy of the Keuffel & Esser and Waltham models. This style of pocket compass would be officially standardized as the Compass, Pocket, Type H-1 under Specification No. 94-27464, on 16 March of 1943.

(Left) One compass adopted by the USAAC for parachute emergency kits was fashioned to resemble a traditional pocket watch. Manufactured by Keuffel & Esser Co., a New York company founded in 1867, this compass had a nickel-plated brass case and spring-loaded cover and was classified as Compass Pocket Type, reference stock number 8300-060930. These compasses included a cotton pouch and were to be included in the F-1, B-1, and B-2 type parachute kits in early production assemblies. This brand used no government property markings. The Type F-1 and B-1 technical data sheets cite Keuffel & Esser to be provided.

(Below) Specifications for USAAF Multi-Person Emergency Sustenance Kits in 1942 and 1943 called for "Waltham or Equal" compasses, supplied as CFE. Waltham Watch Co. supplied the US Army Corps of Engineers with pocket compasses from WWI to 1942 when it was replaced by a lensatic type. This compass was originally manufactured in all brass or nickel-alloy plated models during WWII with a US impression on the front cover.

(Right) Longines-Wittnauer was the primary supplier to the USAAF for parachute and sustenance kits during WWII. Each compass, made primarily of nickel-alloy, bore a US impression on the cover and came in a cotton pouch. Examples can be found with and without the manufacturer's Part No. K1626-2 stamp on the back.

(Right) Significant quantities of pocket compasses needed to be procured in 1942 and early 1943 to supply nearly 500,000 parachute and sustenance kits. Additional quantities were later obtained to supplement stock supply. The previous examples were from early procurements. The Type H-1 compass, as detailed under Specification No. 94-27464, used a radioactive luminous material at each cardinal direction and at each point of the needle. Illustrated is revised pocket compass designated as Type H-1 with luminous materials.

(Right) Unidentified pilot of the 94th FS has what appears to be a pocket compass attached to zipper pull of A-4 flight suit, England. Evasion reports from both Europe and the CBI mention the use and carrying these compasses as personal equipment. In some cases, individuals mentioned carrying two or more in various pockets, one in the jacket and one in the flight suit was common. Having one in the parachute kit was an added bonus. Many extracts from Europe mention the use of "jungle" kits, presumably the B-2 kit. Upon parachute landing, they quickly gather up essential equipment before searching for a place to hide. Often noted was the retrieval of the pocket compass from the kit before abandoning the parachute assembly. The purse with maps and ration box were already on their person. Ultimately, the pocket compass received satisfactory praises except had a tendency to corrode after long durations in jungle environment. (National Archives)

(Below) The pocket compass remained the standard for Types B-1, B-2 and B-4 parachute kits for the duration of the war. In June 1942 it was approved for inclusion in USAAF multi-person life rafts. (Left) Longines-Wittnauer compass in display of accessories for a USAAF life raft, c.1943. (Right) Keuffel & Esser pocket compass in Type B-1 Alaskan parachute kit, c.1943. Each of the companies have distinctive details on the dials. (National Archives)

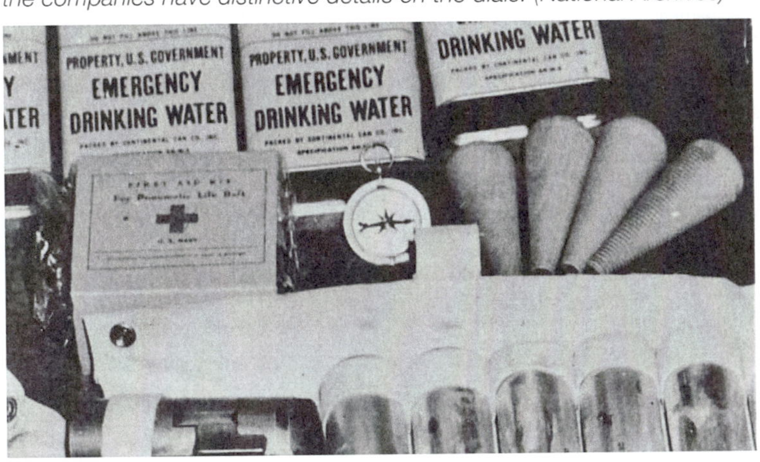

In 1942, a combination compass-match safe was developed, originally as a result of a collaboration between the BuAer and QMC. Both services would adopt the waterproof combination compass-match safe, manufactured of Bakelite with a removable screw top, containing a pivoting compass dial.

The QMC drafted and released Tentative Specification BQD 82, 14 October 1942 with nomenclature Box, Match, Waterproof with Compass, referenced as stock number 74-B-788-50. The BuAer adopted it under NavAer Specification M-554 as the Compass, Match Box, on 3 September 1942, referenced under stock number R37-C-2500 for use in life rafts. The QMC, however, adopted it for use as a jungle warfare item, and upon release of its specification placed a procurement order for 350,000 match safes. In April 1943 the compass was placed on limited standard status, being replaced by a simplified model of match safe without the compass.

All contracts were awarded to the Taylor Instrument Co. Taylor manufactured them in brown for the QMC and red as specified by the BuAer; as shortages developed, the brown ones were also delivered to the BuAer. Each safe was marked on the bottom with a generic US and included a pyrophoric material for fire starting. Quantity contracts for the BuAer are unknown, since most purchases were from life raft and back-pad kit contractors. However, reports from May 1943 indicate there were over half a million of these compass-match safes in service.

(Left) An M-554 match safe in a display of accessories for a BuAer multi-person life raft; it was part of the standard complement from 1943 to 1945. The USAAF requisitioned 55,000 for inclusion in its life rafts to replace nonwaterproof pocket type 8300-060930. (National Archives)

(Below) Red M-554 and brown QMC BQD 82 compass types. The compasses differed only in color.

(Above and left) Elmer Rand ARM2/c of VT-20 aboard the USS Enterprise has modified M-554 match safe secured to his life preserver, March 1945. Prior to availability of the AN-C-101, the M-554 included in the original back pad kit had this modification. (Left) Modification visible with length of cord secured to the safe with friction tape. (National Archives)

(Right) On 28 April 1943, a revised compass match safe was standardized under joint Army-Navy Specification AN-C-101, referenced under Dwg. No. AN5731, and Part No. AN5731-1. The assembly was modified with the addition of an integral molded tab to secure a lanyard, superseding the M-554; it remained under contract with Taylor Instrument Co. By this time, new emergency kits were being developed by the BuAer, in which this combination match safe became a standard component, even sharing stock number R37-C-2500. This model was only fabricated in brown Bakelite. Specification AN-C-101 was cancelled on 24 May 1945 when the new magnetic card compass became available for the BuAer and the USAAF had previously adopted the Engineer wrist compass.

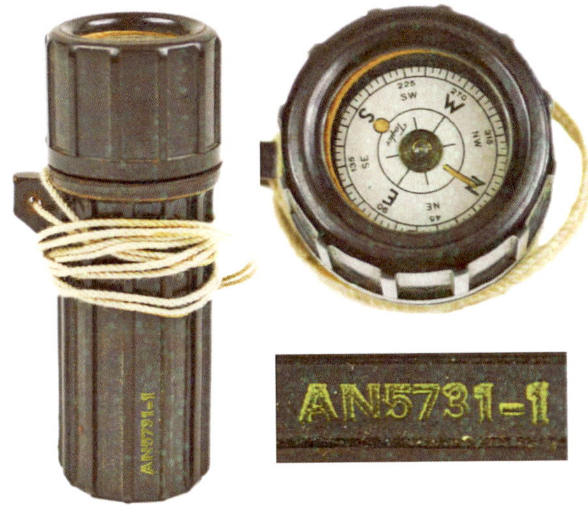

(Left) Pilot in USS Essex ready room prepares for strikes over Luzon, January 1945; he dons an AN-C-101 match safe from his life preserver as a standby compass. (National Archives)

(Below left) Aircrew members of VC-72 aboard the USS Kasaan Bay use AN-C-101 match safes as standby compasses secured to their life preservers, August 1945. (National Archives)

(Below) Gunner Edwin Emmons of VB-1, March 1945. In addition to a whistle, he also carries AN-C-101 match safe from his life preserver. (National Archives)

Combination match safe-compasses remained standard for the BuAer's life rafts and emergency equipment. Through 1943, the USAAF standardized multiple types for its rafts via updated technical orders and specifications. The first to be integrated was the pocket type, followed by the compass match safe BQD 82/AN-C-101. However, match safes had structural integrity issues, particularly with dials, which were known to warp easily.

In June 1943, specification changes were recommended for life rafts to switch to the use of the Compass, Liquid Filled, Wrist Type, US Army Engineer Specification EBP-292-A. 100,000 wrist compasses were requisitioned for life rafts under contract or, if the life rafts were in service, the compasses were available under stock number 6200-061700.

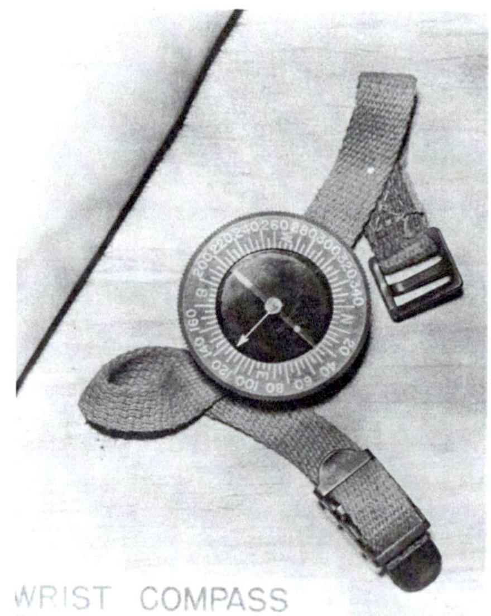

(Above) Wrist compass manufactured by Superior Magneto Corp. (Right) Taylor Instrument Co. compass. These compasses can be found with and without the property marks Corps of Engineers and US Army. It is not clear if those without the property markings were significant to the procuring agency.

(Below) USAAF volunteer demonstrates life raft equipment and an Engineer wrist compass is visible on his left arm. Standard equipment for all multi-person life rafts. One was also included in the E-15 sustenance kit. (USAF)

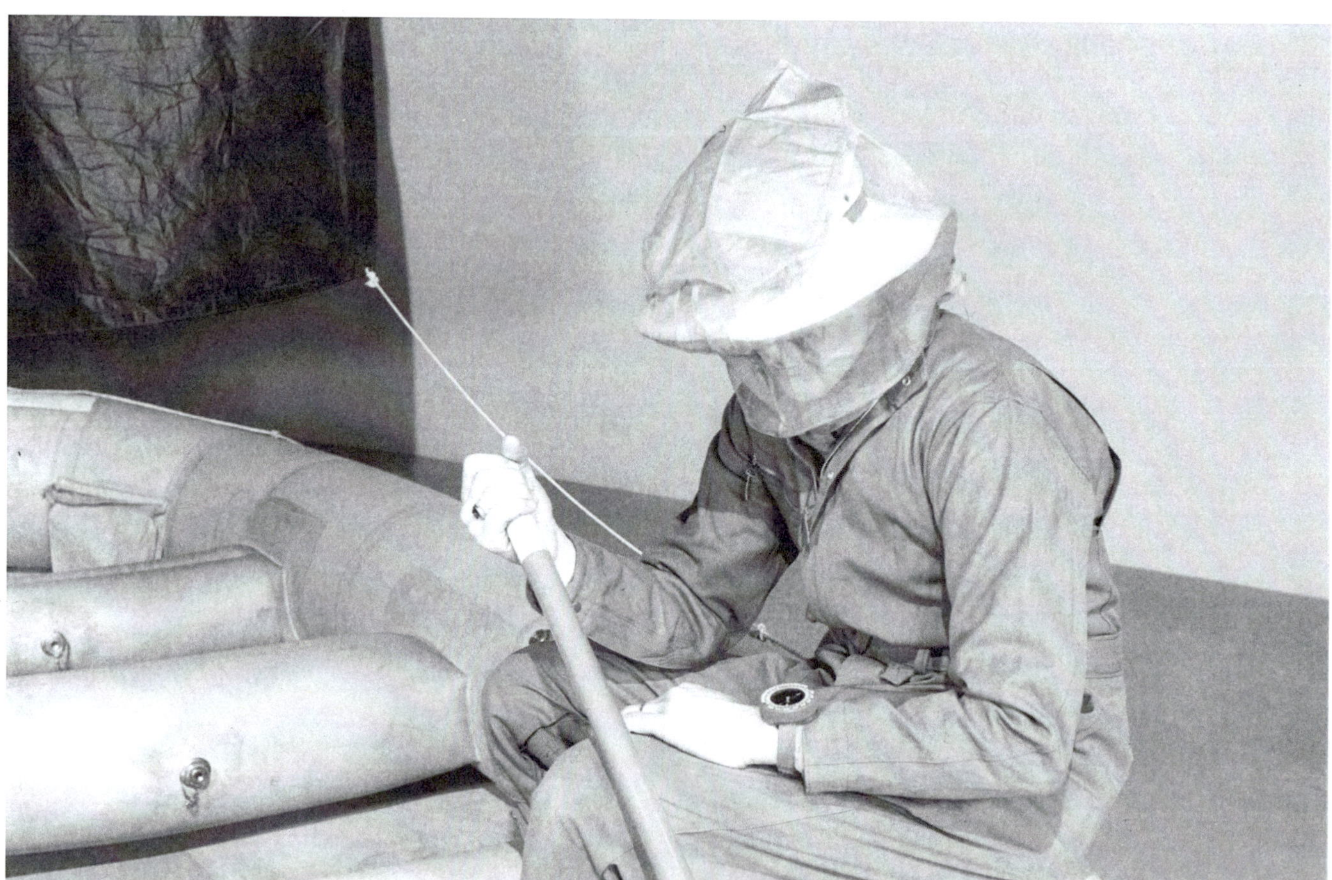

The USAAF had other ideas. It replaced the compass-match safe with the engineer wrist type compass for use in life rafts, while the pocket type remained standard for parachute kits. The Bakelite BQD 82 was utilized in laboratory samples of the C-1 sustenance vest. As the vest approached final approval and standardization, the compass-match safe required revision as well. A design submitted by Livingston Plastics Corp. was approved, constructed of phenolic resin and liquid filled dial like that of the Engineer wrist type.

No specification was drafted for the match safe compass upon standardization of the Type C-1 vest; it was procured directly by the contractors. By late summer-early fall 1944, resident inspectors noted quality control issues on the match safes, which led to formal specifications outlining requirements. Specification No. 94-27610, titled Compass, Liquid Filled, Match Box, Type H-2 referenced under Dwg. No. 45A9033, was effective 3 January 1945.

Due to quality control issues resulting in loss of product, compounded with an overburdened production plant from other contracts, Livingston subcontracted the manufacture of H-2 compasses to DuPage Plastics Co., also located in Chicago. In January 1945, DuPage submitted product samples that deviated from the specification: knurls around neck of cap were eliminated and a pyrophoric fire starter was added. The sample was approved by the Materiel Command; DuPage began manufacturing later that month, taking over all production and supply of the match safe for C-1 vests. The company's compasses had BY DUPAGE embossed on the bottom. DuPage produced just over 221,000, each of which included 56 half-length matches. In the immediate postwar years, DuPage continued to manufacture this match safe in a marbled red mahogany pattern for commercial and Boy Scout markets. US Government safes were specified to be in color OD No.108.

(Right) Livingston liquid-filled match safe compass. Liquid-filled compasses were found to be more accurate and better able to withstand extreme climate conditions than dial types. (USAF)

(Left and below) Variations of Livingston match safes. (Left) First production safes manufactured for test samples of the C-1 vest were unmarked on the bottom. During production, manufacture markings were added; one is embossed while the other has raised manufacturer markings.

(Right) DuPage Plastics example as manufactured under Specification No. 94-27610, Type H-2. Their model eliminated the knurls and added striker. A clear acetate match safe demonstrates the layering method as would be in the H-2 compass.

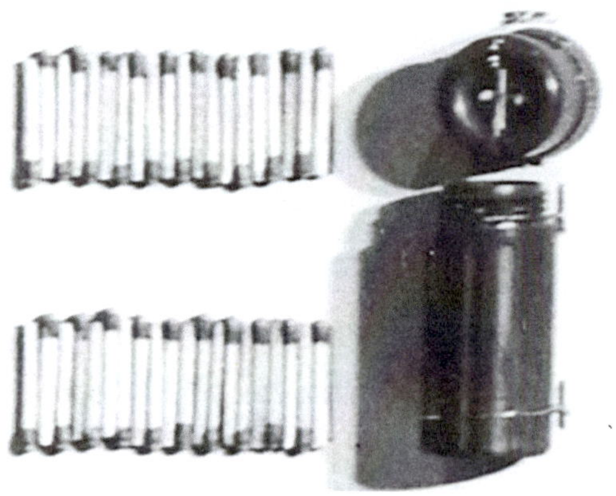

Another commercial compass known as the Gydeway was mass-produced by Taylor Instrument Co. and pressed into service. The case was an art deco Bakelite with a pivoting dial. These had been manufactured since the 1930s and were sold in sporting goods stores and catalogs. Each service used them for differing purposes.

The Gydeway compass served as a standby compass for the US Navy in case the remote electric, magnetic, or gyro compasses were to fail in an aircraft. It was not an unusual occurrence for all three to be nonfunctional due to electrical failures, lack of calibration, and the inherent wobbling and inaccuracy of the gyro compass. It therefore became unofficial standard practice for pilots to wear a standby compass. Compasses became a vital item of personal equipment and were the responsibility of the AESO to requisition before leaving the United States or Pearl Harbor.

(Above) Taylor Instruments Co. Gydeway compass.

(Right) Pilot of Air Group 10, Ens. John Kincaid, has a Gydeway secured to his 23P6 life preserver, USS Enterprise, March 1944. (National Archives)

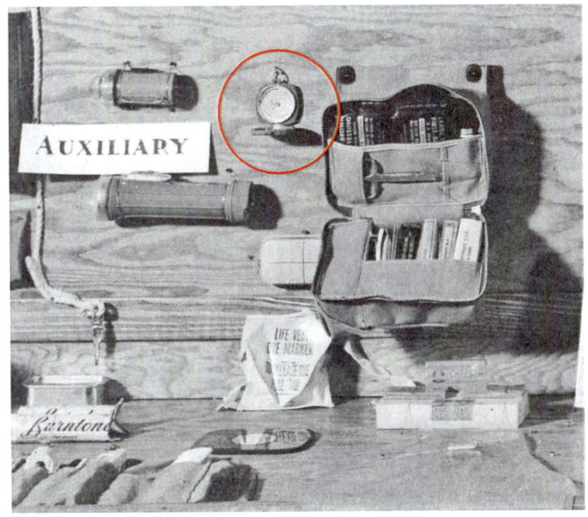

(Above) Accessories for Navy multi-person life raft includes M-554 match safe and Gydeway compass in box, c.1943. Other notables are TB-8 flashlight, police whistle and jackknife. (Right) Equipment display at NAS Norfolk, c.1945. Items categorized as Auxiliary: single and two-cell flashlight, aeronautic first aid kit and Gydeway compass.

(Below) Pilots of VF-14 aboard USS Wasp, July 1944. Second from left, J.D. Stokes has Gydeway secured to sea marker packet. Other assorted equipment in image includes; 23P6A life preserver, pistol lanyard, hunting knives and .38 hip holster. (National Archives)

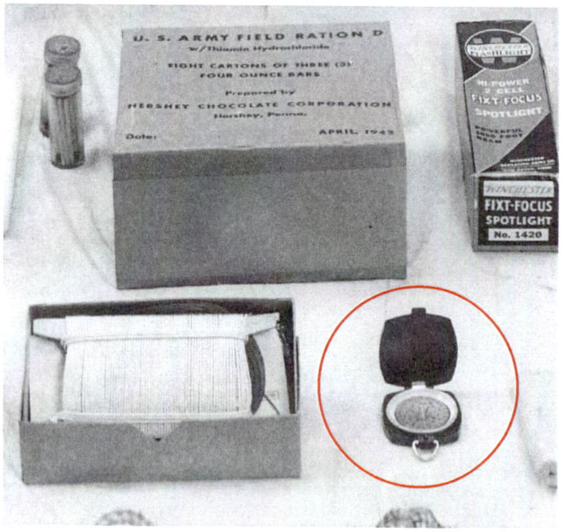

(Above) Gydeway in display of accessories for early generation Type E-5 sustenance kit, USAAF.

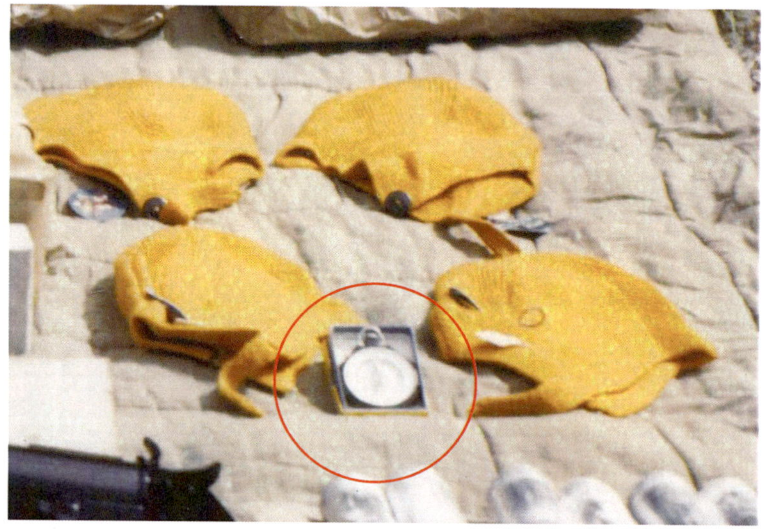

(Above) Another commercial compass by Taylor used was the Leedawl. It was an open face compass with pivoting needle. A Leedawl was a standard component to the USAAF Type E-11 sustenance kit.

(Below) Taylor Leedawl compass.

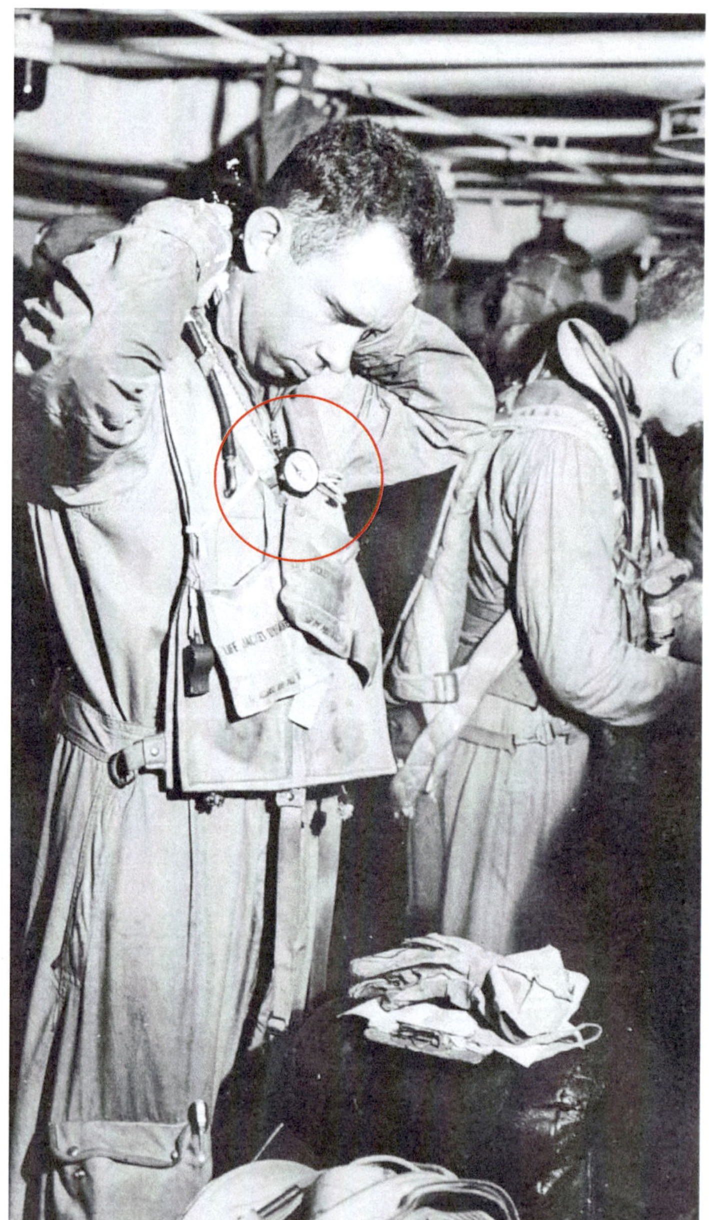

(Right) Pilot Lt. William Griese preparing for air strike in the ready room of the USS Ticonderoga. Attached to flight suit zipper pull is a Leedawl compass, July 1945. (National Archives)

(Below) Taylor compasses were dependable and readily available. A Leedawl compass was locally procured by an unknown naval squadron for the USAAF emergency jungle parachute kit, c.1942. (National Archives)

(Right) Upon employment as USMC officers, Marine aviators were issued a standard complement of 782 gear that included a compass. The model issued to Marine officers was the US Army Engineer lensatic type. The bulk of equipment issued was disposed of as it was nonessential to pilots. However, the compass was often retained and used as a standby compass. W & L.E. Gurley-made compass commonly associated with the USMC issue. Other example manufactured by Superior Magneto Corp., followed by an issued carrying pouch.

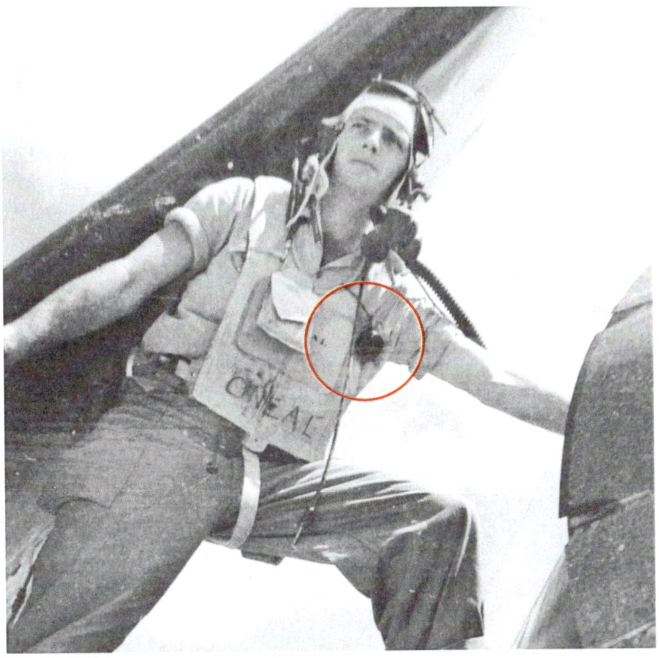

(Left) VMF pilot 2nd Lt. Merlin O'Neil of the 2nd MAW with lensatic compass, without pouch, secured to his life preserver. (National Archives)

(Below left) USMC 1st Lt. George Barlow of VMB-413 wears his lensatic compass pouch secured to pistol belt, also utilizes M2 jungle medical kit, c. 1944. (USMC)

(Below) F6F Hellcat pilot of VC-37 Ens. J.P. McKernan utilizes lensatic compass pouch over right hip, USS Sangamon, September 1944

In January 1944, Waltham Watch Co. submitted to the US Navy Coordinator of Research & Development a sample of a small, dry, watertight compass designated as WTS-40 that could be slipped onto the strap of a wristwatch. This submission resulted in the suggestion that the compass or equivalent be adopted immediately. However, adoption by the BuAer was delayed, pending an investigation regarding a new type of pocket compass in development.

To meet service requirements, BuAer adopted the compass and added it to the list of optional Pilots' Personal Equipment in October 1944. It awarded a quantity contract the same month under NOA-4681 for a production run of 92,000, assigning stock number R88-C-890 to supplement the demand for an emergency standby compass.

The compasses were provided with an OD cloth wrist strap and a printing of WALTHAM on the dial. Early production compasses had the stock number stamped in ink on the back; later production switched to an embossment. This wrist compass would remain a standard option in aviation into the modern era.

(Above) Sample compass submitted to BuAer from Waltham.

BuAer approved and procured Waltham WTS-40 wrist compass.

(Right) Ens. Raymond Lebel of VF-40 wears a Waltham compass on his right wrist, May 1945. Left wrist would be a time piece, a staple for all aviators. (National Archives)

(Below) Lt. jg Ralph Koontz of VF-9 aboard the USS Yorktown, August 1945. A Waltham wrist compass is worn on the right hand. (National Archives)

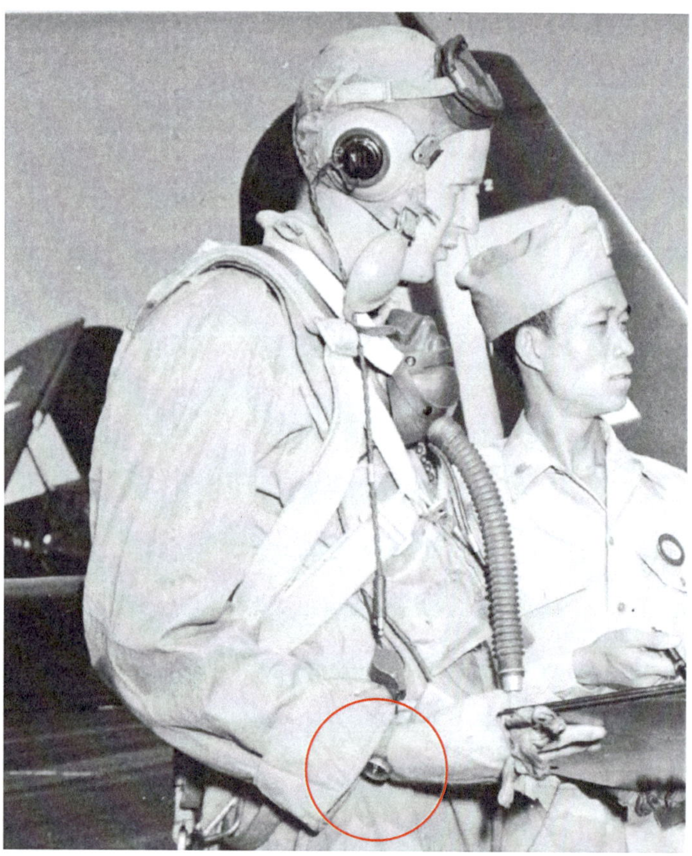

Discussed thus far are the primary compasses used by the aeronautical services. From altering commercial types to drafting specifications for exact service requirements, a substantial amount of trial and error led to the implementation of these compasses.

For life rafts, the USAAF included the US Army Engineer wrist compass, using it to replace all other types when supply permitted. This compass remained the standard until immediate postwar years. The BuAer, too, found the M-554 or AN-C-101 compass-match safes deficient for use in life rafts, especially after long periods of exposure to corrosive saltwater.

By 1944 the Instrument Branch, Engineering Division of the BuAer began development of a new pocket compass. The two primary requirements were that the compass be watertight and accurate for true navigation headings, unlike the AN-C-101 or M-554. The compass developed included a watertight transparent phenolic resin case with a dry card type compass that could be read either from the top or the side, loosely based on standard aircraft compass systems.

The intent was for it to be used in life rafts, but it quickly became an option for use as a standby compass in all service areas. Quantity production commenced in September 1944 through a contract with US Gauge Co. for 50,350 compasses to supply all multi-person life rafts in service under contract NOA-4573. In the following month, Waltham Watch Co. was awarded a contract for 50,000 compasses for additional stock to be used as standbys for pilots under contract NOA-4682. Life raft contracts consumed initial deliveries of this compass, which began in December 1944. Spares were delivered to supply points by the spring of 1945. This compass would influence postwar designs and adoptions such the Type L-1 Compass, Wrist, Magnetic Card, Dry, USAAF Specification No. 94-21051, and MC-1 for the US Navy.

The new compass was referenced as Compass, Pocket Card Type, Magnetic, detailed under NavAer specification SQ-127A, available as stock number R88-C-786. Contract, stock, and contractor part numbers were stamped on reverse. The U.S. Gauge Co. used part number AW-20-1 and Waltham Watch Co. used 91-900. The back also held two clips for lanyards or other securing devices. This compass superseded the AN-C-101 in life rafts.

In the summer of 1945, the USAAF Air Technical Service Command field-tested several fluid-filled compasses and the dry dial type used by the US Navy. Because fluid loss occurred at high altitudes and low temperatures, which resulted in interference with dial movement and loss of accuracy, the dry dial was deemed superior, leading to the adoption of the Type L-1 in 1946. It copied the Navy's design.

(Right) Compass, Pocket Card Type, Magnetic, manufactured by Waltham.

(Below right) VC-20 pilot in the ready room of the USS Kadashan Bay has a card type compass secured to his life preserver resting on the sea marker. (National Archives)

(Below) Stock image of new magnetic card compass.

As part of a revision for the complete overhaul of the USAAF Type E-3 personal aids kit, the BQD 82 compass-match safe was deleted. A new cellulose acetate match safe was instead adopted. Through 1943, Waltham Watch Co. solicited the USAAF for compass contracts, and in December of that year submitted watch compass WTS-40 in hope of receiving contracts for compasses for USAAF life rafts.

The USAAF approved the design but was already in favor of the liquid-filled wrist type. Negotiations with Waltham continued and a revised design was submitted for a compass, eliminating the strap brackets and reducing the diameter from 1-1/8 inches to ¾-inch. The new design, fabricated of a brass casing and painted OD, was approved and adopted in the new Type E-3B sustenance kit in February 1944 supplied by Waltham as part number WTS-43.

This compass would continue to be included in USAAF personal aids kits, to include the E-17 and the US Navy Mark-I emergency flask kit supplied as CFE. This compass was officially standardized as the Compass, Pocket, Type H-4, on 8 August 1945. Because personal aids kits fell under the purview of the military intelligence department, the compass was also obtained for evasion and escape kits for the appropriate intelligence sections operating overseas. A half-inch version was also manufactured at this time, which can be found in some existing examples of the US Navy Mark-I kit. Procurement, requisition and more types of these compasses will be discussed in a forthcoming volume as they are more associated with E&E.

(Above) 3/4 and 1/2-inch Waltham WTS-43 Type H-4 pocket compasses. Type E-3, E-17 and Mark-I sustenance kit contractors were responsible for procuring the compasses under the Waltham part number. Those provided in the E-17 were packaged in a manila envelope. Envelope used when distributed on individual basis. Significant quantities procured by MIS-X for distribution in POA, SWPA, ETO, India-Burma and China.

(Above) In 1945, the USAAF again revised the Type E-3 kit to emulate the flask and match safe included in the Type E-17. A new match safe compass combination was devised to incorporate the WTS-43 compass into the cap of the safe, as detailed under Dwg. No. 45A9185. The new compass match safe arrived toward the end of production run for inclusion into the Type E-17 and E-3C kits. The compasses in preceding match safes were fixed into position, whereas the new 45A9185 was designed so that the WTS-43 could be removed. It was friction-fitted into the cap and backed with a separate removable sheet of neoprene, making the compass easy to retrieve while ensuring protection.

FISHING KITS

For thousands of years, fishing tackle has been a staple for survival. In prewar US, it was primarily used for recreation, with practically no military application until the 20th century. Slowly, fishing implements were integrated into military requirements through the standardization of emergency equipment.

The USAAC was the first service to develop fishing tackle and introduce it into emergency kits. This occurred in the 1930s in overland kits. Tackle was not considered for inclusion in life rafts by the US Navy or USAAC until WWII.

Prior to this, each service branch had its own opinion on what was necessary life raft survival gear: distress signals, first aid, and rations, for example. However, there was no consideration that a fishing kit may prove useful for survivors stranded at sea. Life rafts were stocked with an eye toward self-containment with canned water, canned or boxed provisions; the rations could sustain individuals for two or three days if apportioned properly. There was no thought of gathering food. While this reasoning seemed sound in a peacetime environment, it quickly proved deficient with the ever-increasing episodes of aircrews hitting the drink.

Gifford Pinchot, ex-governor of Pennsylvania, wrote a letter to the Secretary of War, Henry Stimson, on 29 April 1942 stating, "You may be surprised to know that potable fresh water can be squeezed from the flesh of fish, as I discovered by a test last Sunday."

Pinchot, a recreational deep sea fisherman, was influenced to conduct the test after reading an account of three Naval fliers stranded at sea for 34 days without supplies, food, or water. The account was circulated in an article in Life magazine in which a TBD Devastator of VT-6 was forced to ditch with its three-man crew on 16 January 1942. Pinchot conferred with the US Army, Bureau of Medicine & Surgery and the Bureau of Aeronautics, suggesting that fishing tackle be placed in all life boats and life rafts because it could help provide both food and fresh water.

The Navy Department, the Bureau of Ships, and the United States Coast Guard expressed immediate interest. The BuM&S also became interested in this revolutionary shift, as it was responsible for the development of rations and nutrition study, for which it was already conducting tests on suitable emergency sustenance, evaluating both solids and liquids. Due to the increasing number of shipwrecks and aircraft ditchings, the Naval departments began to view fishing tackle as a new and vital element to meet wartime demands, leaving only the USAAF expressing no immediate interest. Following the recommendations by Pinchot and service demands, the BuAer drafted a specification for a fishing kit to be included in all life rafts as NavAer Specification M-552, which was adopted as standard 23 July 1942 and titled Life Raft Fishing Kit.

(Below) The M-552 fishing kit was comprised of a single drawstring cotton bag measuring 8" x 2-1/2"x 1-1/2" that contained 100' of 90 lb. test line in natural or green cotton on a plywood spool, 6" feathered jig, 4" red cedar jig, each with 2 ft. of steel leader, a No. 7 hook, and instruction sheet. This kit became standard for all M-3 series life rafts, available under ASO stock number R37-K-300. Edward K. Tryon Co. assembled and provided 30,000 kits. Below, Variants of drawstring bags.

(Left and above) Expedited version of fishing kit in drawstring bag 14" long, lacks instruction sheet. This type was a local assembly for life rafts. (Left) on display board at NAS Norfolk.

(Below) Due to the volume of M-552s in stock and until sufficient quantities of the M-552A became available, fishing bait was added to standard stock to supplement the M-552. Services requisitioned the bait via stock number R37-B-200. This was an attempt to bring the M-552 to the standards of the M552A.

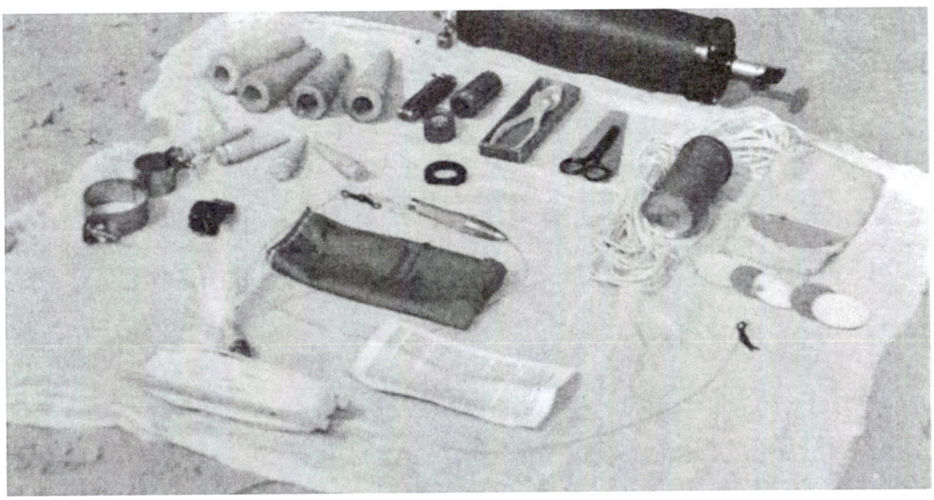

(Below) The Bureau of Ships procured a fishing kit in July 1942 to supplement its lifeboats and floats aboard vessels. It consisted of a cylindrical tin measuring 8" x 3-1/2" diameter. The kit included fishing spear, line, feathered jig with large hook and package of assorted hooks. These were contracted through Edward K. Tryon Co. under the only known contract, NXSS-8537, in a quantity of 33,000. Fishing kit being examined in a life float aboard an LST en route to Africa, April 1943. The typical complement of the abandon-ship ration tin and water breaker is in background. (National Archives)

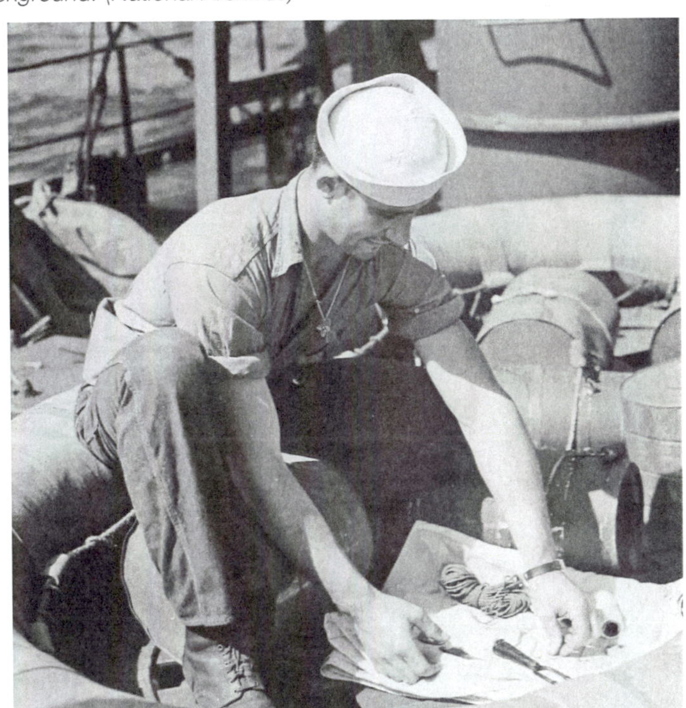

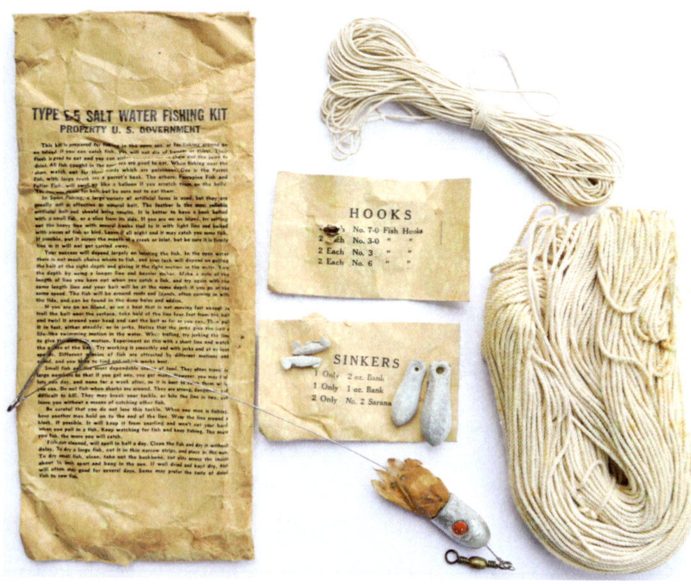

(Left) Following development and procurement of Type E-5 sustenance kit, over water, it was determined that a fishing kit was needed. USAAF parachute emergency kits were provided with fishing kit per Dwg. No. 41A9711. A package of fishing tackle was approved for E-5 sustenance kit, referenced as Packet, Fishing, Liberty Electric. The package contained two size 7-0 hooks, two size 3-0 hooks, two size 3 and size 6 hooks, one 1 oz. and 2 oz. sinker, two no. 2 clincher sinkers, one feather jig, 75 yards of 54 lb. fishing line and 50 ft. of 25 lb. fishing line, in an envelope with instructions. Future assemblies requiring fishing kits used Type-5/AN-K-2. Specification for E-5 was not revised to adopted type. Example of Liberty Electric model fishing kit provided for E-5 sustenance kit.

Pinchot, having piqued the interest of the US military, collaborated with big game fisherman and explorer Michael Lerner to develop a suitable fishing kit for the war effort. The two assembled a team of knowledgeable experts in the field of deep sea fishing, including Kip Farrington, William Hatch, and Julian Crandall of the Ashaway Line and Twine Co. This team developed a fishing kit that could be worn as a bib apron and was packed in a waterproof key-opening can. It weighed three pounds, contained 12 items and was named the No. 10 fishing kit. The unit was presented to all interested parties, and adopted by the Life Saving Board of the US Navy and USCG in December 1942 as standard equipment for life boats and rafts, to be provided on vessels over 3,000 gross tons.

(Below)12 unit fishing kit as developed by Pinchot, Lerner, Farrington, Hatch and Crandall. US Navy contracts were first awarded in January 1943 under contract NXSS-22619 to Ashaway Line and Twine Co. Additional contracts were awarded in May, October, and December 1943 under contracts NXSS-28824, NXSS–38922 and NXSS-45761; contract quantities unknown. These kits were available under stock number 23K45 titled Kit, Fishing, US Navy, Emergency, referenced under Bureau of Ships Specification 37K1 (Interim), adopted as standard 1 July 1943. (National Archives)

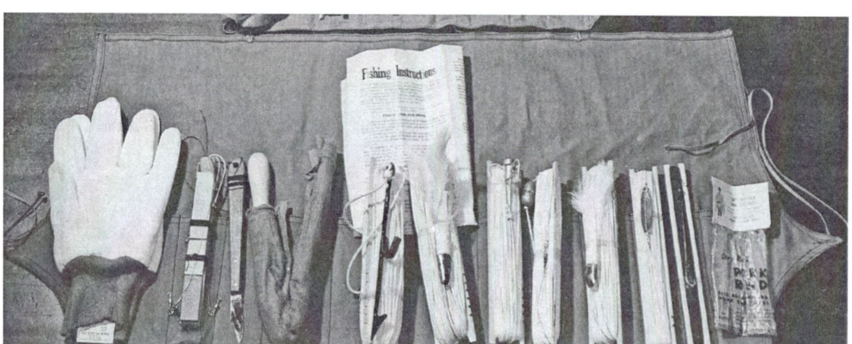

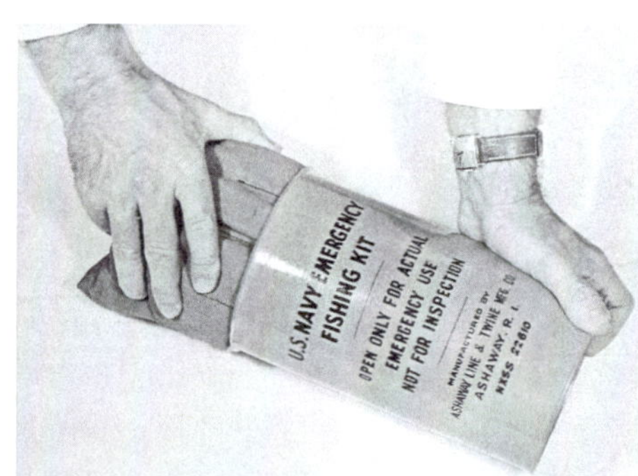

(Above)12 unit fishing kit as developed seen field tested at Cape Fear, North Carolina, June 1943. One can provided per life float aboard Navy vessels. (National Archives)

(Right) Apron fishing kit being demonstrated. Designed to be secured over standard kapok life jacket.

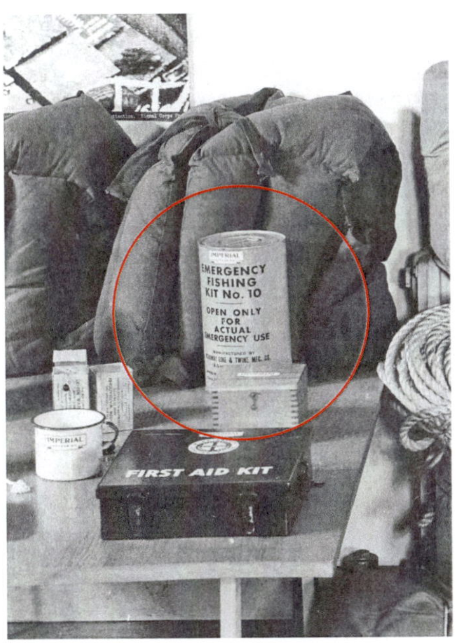

(Left) Fishing tins from manufacturers approved by the USCG. The 12 unit fishing kit was manufactured by Ashaway as part No. 10; however, this was not reflected on Navy contracted kits. These were USCG approved types available for maritime and Merchant Marine services. Pachner & Koller Inc. was another supplier. Why the unit on the right is labeled Kit No. 4 is unknown, because the apron, components and tin are identical to Ashaway's No. 10 unit. No. 4 only referenced an item contracted by Transportation Corps (TC) with Edward K. Tryon Co. Each manufacturer had a different USCG approval date stamped on the can, which reflected when the company was qualified to bid.

Upon review of the No. 10 fishing kit, the BuAer approved the concept but not the bulk. It requested the kit be altered in consideration of space limitations aboard aircraft. The Ashaway Line and Twine Co. reduced the kit by four units, which dropped the weight to one pound. This modification was approved by the BuAer, prompting a revision to Specification M-552 that reflected the changes. The new unit was referenced under NavAer Specification M-552A effective 13 February 1943. Procurement, however, was initiated in January prior to approval. Acquisition for the BuAer was conducted by the life raft manufacturer as CFE on existing contracts. The ASO awarded contracts for stock and to supply life rafts already in service with kits available under stock number R37-K-300. The M-552A superseded the M-552 but remained in service until fully replaced; they shared the same stock number. Upon requisition, either kit was supplied.

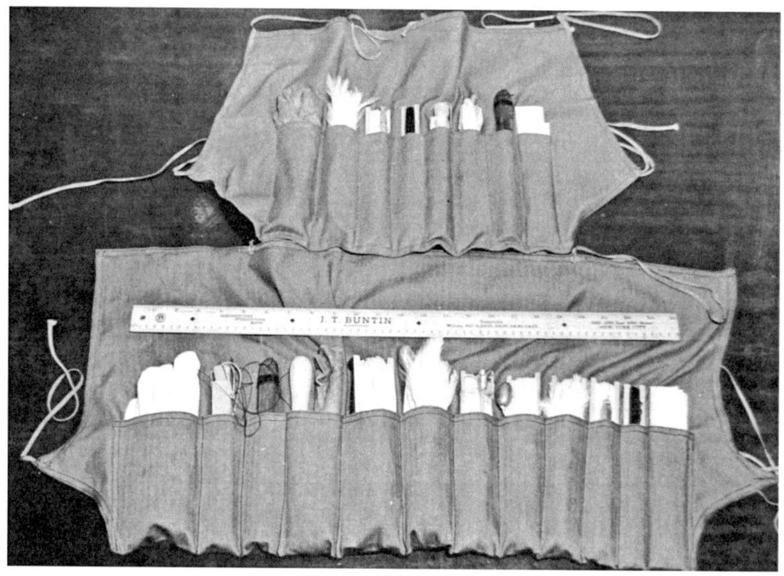

(Above) Size comparison between the M-552A and the No. 10 units. Unlike No. 10, the fishing kit developed for aircraft use was not stowed in a tin container for weight conservation. Aprons were rolled, tied closed, and wrapped in Kraft paper before being packed for shipping. Once removed from shipping crate, kit was packed in life raft or other emergency kit, with or without the wrapper. (National Archives)

(Below) Revised NavAer Specification M-552A life raft fishing kit assembled by Ashaway Line and Twine Co. Nomenclature is stamped on the outside of the HBT apron. The components were 100 ft. of 15 lb. test nylon line with #6 O'Shaughnessy hook, 100 ft. of #12 Ashaway white cotton shoreline with stubbed mackerel squid jig, 100 ft. of #24 Ashaway white shoreline with 1 oz. feather jig, 12 pork rinds, instruction sheet, wooden block with abrasive stone that included three 6" lengths and one 3" length of stainless steel leader, one of each having a #1, #2, #3, and #6 O'Shaughnessy hook and swivel, fishing knife, dip net measuring 12" x 16", and an 8" barbed spear.

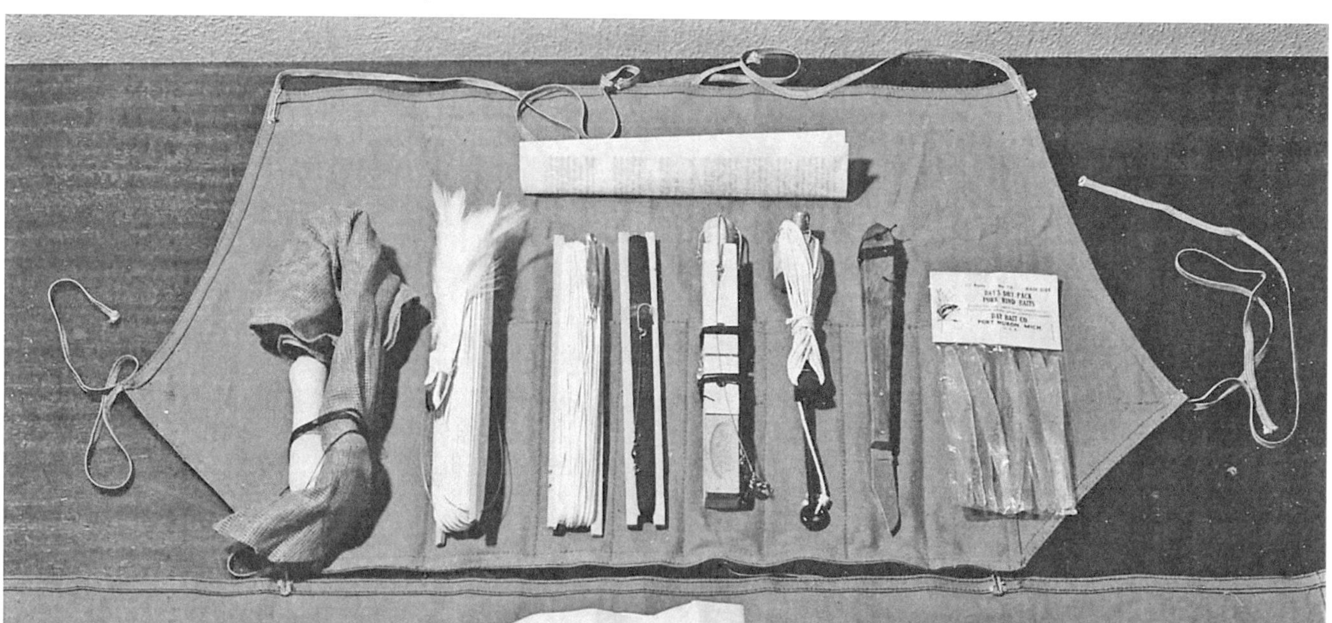

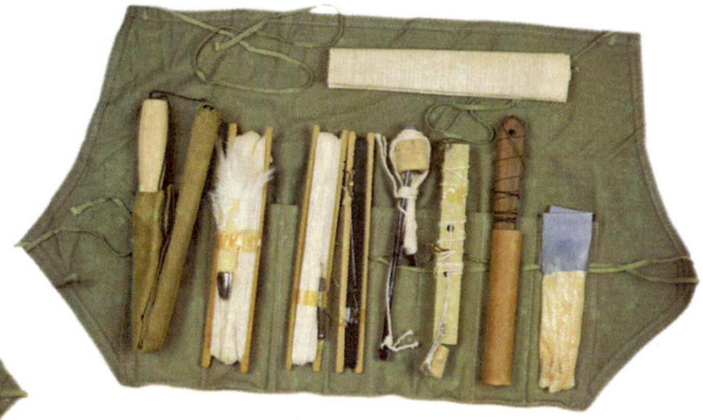

(Left and above) BuAer M552A contract life raft fishing kit.

The USAAF also adopted the fishing kit shortly after an interview with Captain William Cherry on 11 December 1942. Cherry was the pilot of the B-17 that ditched in October carrying Captain Eddie Rickenbacker. Soon after, Rickenbacker submitted the report of his experience adrift for 21 days to the Secretary of War, expressing the need for fishing kits and other life raft provisions. The BuAer revised the fishing kit, now known by the manufacturer as part No. 5, and submitted it to the USAAF for review. On 24 December it was approved for adoption as the Pinchot-Lerner Emergency Fishing Kit No. 5 as a joint Army/Navy item manufactured at an average cost of $2.80 each. Procurement by the USAAF was expedited on 5 January 1943, executed under NavAer Specification M-552A to manufacture 15,000 Type-5 fishing kits, awarded to the Ashaway Line and Twine Co. under Purchase Order 43-19275P. Two additional contracts were awarded in February to Sports Products, Inc. for 64,633 kits under P.O. 43-22408P and to the Horrocks Ibbotson Co. for 64,634 under P.O. 43-22409P. This would be the sole contract awarded to the Ashaway Line and Twine Co. by the USAAF. The designator P at the end of the purchase order signifies a Priority Procurement.

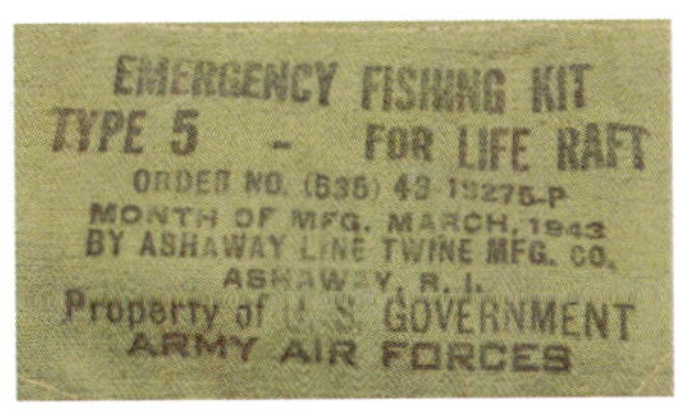

(Left) First USAAF contract for the Type-5 or No. 5 fishing kit. Classification and contract data are stamped on outside of apron. Kit was available under stock number 8300-559020. NavAer specifications were made for initial procurements until a formal joint Army-Navy specification could be drafted.

(Right and below) February 1943 USAAF contract to Sports Products Inc. referencing it as No. 5. Ashaway used both Type-5 and No. 5 designations. The pattern of the apron was an approved deviation.

(Left) Joint Army-Navy standards went into effect 24 June 1943 under Specification AN-K-2, Dwg. No. AN-6522 and Part No. AN-6522-1. Further procurements by both services were made under this specification and part number. Due to the volume ordered by the USAAF in January and February, a limited amount of 4,326 AN-K-2 kits were procured under P.O. 44-10065 to Sports Products, Inc. dated 18 October 1943. Overall production of the AN-K-2 was claimed to have reached a quantity of 160,000 with the bulk allocated to the BuAer and the Air Transport Command to supply life rafts in the vast network of cargo aircraft. These acquisitions were executed under procurement specification AN-K-2.

(Below) Wartime image of the components in roll type fishing kit No. 5. They remained consistent from the M-552A to the AN-K-2, although minor variations were attributed to the multiple suppliers involved in the final assembly. AN-K-2 integrated new fish knife. (National Archives)

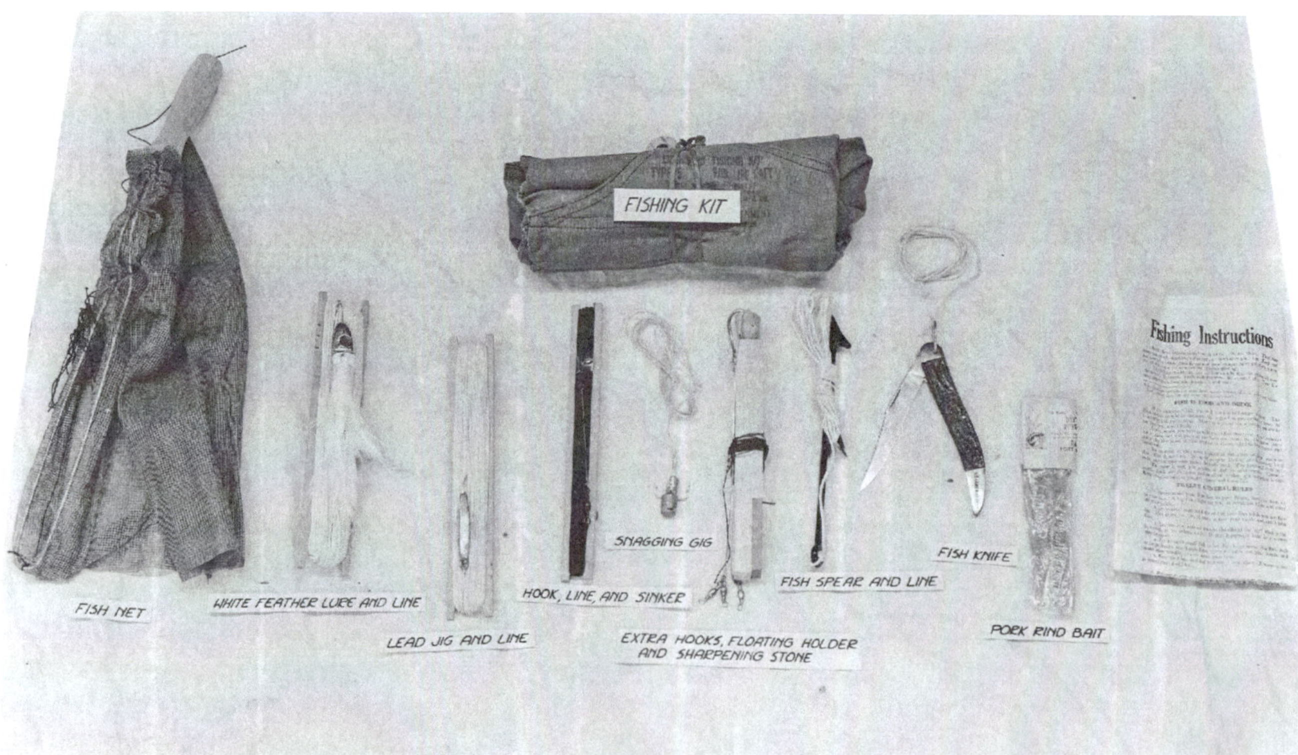

By December 1943, the USAAF was working on a revision for the AN-K-2 fishing kit to address complaints of damage to the pork rinds by rats and items missing after prolonged storage. The proposed revision presented to the Joint Army-Navy Board in February 1944 included a sealed waterproof can. As with any item under joint specification, all parties involved had to approve the changes. BuAer waived its approval authority since it had no further interest in the AN-K-2. Following receipt of that waiver, USAAF awarded two contracts on 22 March 1944 for the improved AN-K-2 fishing kit to Sports Products, Inc. under contract (30-053) 44-12103 and to Horrocks Ibbotson Co. under contract (30-053) 44-12104 for 41,340 kits to each. The kits were procured under the Army-Navy Part No. AN-6522-1, which was reflected on the can and apron. Deliveries began reaching supply points and life raft contractors by summer 1944. The only change was the packing of the apron fishing kit in a key-opening tin can. Components remained unchanged. Specification AN-K-2 was cancelled on 20 August 1947 upon adoption of a new MILSPEC fishing kit.

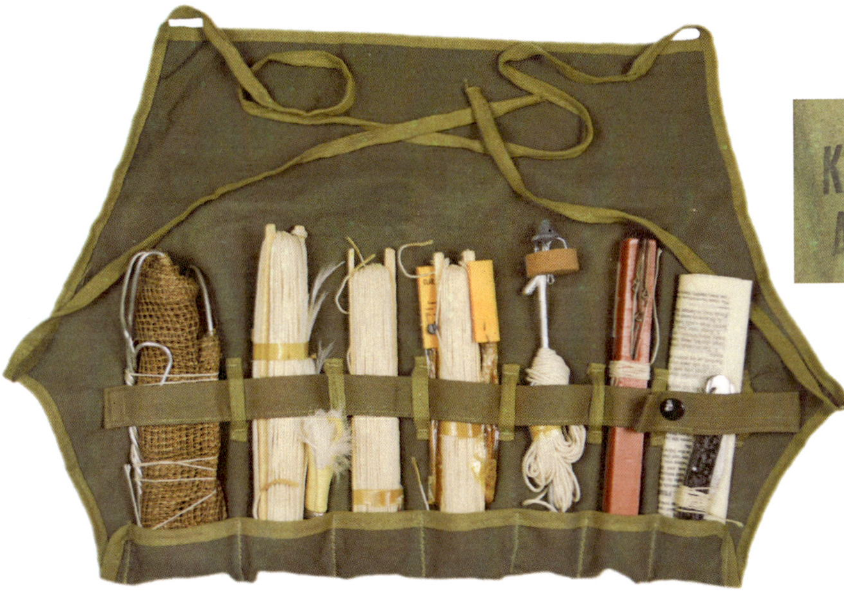

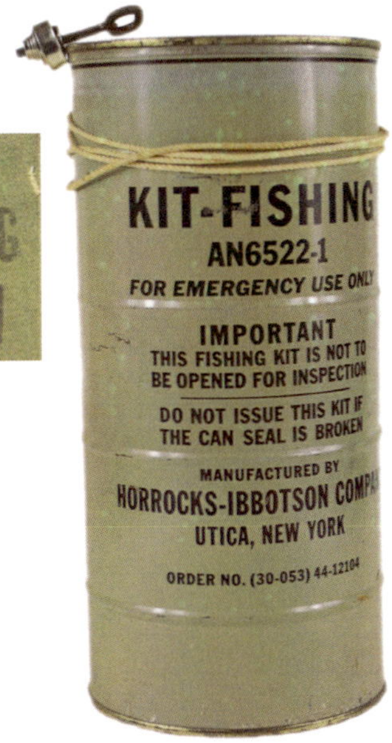

(Above and below) Revised AN-K-2 fishing kit with sealed tin container. Now referenced under Part No. AN6522-1.

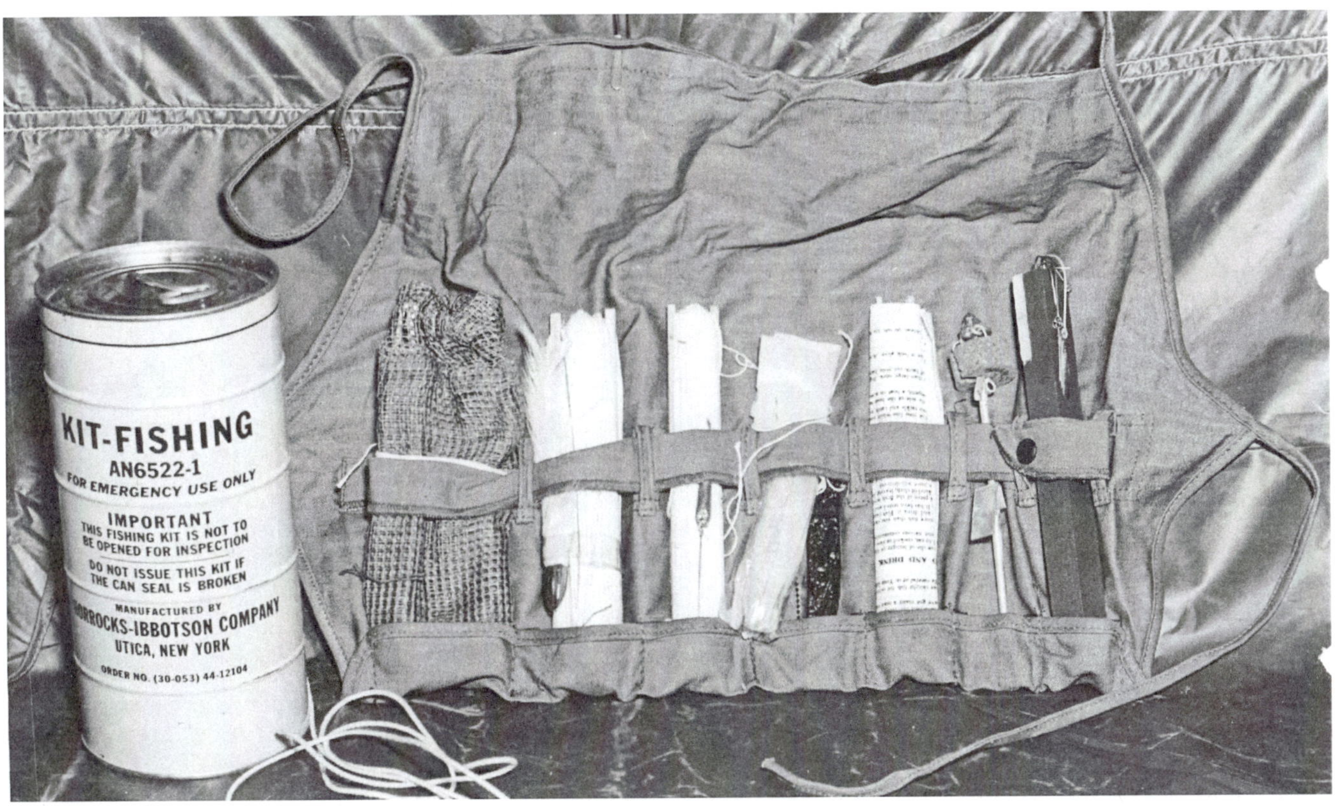

In early January 1943, Camillus Cutlery Co. contacted both the USAAF and BuAer with a marketing pitch regarding a standard commercial fishing knife. In a presentation, the company's representatives pointed out the deficiencies of the floating knife and promoted the benefits of their "real" fishing knife that boasted a specially designed four inch locking blade with fish-scaling abilities and a hook disgorger opposite the blade. They noted a clevis could easily be included with a lanyard to prevent loss to compensate for the lack of floating capabilities, although the commercial model did not include this feature.

The knife presented was approved and added to approved draft of specification M-552A under which the No. 5 fishing kits were tentatively being procured. Camillus referenced the life raft fishing knife under part number 5369 L67B.

Ashaway Line and Twine Co. was the initial contractor for the Type-5 kits and used the floating type until the Camillus model became available. All additional contractors included the 5369 L67B for No. 5 fishing kits. Due to volume demands, manufacturing of the fishing knife was extended to Imperial Knife Co. subsidiary Kingston and marked MADE IN USA. The closed knife measured five inches and contained a four-inch blade. Production of this fishing knife exceeded 150,000.

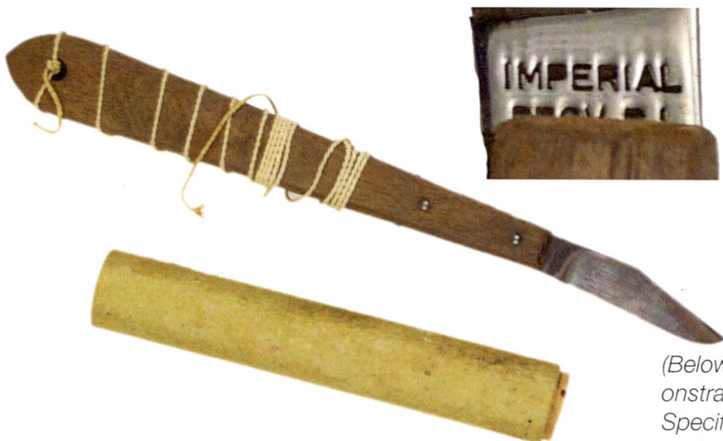

(Left) No. 10 fishing kit was designed to comply with USCG regulations prior to being submitted to the BuAer. This meant it was more important that the incorporated knife float. The contracts awarded in January 1943 from the USAAF and BuAer included this floating knife in kits from the Ashaway Line and Twine Co. It had a 6" wood handle with a 2" blade, and a fiber tube with a cork stopper at one end to prevent the knife from puncturing through the apron and the bladder of life rafts when stowed. Imperial Knife Co. is the only known knife supplier for the No. 10 and No. 5 fishing kits.

(Below) Camillus pattern 5369 L67B fishing knife. Also made by Kingston (Imperial). Knife applied to all AN-K-2 and AN-6522-1 contracts. Provided with 3' lanyard. Clevis is same application as provided on USAAF three-bladed life raft knife.

(Below) Continental Can Co. employee, Mrs. Ethel Fischer, demonstrates streamlined fishing kit for use in aircraft, procured under Specifications M-552A and AN-K-2. Unit illustrated was acquired under A/N Part No. AN-6522-1. Other uses for can include catching or storing water and bailing.

(Left and above) From experiences in 1942, survivors reported that fish were consistently swimming around the raft and they wished they had the means to spear them effectively. The 8" barbed spear included in the No. 5 and No. 10 fishing kits was designed to attach to the end of wood or aluminum oars. It was used as a harpoon, releasing from the oar once impaled in the fish, then dragged to the boat by the line provided. (Left) Volunteer attempts to spear a fish from a life float during sea trials off Cape Fear, North Carolina, June 1943. (National Archives)

By January 1943, the BuAer established a directive to equip all life rafts with a fish spear to supplement fishing kit M-552; bait was made available from stock. A spear design submitted by Edward Vom Hofe & Co. was approved for production in a quantity of 100,000 9-inch single barb steel fish spears. It was designed to be integrated into the standard aluminum sectional oar Specification M-162 for storage and use. Following attainment, NavAer Specification S-146 was drafted; it was adopted 8 March 1943.

Technical Note 9-43 was distributed to services 23 March 1943, instructing that all life rafts be equipped with the fishing spear, illustrating how the oars are to be modified for storage and its use. These were available under ASO stock number R83-S-648150.

Development of a fishing spear or gig was reported to have been inspired by the 34 day account of three standard Navy fliers: Harold Dixon, Gene Aldrich and Tony Pastula. In their account, the raft was noted to oftentimes be surrounded by plenty of fish but spearing them with their Scout knives was next to impossible. Prior to official adoption by the BuAer, squadrons developed gigs to supply life rafts on aircraft. Such developments were reported by VB-19 and VT-9. Their designs principally incorporated a three-pronged gig that could attach to a standard life raft oar.

(Right) Fishing spears developed by VB-19 personnel, based on a report submitted in March 1943. Members of VB-19 test new life raft equipment in New York Harbor using the single gig spear; the principle was applied to the standardized S-146 type. (National Archives)

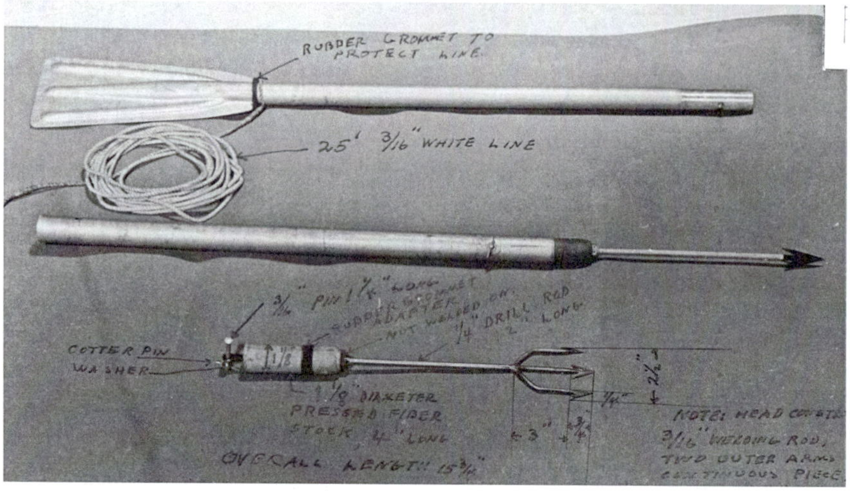

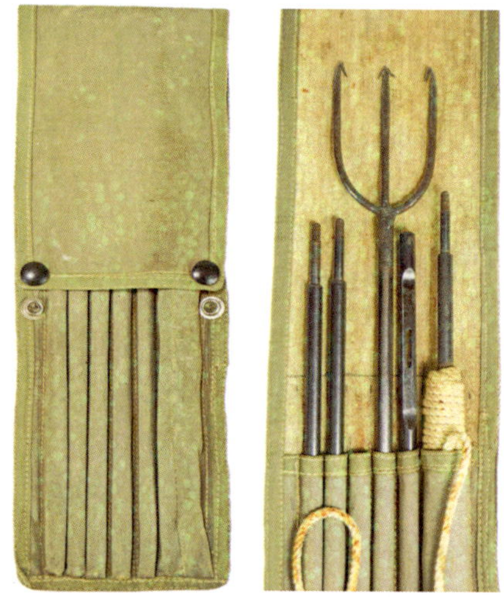

(Above) Fishing spears developed by VB-19 personnel; single and triple gig spears. (Right) Locally improvised fish gig adapted from a modified .50 cal. cleaning rod set; the container was fabricated from waterproof airplane cloth.

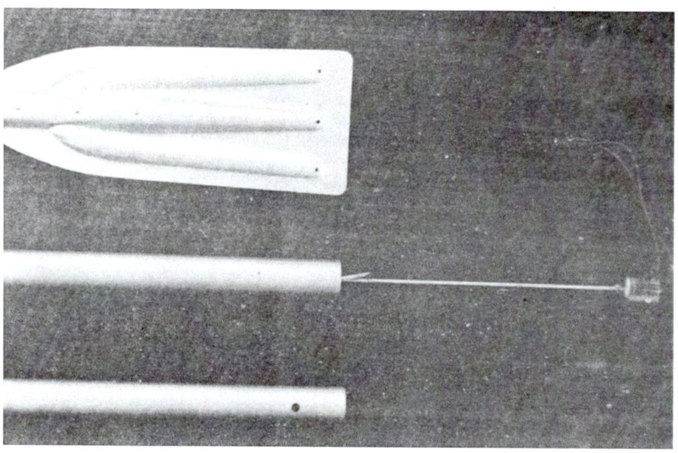

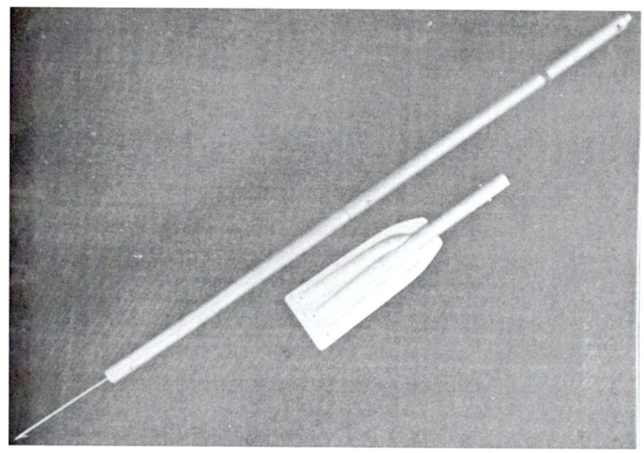

(Above) NavAer Specification M-162 was revised 22 March 1944, Amendment-D, to make the fishing spear standard. Following this date, oar manufacturers supplied the fishing spear and a label that read FISH SPEAR STORAGE SPACE THIS END added to one oar section. The spear was secured by two spring-loaded locking pins. To release the spear from storage, pins were squeezed and pulled from tube by the cord provided. To use the spear, process was reversed: the back end was slid in until it locked into place. Images are from Technical Note 9-43.

By the end of 1943, the BuAer was placing greater emphasis on signaling devices oriented to rescue, moving beyond the concept of mere survival, for which the AN-K-2 fishing kit was categorized.

A great deal of data had been accumulated by laboratory research and combat experience in all aspects of equipment and personal health. It was widely understood that water was the primary factor in preservation of life. It was calculated that, with the pending introduction of a desalting kit, one can of water and one desalting kit could provide adequate amounts of water for one person for a minimum of six days. The fishing kit had been adopted largely because water could be squeezed from fish, according to Gifford Pinchot's one-time event. However, formal experiments conducted at the Naval Medical Research Institute and Yale University later found this false, contradicting Pinchot's finding. Lab results showed that dehydration was not prevented by eating fish. In fact, protein from fish excreted body fluid to digest and to absorb the salts. In conclusion, Yale University's experiment showed that men should be cautioned not to eat fish and gulls unless they had drinking water. Another important factor was the limited number of reports by survivors successfully catching fish.

Based on these findings and reports, the importance of a fishing kit was greatly reduced (especially in comparison to a desalting kit), but it was still considered a viable piece of emergency equipment. The BuAer wanted to reduce the bulk, thereby decreasing the space required in multi-person life rafts. Its solution: replace the fishing kit in the back-pad kit and transfer it to an item of personal issue for those not equipped with the back-pad kit. The BuAer was initially influenced by a small compact fishing kit utilized by the Royal Canadian Air Force measuring 3-1/2 x 5 x 5/8 inches and containing a basic selection of hooks, line, and spinners. Other samples were obtained from various companies soliciting bids for the fishing kit required in the USAAF C-1 emergency vest for evaluation.

The Edward K. Tryon US Govt. Emergency Fishing Kit was revised and formalized under NavAer Specification M-627 14 July 1944 to replace the AN-K-2 fishing kit. The BuAer was contemplating this streamlined unit upon submission of the revised specifications for the AN-K-2 presented by the USAAF and, with full intention of standardizing it, declined any further interest to the Joint Army-Navy Board.

The M-627 fishing kit was placed into production 19 September 1944 under contract 288-25897 for a quantity of 150,000 units. Deliveries were expected in December 1944 with stock available under ASO number R37-K-110. The M-627 was to replace the AN-K-2 for multi-person life rafts, specifying one for the MK-II and two for the MK-IV and MK-VII. This was first officially detailed in Technical Order 119-44 for inspection and maintenance of multi-place life rafts and equipment in September 1944. Technical Note 13-45 dated 23 February 1945, specifically addressing the M-627, echoed the Technical Order and prescribed this kit as optional personal equipment for pilots and aircrew, to be carried in flight suit pockets. The fishing kit was eliminated for inclusion in PK-1 pararaft. An additional contract was awarded for 150,000 more kits in May 1945 under contract 288-31916.

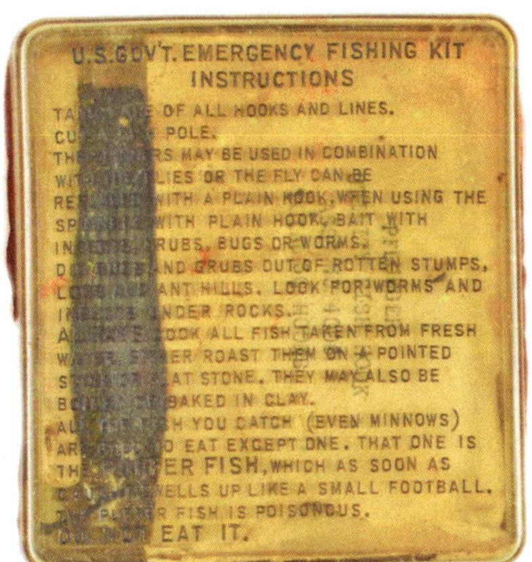

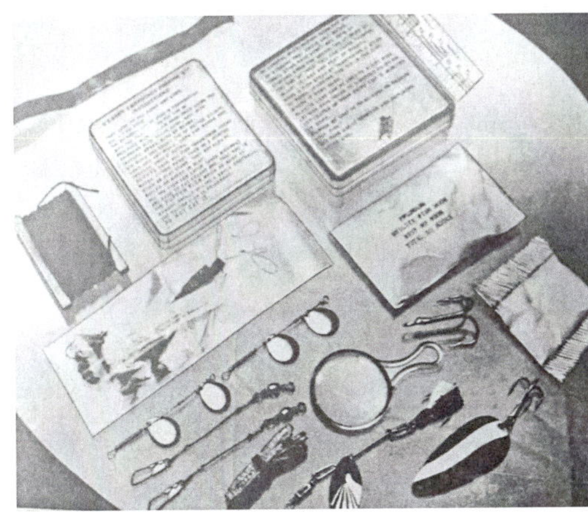

(Above) The fishing unit of greatest importance to the BuAer was obtained from the Edward K. Tryon Co. and designed by Canadian outdoorsman Roy Teller. It was specifically marketed to military services. The kit contained fishing tackle in a cellulose plastic container 3" x 3" x 1", with printed instructions on both sides. Limited quantities were procured by the BuAer for use in life rafts and issued as personal equipment by early spring of 1944.

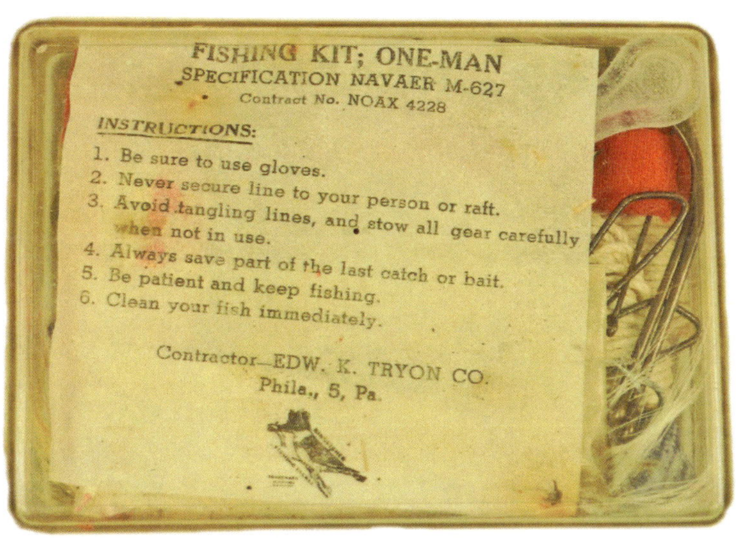

FISHING KIT; ONE-MAN
SPECIFICATION NAVAER M-627
Contract No. NOAX 4228

INSTRUCTIONS:

1. Be sure to use gloves.
2. Never secure line to your person or raft.
3. Avoid tangling lines, and stow all gear carefully when not in use.
4. Always save part of the last catch or bait.
5. Be patient and keep fishing.
6. Clean your fish immediately.

Contractor—EDW. K. TRYON CO.
Phila., 5, Pa.

(Left) Upon standardization in July 1944, an unknown limited service test quantity of fishing kits was procured under contract NOA-4228. This expedited manufactured kit, called Fishing Kit; One-Man, measured 4" x 3" x 1" and included a rayon instruction sheet. (Robin Bartel Collection)

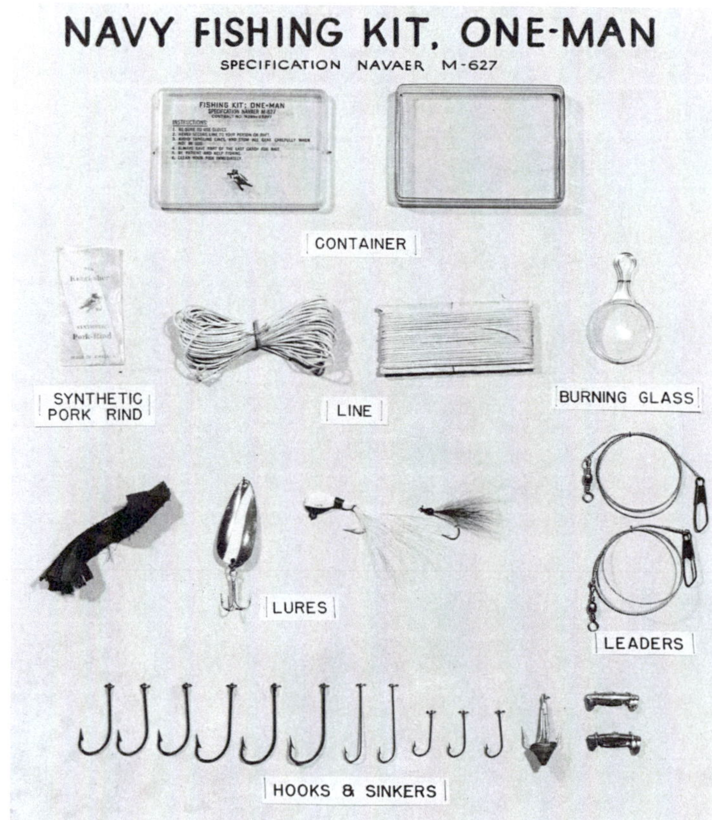

(Above) Contracts 288-25897 and 288-31916 applied nomenclature and instructions to external surface of acetate container. Image on right illustrates the components included in M-627 fishing kit.

Fishing was a common recreation for units operating in the Pacific. The isle reefs provided a variety of species for fishing enthusiasts. Fishing kits were procured by the Quartermaster for the Special Services Division, available to servicemen for recreation. Ashaway was the primary supplier of these streamlined drawstring fishing kits with the same type components as the No. 10 and No. 5 units. They were also available through the Red Cross, donated by the International Game Fish Association, American Museum of Natural History. Most large installations had recreation facilities. Fishing kits and other athletic items could be checked out by service members.

(Below) Example of the type procured by the QMC for recreational purposes.

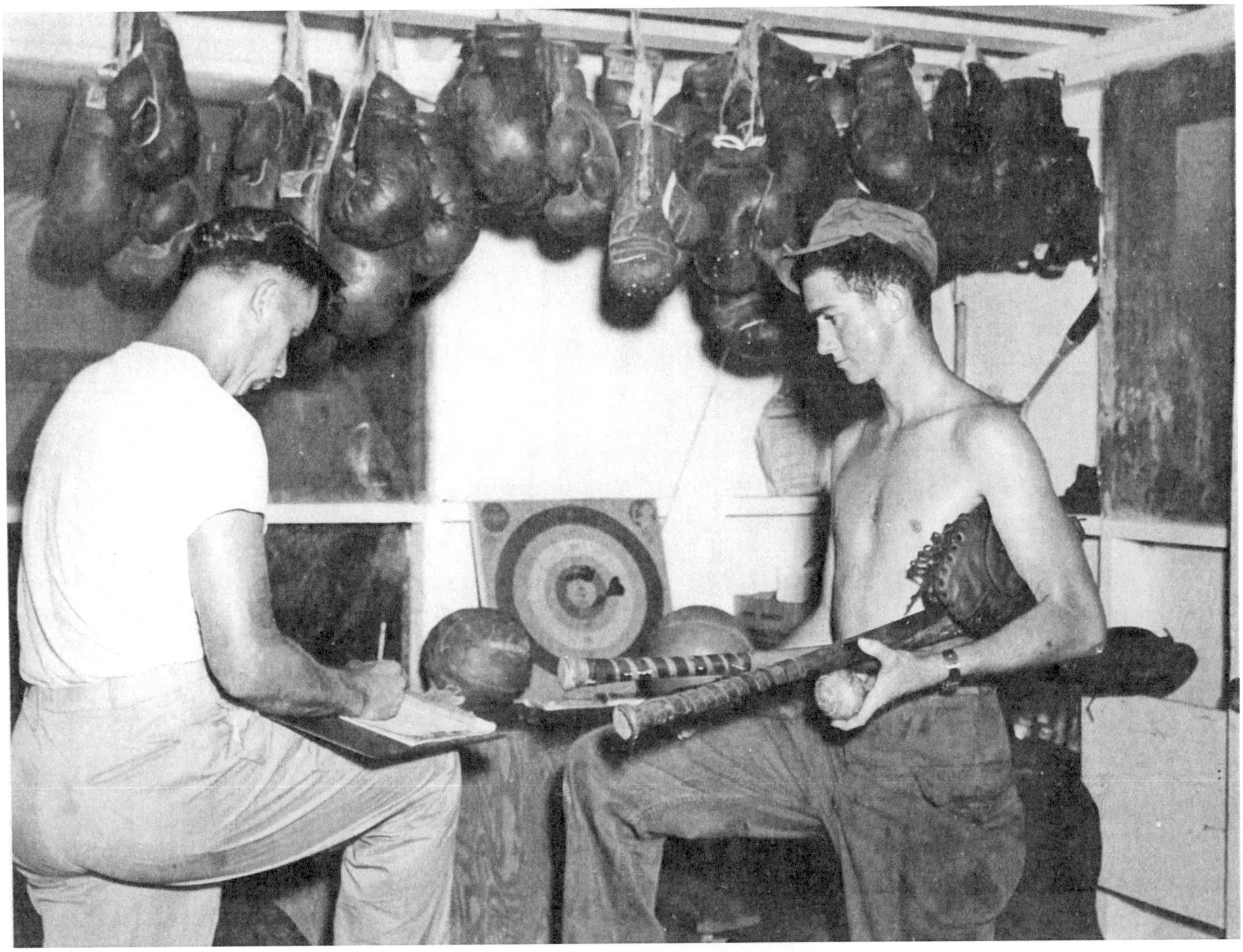

(Above) This is typical view of a Special Services Division recreation room, baseball equipment being checked out, c.1944. (National Archives)

(Left) Pfc. Robert Sloop about to launch a wooden float from shoreline in the Marianas with modified grenade launcher on his 1903 Springfield rifle, with assistance from M/Sgt. Thomas Whittemore, 7th AF, c.1944. Both enjoying some recreational fishing. (National Archives)

SURVIVAL LITERATURE

With the outbreak of war, many agencies expressed interest in providing handbooks, manuals, and pamphlets for survival guidance. They wished to preserve life in emergencies in which men were separated from their bases without ground force, naval activity, and air operation supplies. To answer the call, new offices and agencies were created within the Army and Navy that supplied general, technical, geographical, and environmental information to cope with the various problems survivors faced.

Fundamentals of survival became a major factor for all service members destined for combat area in all regions of the world. However, the most significant emphasis was placed on the men having a working understanding of the environment they might encounter. Literature was prepared by military and civilian organizations, distributed through special publications, and eventually integrated into survival training programs. Professionals and experts in the fields of anthropology, geography, exploration, biology, geology, astrophysics, and botany were sought out and assembled for consultation. The primary institutions established as central clearinghouses for information included the Smithsonian Institute's Ethnogeographic Board, The Arctic, Desert, and Tropic Information Center (USAAF), Aviation Training Division of the Bureau of Aeronautics, and the US Navy Bureau of Medicine & Surgery. Other organizations oriented toward all aspects of air-sea rescue that were consulted included The Liaison Committee on Emergency Rescue Equipment (est. April 1943), Emergency Rescue Equipment Section under the US Navy Office of Scientific Research and Development (est. April 1943), USAAF Emergency Rescue Branch (est. August 1943), and the Air Sea Rescue Agency under the direction of the USCG (est. February 1944). Combined, these entities provided hundreds of publications for the education of service members for the war effort.

The concept of providing technical information for aviation units can be traced to 1935 when, upon request of the USAAC, the world-renowned explorer Vilhjalmur Stefansson prepared a report on the problems of living and operating in the arctic. Only a limited number of copies were made available. In 1940, due to many requests, the Army granted permission to print a commercial version in two volumes titled The Arctic Manual. It was revised, enlarged, and republished in one volume in 1944 consisting of 556 pages on native food, shelter, clothing, climate, fauna, transportation, and medical complications and treatment.

The demand for these types of educational tools increased upon US entry to WWII. The initial literature, efficiently compiled in compact booklets, was intended for general distribution to all service members. A condensed and revised Arctic Manual was prepared and issued as The Arctic Manual TM 1-240 on 1 April 1942. This technical manual, assembled by the War Department, was comprised of 131 pages that gave a brief overview of arctic geography, climate, vegetation, shelter, travel, and procedural outlines for aircrews in the event they were forced down.

(Below) The Arctic Manual, reprinted in 1944, was a textbook geared to classroom use. The revised version, TM 1-240, was printed in a form more fit for field use. A related item was FM 31-15, Operations in Snow and Extreme Cold, published earlier, on 18 September 1941. This manual was included in the USAAF Types E-1 and E-2 sustenance kits. Its 82 pages covered many of the subjects in The Arctic Manual. A similar manual was titled Desert Operations FM 31-25, published on 14 March 1942. Dates on manuals reflect when the material was released for print, does not reflect its printing date.

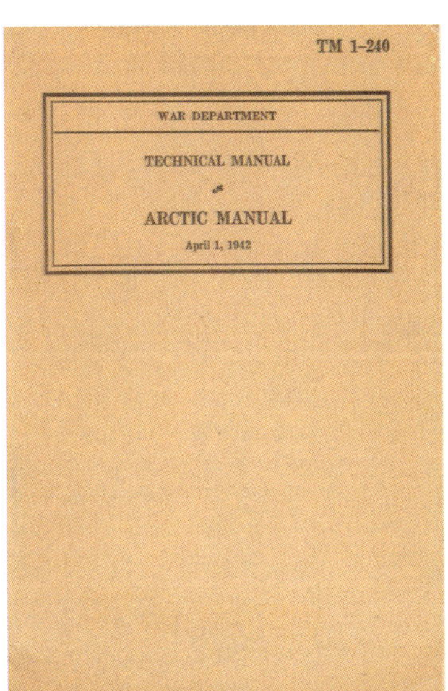

The Arctic Manual, reprinted in 1944.

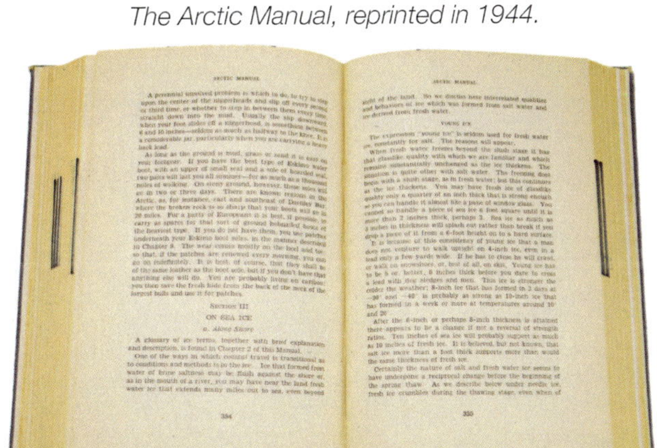

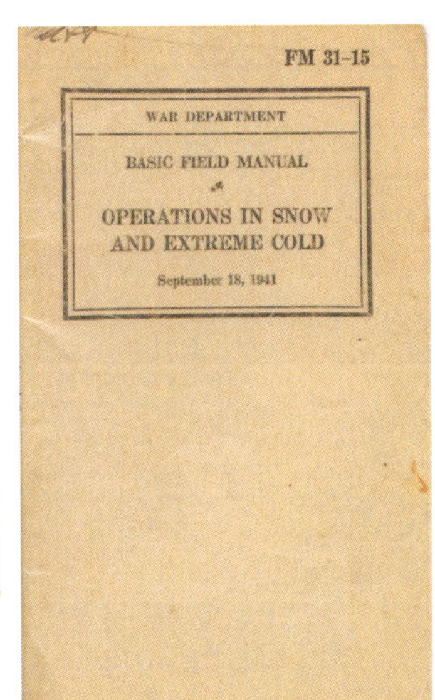

By the spring of 1942, several dozen individuals and agencies representing various branches of service commenced preparing literature for the benefit of servicemen. There was no interdepartmental coordination, and, too often, individuals with neither much knowledge of the subject nor any arctic or tropical experience were assigned the tasks of compiling data. Due to this oversight, a meeting of departmental representatives was called under the auspices of the National Research Council in September 1942.

The ADTIC and the Ethnogeographic Board were officially formed after this conference to help consolidate and coordinate efforts. The War Department Technical Manual TM 10-420 titled Emergency Food Plants and Poisonous Plants of the Islands of the Pacific was sourced from this meeting, for which printing was approved on 15 April 1943. Illustrations in this manual were then copied into coinciding and future survival manuals.

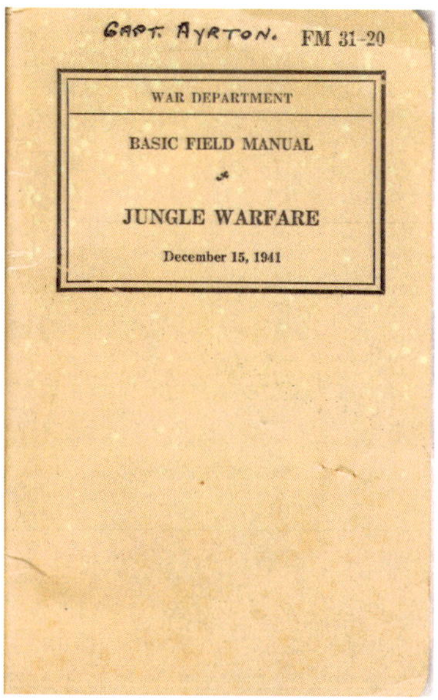

(Left) For an expedited response to service needs, the War Department published a basic field manual titled Jungle Warfare FM 31-30, 15 December 1941. This was an 87 page manual discussing personal care and jungle tactics. It and The Arctic Manual were the principal reference aids for USAAF aviation personnel and the BuAer during initial stages of war.

(Right) TM 10-420 had 149 pages with 113 illustrations of flora and their uses. This manual was considered such a vital asset that the BuM&S distributed an All Pacific Activity Circular, 11 August 1943, stating it should be placed in the first aid kit of each life raft and life boat on naval vessels and aircraft operating in the Pacific area. The manual was maintained in standard stock, number S16-2004.

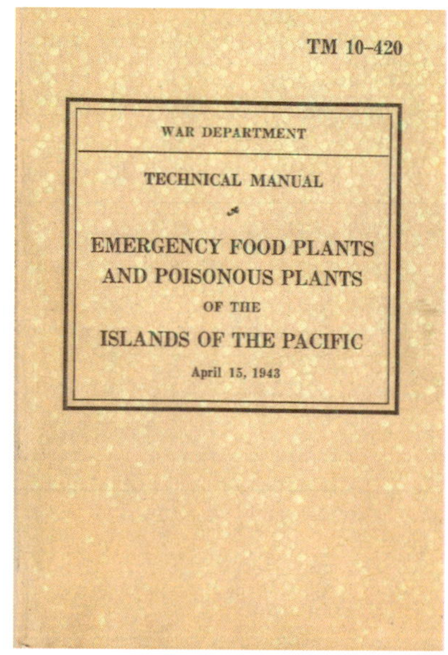

(Right) The first emergency booklet prepared for US armed forces during the war was created by Capt. Ammon Godshall and published in Panama, 20 March 1942. This pocket-sized booklet of 47 pages contained 45 illustrations of vegetation in the Canal Zone and was titled Edible, Poisonous, and Medicinal Fruits of Central America. Two subsequent booklets influenced by Godshall's were Edible Plants of the Arctic Region (1943) and Edible and Poisonous Plants of the Caribbean Region (1944). These were printed and distributed by the US Navy Bureau of Medicine & Surgery and available for purchase through the Government Printing Office. The Godshall manual was compact, measuring 4"x 6", and could be obtained only by direct requisition from the 6th Air Force HQ.

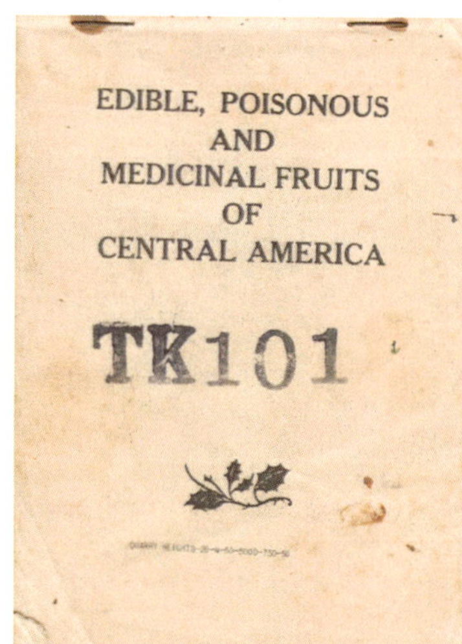

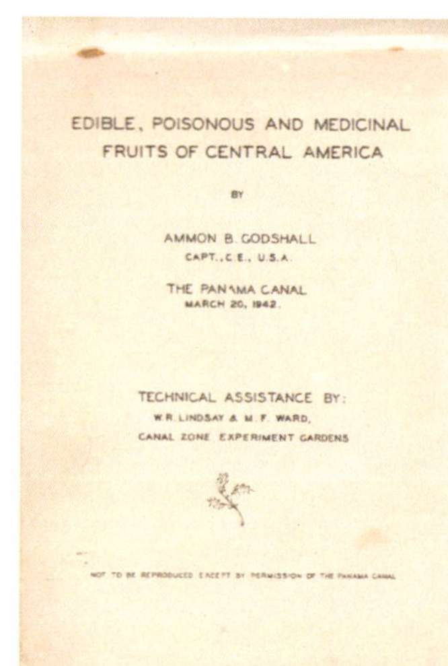

Through 1942, there was no central agency for the distribution of survival literature; however, by that year's close, the deficiency was being addressed. As an interim until service requirements could be met, the US Navy Objective Data Section, Air Intelligence Section, Pacific Ocean Area requested a manual to be compiled by the Bernice P. Bishop Museum located in Honolulu, Hawaii. This booklet was intended for general distribution to all interested parties operating in the Pacific Ocean Area but was oriented toward Naval aviators. Booklet preparation was assigned to Kenneth P. Emory, who compiled notes from personal experiences on geography, vegetation, animals, and insects on Pacific tropical islands. He included instructions on native foods, plant usage, and life raft navigation. This 63 page booklet went into print on December 1942 and was titled Castaway's Baedeker to the South Seas.

The Bernice P. Bishop Museum quickly grew to be an integral entity in survival indoctrination by providing survival training and literature for all units stationed in Hawaii. Following the compilation of the Baedeker, Emory continued assembling educational materials for servicewide distribution endorsed by the museum.

His next publication was titled Native Lore for Castaways in the South Pacific, printed in January 1943. This was a 20-page pamphlet illustrating plant and animal food, and shelter and clothing improvised from natural materials.

In September 1943, the museum published a booklet that combined attributes from Baedeker and Native Lore titled South Sea Lore Special Publication #36. The first printing, green cover, consisted of 75 pages. The second printing, red cover, 79 pages, was published in September 1944. It is an extended and revised edition adding information for the Philippines and Dutch or Netherlands East Indies.

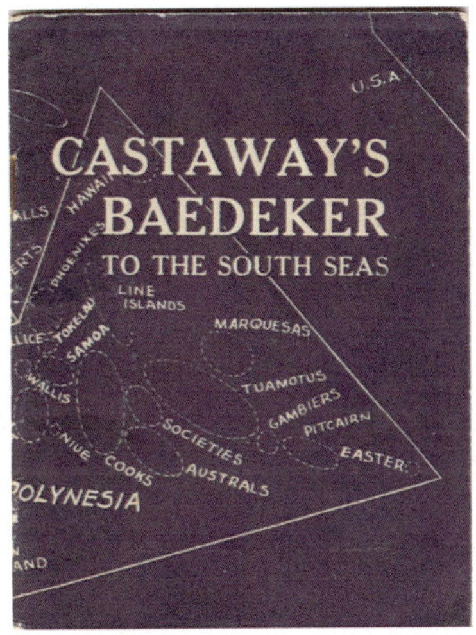

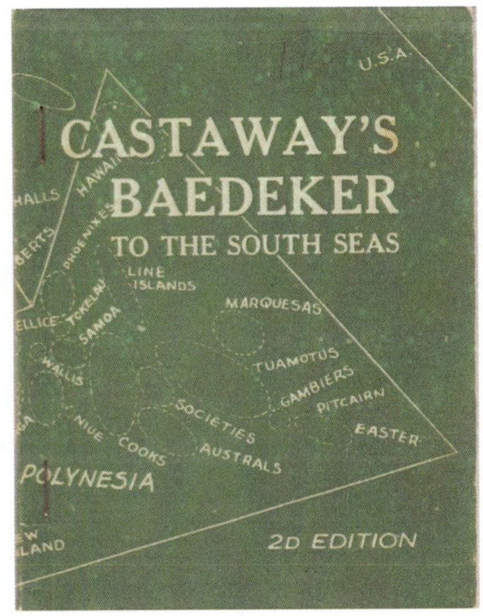

(Above) The first printing, December 1942, of Castaway's Baedeker had a blue cover. The second edition (September 1943) had a green cover. It was not just reprinted; it was revised and enlarged to 100 pages, with a glossary of phrases in Pidgin-English. Both booklets measured 4-1/2"x 6". Castaway's Baedeker was the first all-encompassing survival manual of the Pacific region for general distribution for US Armed Forces. (Left) Baedeker was often included as a supplementary item in emergency kits as seen in this Original back-pad kit with 2nd Edition.

(Left) Native Lore for Castaways in the South Pacific, printed in January 1943 was 8-1/2"x 11". This was a 20-page pamphlet illustrating plant and animal food, and shelter and clothing improvised from natural materials. South Sea Lore Special Publication #36, first printing, green cover, consisted of 75 pages. The second printing, red cover, 79 pages, was published in September 1944. Each booklet measures 4-1/2"x 6".

Before arriving in New Guinea 11 December 1942, Maj. John Small, USAAF had determined to investigate the rescue situation. He found a universal interest in rescue, but no organization and no coordination of effort. Each squadron or group was largely dependent upon its own enterprise. Other difficulties were lack of knowledge of New Guinea emergency landing areas, insufficient jungle kits, insufficient knowledge of the jungle, and primarily, a complete lack of rescue materiel in the Army Air Forces. Small began immediately a collection of material for a jungle book and on 5 February 1943, issued the first edition for limited distribution in the 5th Fighter Command. He also began a collection of all available data on New Guinea air strips, of which about 150 were listed on aeronautical charts, but of which little was known and that conflicting. It was apparent that more adequate steps had to be taken to promote flying safety and rescue. Small developed and organized a coordinated rescue program between Australian and American resources based from Port Moresby.

From Australian sources, from Allied Geological Section, and from other material, a rescue book Escape from the Jungle was written and published 24 July 1943. It was distributed throughout the 5th Fighter Command, 5th Bomber Command and Troop Carrier Wing.

In November 1943, MIS-X, SWPA began a draft of a new survival booklet due to the viewed shortcomings of the manual Instructions to Air Crews, which was widely distributed in the SWPA at the time. By February 1944 it was in its final draft form and in the hands of the printer by March for a first printing run of 10,000 copies being assigned the title Survival Hints. Expected deliveries were projected to commence in April. Each booklet was printed on waterproof paper and enclosed in a cellophane envelope. One booklet was to be included with each E-3J kit, and distribution was also to be made to all combat crewman of Allied Air Forces, one to each person.

As of November 1944, MIS-X reported that approximately 12,000 copies had been received from the printer, 9,000 of which had already been distributed. Distribution was to all forces stationed in the SWPA. As of 1 January 1945, 5,385 had been received by the FEAF, 1,000 to the US Navy, 2,200 for Marine units and 2,349 categorized under Miscellaneous. As of the end of May 1945, the FEAF had received a total of 8,395 with an increase to 1,916 for the US Navy. Marine accountability remained at 2,200 in total. This does not account for the booklets accompanying the E-3J.

(Left) Escape from the Jungle, 4-1/4" x 5-1/4". A 135 page booklet covering all aspects of surviving on New Guinea. It details equipment, travel, living off the land and contact with native populations. The back includes a glossary to Pidgin English.

(Right) Booklet Survival Hints drafted by MIS-X, SWPA, 4"x 5". First draft was 50 pages, increased to 79 pages when it was released for print. Focused on aspects of New Guinea, but transferable to Solomon Islands area. Difference in this booklet from that of Escape from the Jungle and Instructions to Air Crews is that it includes more depth on obtaining food to include illustrations.

The gathering and dissemination of information concerning survival was a high priority for the US Government during the war. Successful military operations were dependent on geographical and anthropological knowledge of regions, knowledge ranging from large scale attacks down to individuals who might find themselves isolated in unfamiliar territory.

With the increased frequency of aircraft accidents, education was vital for preservation of life. To supplement resources, survival reference aids were compiled and printed at overseas stations. Offices and agencies in the continental United States were also diligently working on providing technical information for all services regarding living abroad.

At the USAAF, prior to the establishment of the ADTIC, responsibility for the preparation of survival literature was relegated to the Directorates of Air Traffic and Safety, and of Safety Education. These offices were established in May 1942. By the end of that year, they had compiled an 88 page booklet for inclusion in emergency kits titled Jungle and Desert Emergencies. The booklet included notes on signaling, shelter, edible and poisonous plants and animals, trail craft, and health. In March of 1943, these offices were re-designated as the Flight Control Command, Safety Education Division, still responsible for all aspects of flying safety and the preparation of educational training materials.

Continuing the previous work, the new office prepared a 65 page booklet titled Arctic Emergencies. Both the Jungle/Desert and Arctic manuals became available for distribution by spring 1943. Due to the high demand for such literature, the two booklets were combined and titled Jungle, Desert, Arctic Emergencies, consisting of 149 pages. Since previous printings included the same technical information, it was more efficient to provide a singular universal manual for all requisitioning departments, since by this time all sustenance kits were directed to include this manual. In October 1943, this office, again, was renamed the Office of Flying Safety, Safety Education Division. In line with the development of streamlined selection of emergency sustenance and parachute kits, it compiled and printed a single universal manual for all climate zones titled Jungle, Desert, Arctic, Ocean Emergencies. This 174 page booklet superseded the previous printings and was directed to be included in all emergency kits.

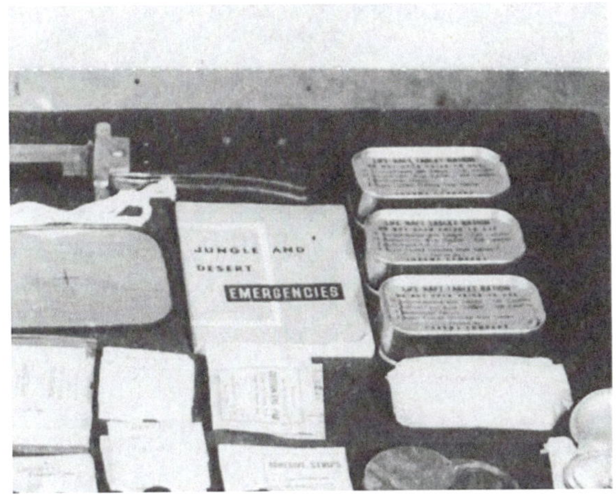

(Above) USAAF Jungle and Desert Emergencies, 4-1/2" x 5-5/8". The initial printings of Arctic and Jungle and Desert Emergencies included a snap cover. With an increased demand for this manual, a second printing removed the snap cover feature.

(Right) USAAF Arctic Emergencies manual 4-1/2" x 5-5/8". Only printed with snap closure.

(Far right) Manual included in locally assembled arctic aerial dropping kit by the Air Transport Command, April 1944.

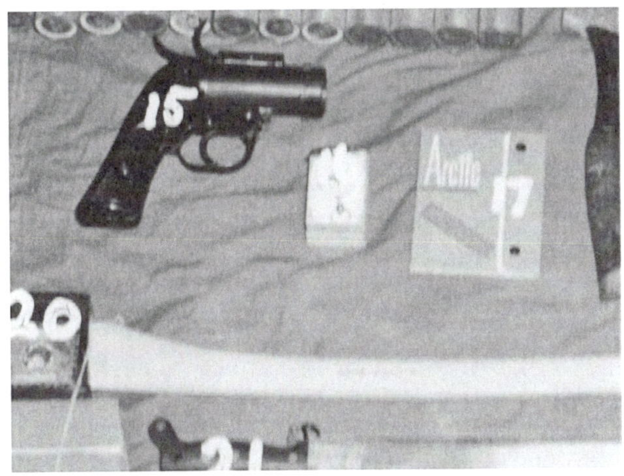

(Above) Consolidated manual Jungle, Desert, Arctic Emergencies. Intended to simplify distribution for requisitioning activities.

(Below and right) Final revision of the USAAF "emergencies" manuals with the addition of ocean survival. This manual with the previous printings would be replaced by the AAF Manual No. 21. Emergencies manuals were provided at local commands decision, as seen in the B-2 jungle kit to the right.

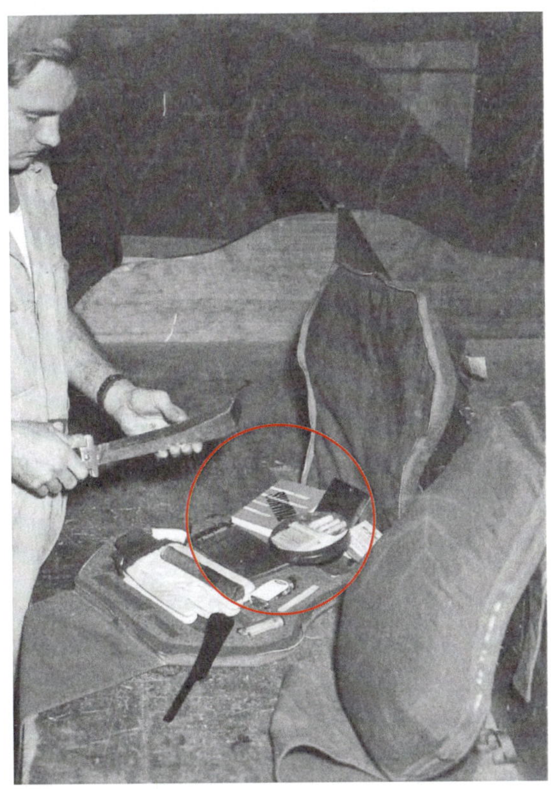

Eventually, dozens of organizations, especially those operating overseas or responsible for training applications, printed materials addressing personal care. One was the Airlines War Training Institute, established 12 August 1942. This organization trained civilian and military pilots. In 1943 it prepared a series of booklets with such titles as Survival (1943), Survival in the Arctic (1943), Survival in the Desert (1943), and Out of the Jungle (1943). Survival was a 116 page booklet for aircraft crews forced down in jungle, arctic, desert, or ocean environments. Its content was similar to the USAAF's Emergencies manuals. It was subsequently printed in an extended format in the Infantry Journal.

The first printing of Infantry Journal was issued in 1944 in hardcover and consist of 376 pages. In 1945, a soft-cover version became available for sale to the general public as part of a military book series. The Airlines War Training Institute also printed a book on plant life of the Pacific area.

(Above) Airlines War Training Institute Survival booklet, 116 pages measuring 5"x 7". Often utilized by the USAAF and airlines under direction of the ATC. Survival with USAAF life raft accessories.

(Above) Infantry Journal prints sold to the public and military personnel for $.25 each.

(Below) Just in Case was a 4-1/2"x 6-3/4" 10-page pamphlet with notes on ditching and life raft operation compiled and printed by the AAF Air Transport Command (ATC). The tinted example was reproduced by the 942nd Engineer Aviation Topographic Battalion. ATC provided other literature, including Guide to Arctic Survival (1943), Read and Live (Northern Regions in Canada), and Pilot's Guide (1943). The literature was for distribution to the vast network of commercial airlines under its purview.

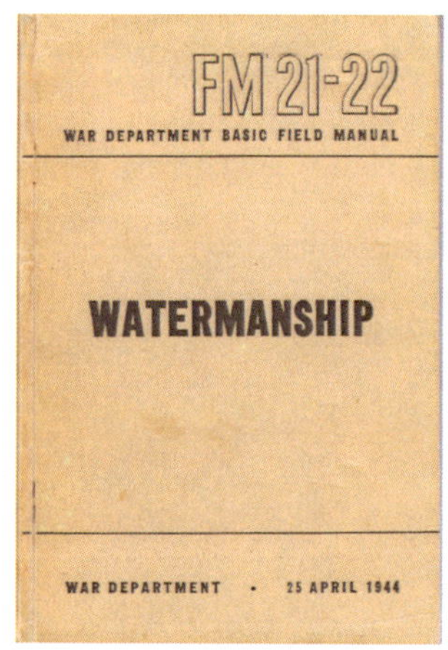

(Above) Survival guides were not pertinent solely to aviation activities; survival was important for all services. Here, an 8-page pamphlet written by Lt. Commander William Chambliss, USNR, titled Recipe for Survival. It first appeared in US Naval Institute Proceedings in July 1943. Due to the clear and comprehensive nature of the article as abandon-ship protocol instructions, it was reprinted in pamphlet form. It later appeared in a March 1944 issue of Colliers Magazine. The pamphlet cover states it was issued through the Training Division of the BuAer. This division had the responsibility of distributing training material for all US Navy departments. A related manual that addressed abandoning ship and survival at sea was distributed by the War Department as FM 21-22. In 151 pages it detailed swimming, abandoning ship, life rafts, and other safety measures for personnel at sea.

On 20 September 1942, a directive was issued by O/C&R to establish a centralized collection and distribution office for educational material. In November, the Arctic, Desert and Tropic Information Center (ADTIC) was founded at the AAF Proving Ground in Eglin Field, Florida. Its mission was to collect, evaluate, disseminate, and prepare information for areas of military interest in the form of flight data sheets and water/land surveys. A collective of at least 30 scientists supplied technical and intelligence information on topics ranging from climate and weather data to the attitudes of native inhabitants in all regions of the world. The ADTIC was one of the first offices to supply general geographical and environmental information to all branches of the military. In October 1943, this office was transferred to the Office of Assistant Chief of Air Staff, Intelligence and relocated to New York City. It was again transferred in April 1944 to AAFSAT in Orlando, Florida. This last transfer was part of a consolidation effort within the USAAF command structure that included the AAF Proving Ground, AAF School, AAF Board, and the ADTIC. At this time responsibilities were divided with the following functions: the AAF Board tested new material for operational and tactical combat conditions, AAF School tested aerial warfare tactics and instructed Army and Navy personnel, and the ADTIC worked on the distribution of information concerning climatic zones. In June 1945, this coalition was renamed from AAFSAT to the AAF Tactical Center and the ADTIC renamed the Arctic, Desert and Tropic Center.

(Above) Literature prepared by the ADTIC. 8" x 10" information bulletins were compiled and printed under the direction of the ADTIC for servicewide distribution. A series of at least 24 bulletins was prepared from 1943 to 1945 discussing a variety of topics from living off the land to poisonous animals, insects, and personal care in all climates. The full-size examples include (Left to right): No. 1 Emergency Living in the Desert, No. 5 Living Off the Southwest Pacific Tropics. Smaller renderings of these bulletins were printed 4-1/2" x 6". With the assistance of the ADTIC, AAFSAT prepared a 2-part information bulletin titled Living Off the Land. Part 1 detailed living in the Southwest Pacific tropics. Part 2 discussed surviving in the Arctic. The ADTIC also published a Medical Series and a Survival Lecture Series utilized by Personal Equipment Officers and other agencies.

(Right) With an increase in air activities, education and training became vital. Knowledge gained from previous ordeals was applied to curriculum development, with emphasis on the best emergency procedures in the arctic, jungle, desert, and ocean. AAF Manual No. 5 was prepared for life raft procedures on 1 April 1944 and titled Ditching Dope. It was 16 pages adapted from the RCAF manual Ditching Gen, discussing ditching and life on a raft. Dinghy, Dinghy, Prepare for Ditching was a manual prepared by the 396th BG at Drew Field, Florida, 15 June 1944. Dinghy, Dinghy, Prepare for Ditching was designed to educate aircrews on ditching procedures, emergency equipment, and life at sea.

Continual publication of specialized educational aids was requested by returning combat personnel and commands stationed overseas. The curriculum shifted to promoting knowledge before an emergency occurred, expressed by "Know before you go!" It wasn't until the latter part of 1943 that this took effect. To supplement standard flight training, the Aviation Training Division of the BuAer produced Sense manuals. Approximately 23 manuals were prepared through the war relative to safety and survival. Interestingly, no Sense manual was prepared for the jungle or tropics. Sense publications began being released in the spring of 1943; each manual contained basic information and Dilbert the Pilot cartoon illustrations. The manuals were available aboard ship libraries and at air stations.

(Below) Some examples of the many Sense manuals published by BuAer's Aviation Training Division, 7-1/4"x 9-1/2". (right) Compilation of literature for naval aviation training, including Sense manuals, ADTIC bulletins, training aid posters on ditching, and the pamphlet Recipe for Survival.

(Below) Collage of literature addressing preservation of life in the event of an emergency. Such literature was available to all aviation personnel c.1944 and included ADTIC bulletins, Sense manuals, survival booklets for emergency kits, and textbooks. The two books written in 1943 detailing the 24 day life raft experience of Eddie Rickenbacker and party, Seven Came Through and We Thought We Heard the Angels Sing were especially recommended. Another suggested reference was the book The Raft by Robert Trumbull detailing the 34 day experience of three TBD Devastator aircrew members at sea. .

(Above and below) Libraries were established at most all air bases or stations, available for recreation and education where many of these survival type booklets and manuals could be found. (Above) Ready room and intelligence office at NAS Amichitka, Alaska, August 1943. (Below) Library established by US Army Special Services, Ascension Island, Atlantic Ocean area.

Survival literature was provided in three categories: training textbooks, operational duty education, and emergency kit manuals. For a time, survival literature was only available for requisition upon request. By the close of 1943, the opinion developed that some form of a manual should be provided in all emergency kits and life rafts. The most efficient method was to provide a pocket-booklet that could be conveniently carried both in equipment and on the person.

With the development of the emergency vest kit, the USAAF began to prepare a compact concise manual with survival information. With assistance from the ADTIC, the Materiel Command printed a manual titled Survival. It was to be included in the vest kit. Following a decision to include a survival manual in all emergency kits, it was approved for mass printing. After adoption, it was reformatted and titled AAF Manual No. 21. In July 1944, the Command approved 500,000 copies for printing, 300,000 on waterproof paper designated 21W, and 200,000 on ordinary paper.

Until these became available, Materiel Command continued printing and circulating its version. In May 1945, a second order of 100,000 copies of AAF Manual No. 21W was approved. In June, a revision was approved, titled AAF Manual 64-0-1 with 300,000 copies printed. AAF Manual No. 21 included 77 pages and 64-0-1 had 84 pages. The principle additions to 64-0-1 were religious (Protestant, Jewish, and Catholic) scriptures and more log pages. All edition dimensions were 3-3/4 x 8 inches. These manuals remained in service until superseded by AF Manual 64-5 in December 1952.

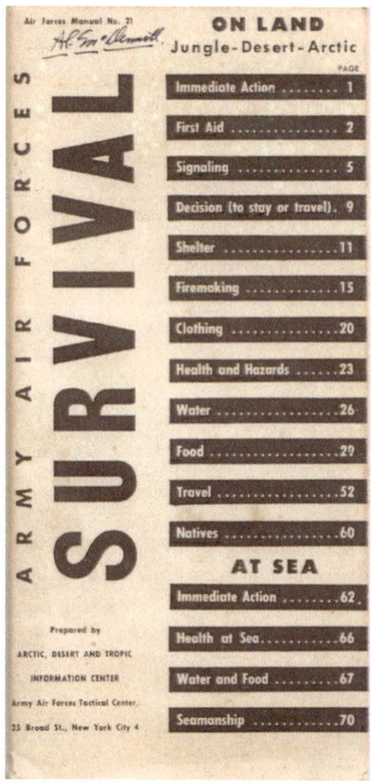

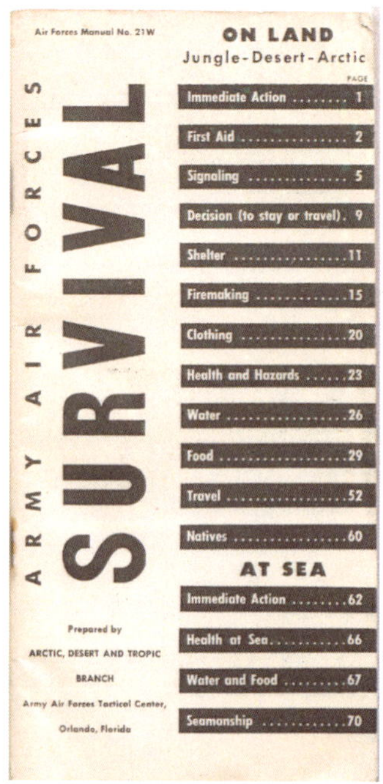

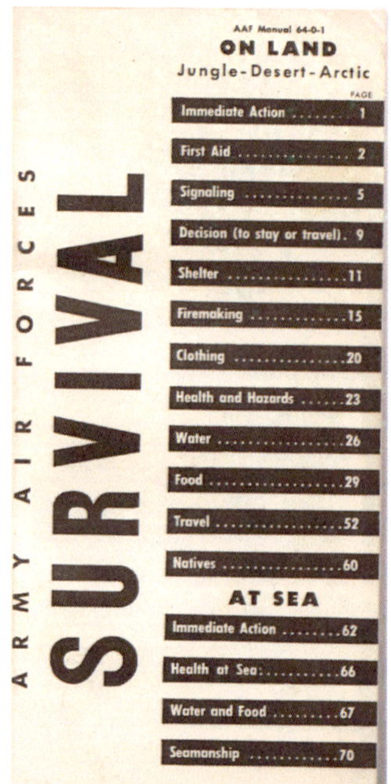

(Above) USAAF Survival booklet directed to be included in all emergency equipment. Left to right; Materiel Command print, contract printing Manual No. 21 (ordinary paper), Manual No. 21W (waterproof paper), revised Manual 64-0-1.

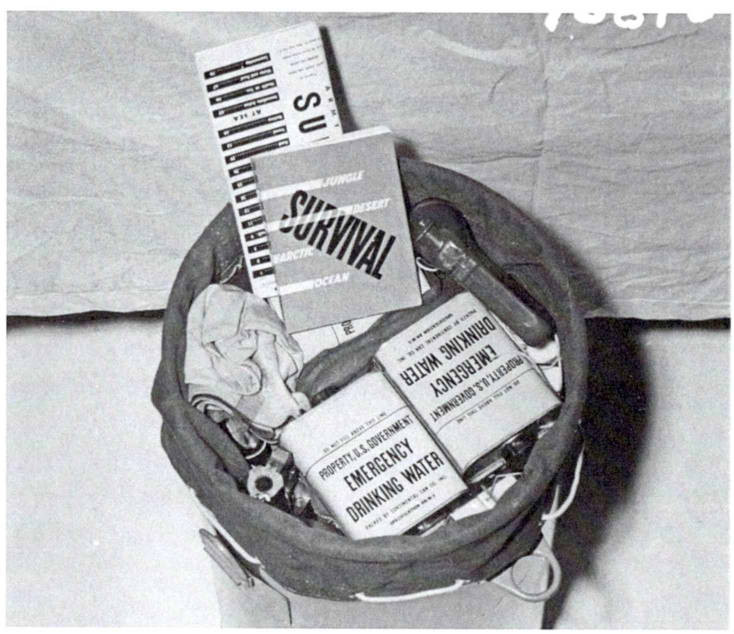

(Right) Both the Survival and equivalent Emergencies manual are displayed with a Type E-15 sustenance kit. Manual No. 21 was a reformatted Emergencies manual of the four climate zones. No. 21 was designed to reduce bulk when used with sustenance vest.

(Below) Printer, James Gray Inc. provided quantity and approval date of printing on the back of 21W.

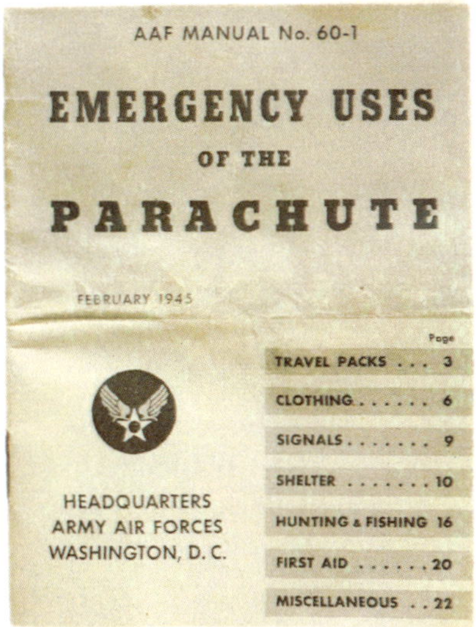

(Left) In February 1945 a booklet was prepared for inclusion in all parachutes with the intent to provide emergency information on all emergency kits, life rafts, and the individual. This booklet, AAF Manual No.60-1, Emergency Uses of Parachutes, consisted of 24 pages and measured 3-1/4" x 4-1/4". It detailed a multitude of parachute uses in emergency situations and was designed to be folded and placed in the inspection and data pocket on parachute packs. Its required installation was detailed in Technical Order 13-5-55, 26 May 1945.

(Left) Another instructional pamphlet. Printed in April 1945, compiled by the USAAF and titled "Going Down" - AAF Manual No. 70. It had 16 pages, 5" x 8" format and contained cartoons walking the reader through the do's and don'ts of parachutes and bail-out procedure.

In September 1943 the BuAer recognized an urgent need for a suitable survival booklet for inclusion in life rafts and emergency equipment carried on naval aircraft. Until this time, Castaway's Baedeker was the principle literature used. Revisions were soon approved by the Office of Naval Intelligence and War Department Military Intelligence, redesignated as the Joint Intelligence Center, Pacific Ocean Areas.

The Publications Branch, Office of Naval Intelligence, U.S. Navy, BuAer, BuM&S, and the Smithsonian's Ethnogeographic Board coordinated their efforts to prepare a manual titled Survival on Land and Sea. This publication, intended for servicewide distribution, was printed in December 1943 at 187 pages, 4-1/2 x 5-1/4 inches. It highlighted the importance of drill, equipment, knowledge of survival techniques, geographic features, and natives. A 1944 edition added an aviation section in the front, provided by the Aviation Training Division of the BuAer, detailing 16 pages relating to bailing out, ditching, crash landing, and water safety. Upon its release, 14,000 copies were distributed by Air Combat Intelligence. A further 100,000 copies were requisitioned by the ASO; no stock number was assigned. Contractors of life rafts and emergency kits were directed to supply the booklet, and specifications were revised for its inclusion.

Survival on Land and Sea was eventually designated as NavAer Publication 13-1-501. By the close of 1944, nearly one million copies had been printed and distributed to armed forces in the Pacific Theater. With its introduction, the USAAF and BuAer each adopted a single universal survival manual for inclusion in all emergency equipment. Overseas stations were no longer pressed to provide relevant survival information.

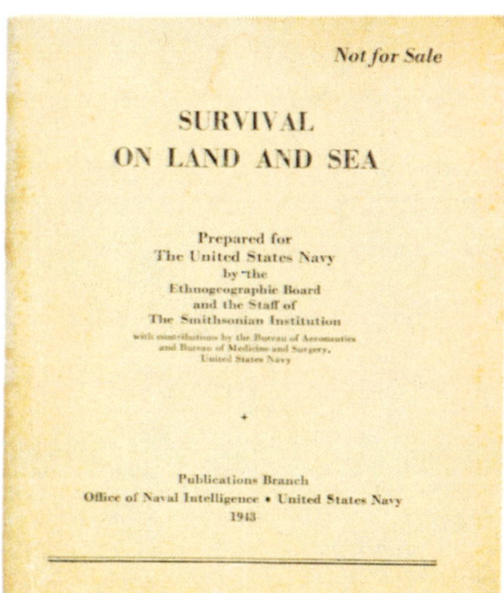

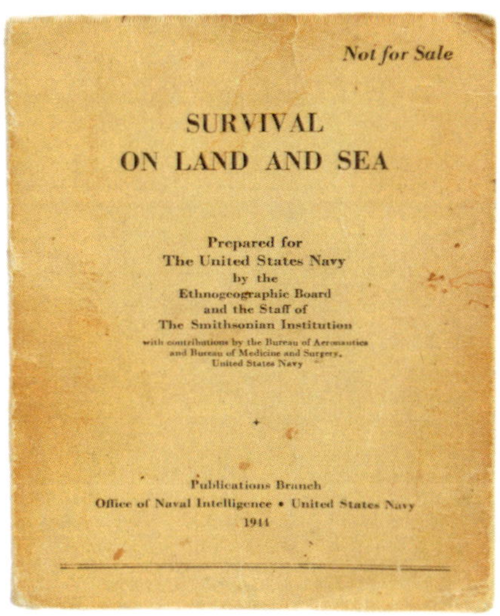

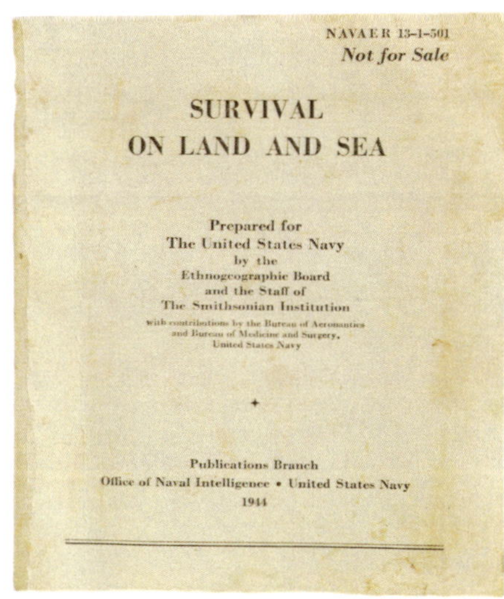

(Above) Three variations of Survival on Land and Sea. Left to right, 1943 printing, 2nd printing with added BuAer content (1944), and 3rd printing (1944) with publication number.

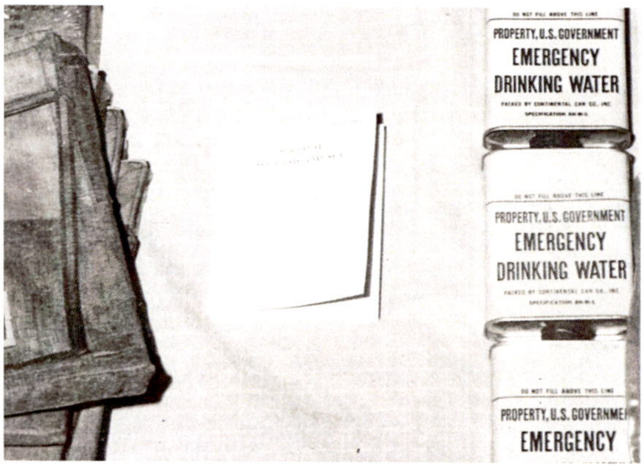

(Above) Survival on Land and Sea would be supplied by contractors of life raft and sustenance kits. From the factory, they would be provided in the M-3R (as seen above), PK-1 and Shipwreck kit M-594B. Further inclusion and distribution conducted at overseas stations.

(Right) To supplement orientation for travelers to the Pacific Ocean Area, Commander Air Force Pacific Fleet prepared a 23-page pamphlet, So You Are Going to the South Pacific? It provided notes on clothing and life at island bases and was reprinted by the Aviation Training Division in a 4-3/8" x 6" format in August 1943, with further printings in 1944.

(Left) To further educate military personnel, Guide to the Western Pacific was printed in August 1944. This 124 page booklet discussed the Pacific island groups, including their history, settlers, native customs, languages, weather, and climate.

(Below) Cartoon by unidentified artist from VPB-18 depicting a well equipped crew stranded at sea with all the necessities.

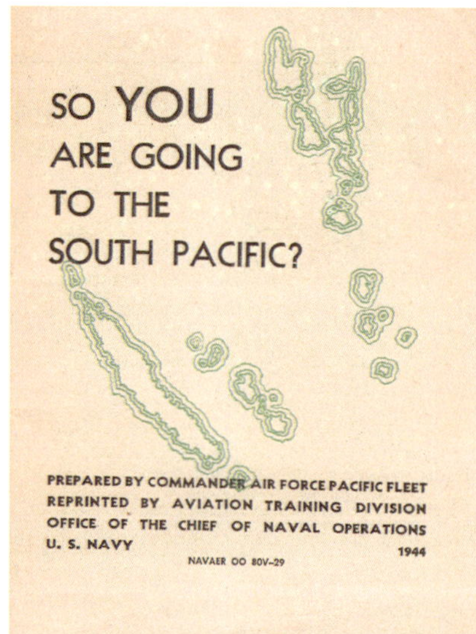

With an increase in operations and ditching events in the Southwest Pacific, South Pacific, and Central Pacific, the ability to navigate those waters became imperative. In 1943, the US Navy approached Harold Gatty, who had previously worked with the USAAF in the South Pacific, to write a book detailing navigational instruction in the Pacific. He compiled a series of charts and literature for navigation by the stars and wind/sea currents, and detailed marine signs such as birds, seaweed, water color, and clouds. This work was privately published in July 1943 as The Raft Book: Lore of the Sea and Sky and available at $2.50. It was a 152 page booklet with several waterproof charts and a folding tape housed in a hard slipcover, measuring 5-3/4 x 8 x 7/8 inches. The U.S. Navy, USCG and USAAF believed it was too bulky and covered too much to warrant a place in a lifeboat or life raft.

With modifications to streamline the work to make it lighter, smaller, and waterproof, the USAAF recommended it for procurement in August 1943, when it also garnered the approval of the US Navy. A new edition, streamlined to the essential information and charts contained in a waterproof asphalt wrapper, went into print in May of 1944. This edition was available for commercial use at $2.00 each. The USAAF, U.S. Navy and USCG all procured unknown quantities of the waterproof edition for inclusion in life rafts and lifeboats. However, because stock catalogs do not reflect their existence, it is not clear exactly how they were distributed.

(Above) First assembly of The Raft Book: Lore of the Sea and Sky, by Harold Gatty.

(Right) Waterproof addition of The Raft Book.

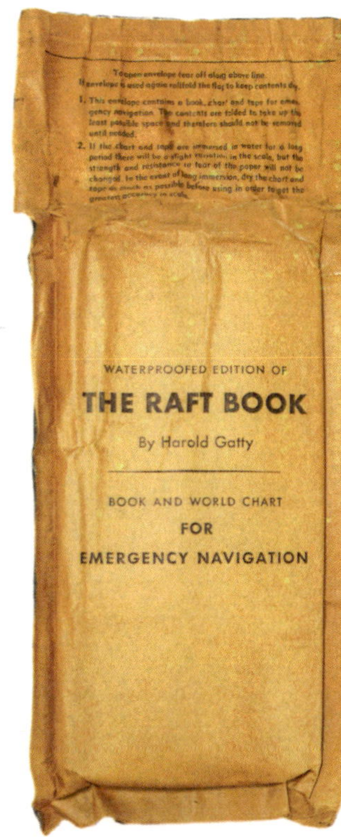

(Below) Display of charts and instructions in the waterproof addition of The Raft Book. This addition can be observed in life raft equipment layouts provided in Volume-I of this series for both the USAAF and US Navy.

(Below) Three religious booklets in a waterproof wrapper were deemed suitable for inclusion on life rafts. They were included in revised USAAF life raft specifications in September 1943. Each booklet contained about 100 pages of select readings from three Biblical versions: Protestant, Catholic, and Jewish. These were replaced by AAF Manual 64-0-1 that integrated religious scriptures. Each booklet measured 3" x 5".

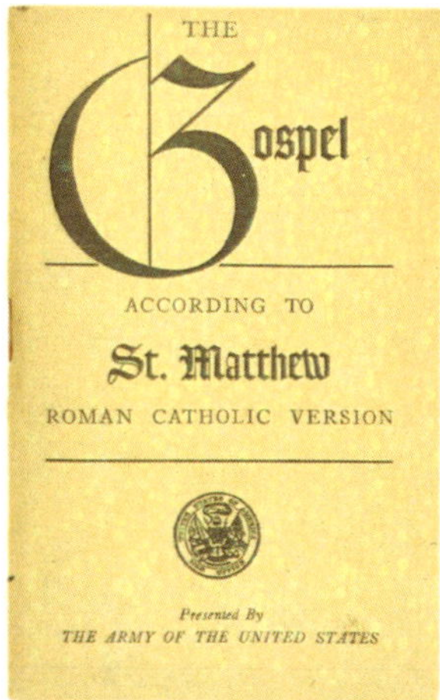

Most literature educated service personnel on personal care and operations overseas. However, there was also literature that aided in emergency survival situations. Publications printed by allied governments were also used by US Armed Forces. Conversely, allied governments prepared literature for their service members that mimicked that of the US Government relative to their regions of operation, and this effort greatly benefited US troops arriving in those areas.

In March 1943, Military Training Pamphlet #9: The Jungle Book, was issued by British Army Staff to personnel stationed in India. Other booklets include, Under the Greenwood Tree and The Jungle Hiker. Under the Greenwood Tree was distributed by the RAF in India for operations for Burma and Assam. The Jungle Hiker referenced Malaya, Sumatra and Ceylon. Since either of the books were not secret they were made available for issue to ships, depots, formations and squadrons.

G.S.I.(e), and later E-Group, distributed these two booklets to the 10th and 14th Air Forces to accompany briefings and lectures on jungle survival. AGAS, China had acknowledged receipt of 800 copies of Under the Greenwood Tree in January 1944. Copies were in circulation amongst the 10th AF by August of 1943.

(Right) Two renderings of The Jungle Hiker. Royal Air Force Welfare Publication. Top example copyrighted in 1942 published for sale at 75 cents. 77 pages, 4-3/8"x 5-1/2". Bottom print is 160 pages covering jungle survival measuring 5-5/8"x 6-1/2". Copyrighted in 1944. Sold to the public for one Rupee. Both printed at Fort Colombo, Ceylon. Each booklet contains a language guide in Tamil and Sinhalese.

(Above) Under the Greenwood Tree consists of 43 pages measuring 5"x 6-3/4". Written as a guide to jungle living.

(Below) Another booklet distributed by the RAF and E-Group for American personnel was The Edible Jungle Fruits, 4-1/4"x 5-1/4". It is 26 pages of color illustrations of edible fruits for Burma, Siam (Thailand) and Malaya. It was printed in Bombay, India with reports of circulation by early 1944.

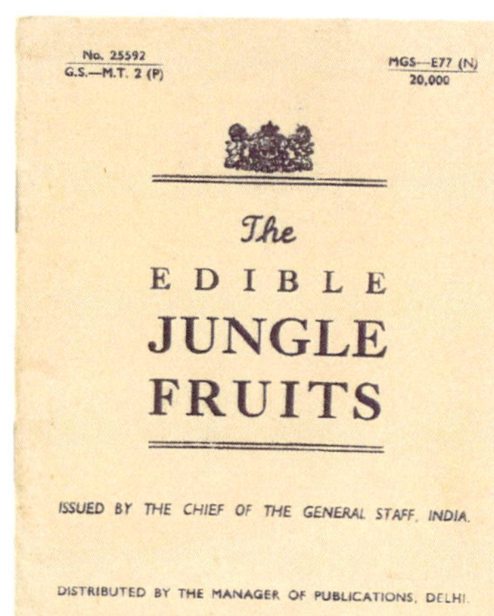

(Left) Survival booklet; How To Stay Alive In The Jungle was also distributed to flying personnel in the India-Burma and China theaters. Most of the content first appeared as an article published in Yank magazine. It was compiled and produced by the Training Branch, India-Burma Theater Signal Personnel and Training Division. For Yank magazine, the printing was contracted for and supervised by the AG Publications Section, Headquarters, India. MIS-X, India made a canvas of the organizations that it supplied with E&E materials to determine whether they could use that information in their escape kits. The findings of that canvas resulted in MIS-X requesting AG Publications to reprint 15,000 of these in booklet form for distribution by MIS-X, India or E-Group. Printing and delivery of the 15,000 booklets was completed by February 1945. How To Stay Alive In The Jungle, 49 pages, 5-1/4"x 4".

Additional helpful material originated in Australia and New Zealand, some examples of which include a 71 page booklet titled Friendly Fruits and Vegetables, May 1943, and a 48 page booklet titled Soldiering in the Tropics, August 1942. Both were prepared by the General Staff, Land Headquarters, Australia and distributed under the authority of the Commander Allied Land Forces SWPA.

Other works from Australia include a 17-page booklet, You and the Native, about relations with New Guinea natives, February 1943. The other, Getting About in New Guinea, is a 29 page booklet discussing the hard jungle terrain and the dangers therein, April 1943. Both were prepared by the Allied Geographical Section, SWPA. Prior to additional material printed by the US Government, these were available for American servicemen arriving in this part of the world for the first time.

(Below and right) Examples of the above mentioned booklets printed in Australia.

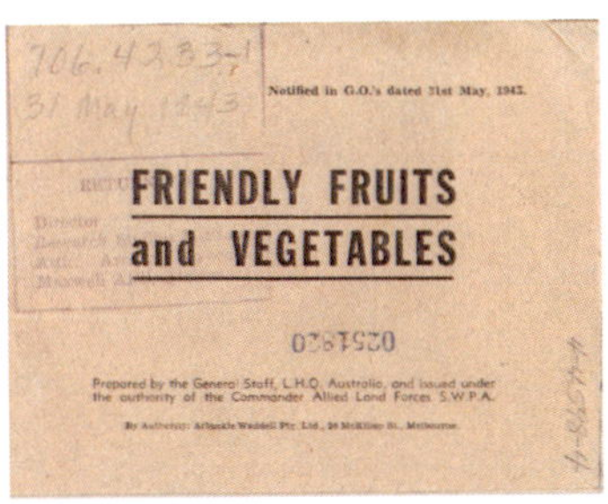

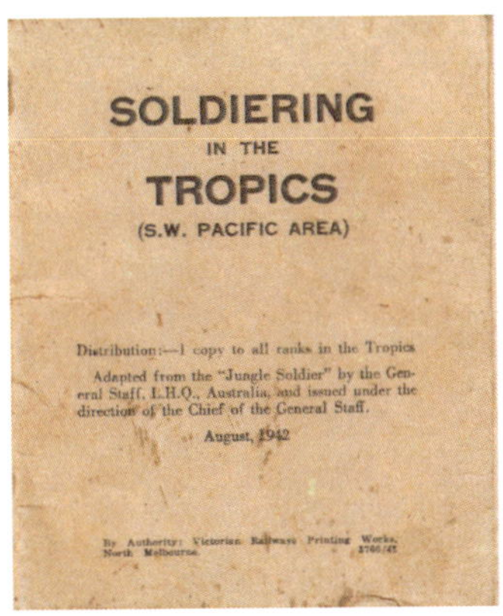

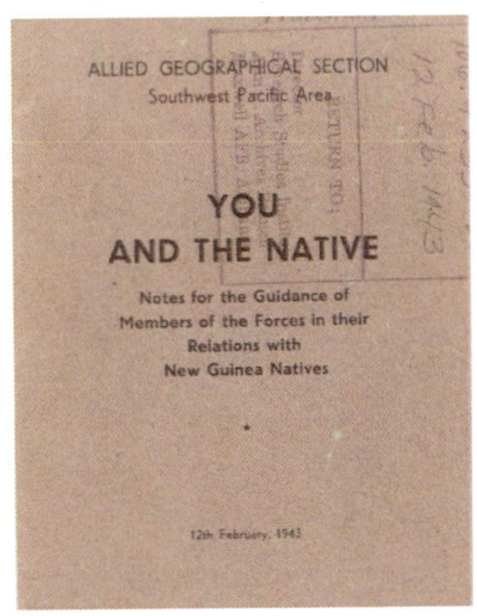

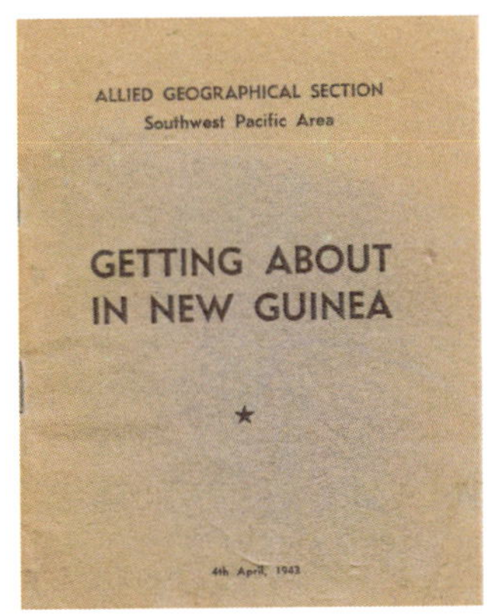

Prior to the printing of Castaway's Baedeker, RAF, RNZAF, and RAAF aircrew and emergency equipment kits were provided with a 29 page booklet titled Instructions to Air Crews, 1942. It was published by the Directorate of Intelligence, HQ Allied Air Forces, SWPA and provided notes on living and traveling in Papua New Guinea, New Britain, the Solomon Islands, and Northern Australia; the end pages included a list of phrases in Pidgin-English. This booklet went through several printings and was typically included in Commonwealth survival kits.

The first printing was complied by the RAAF Air Intelligence Section of the North-Eastern Area Command in 1942. Into 1943 it was primarily only distributed to RAAF aircrew; the 5th AF compiled its own survival manual. There was some anecdotal evidence that Instructions to Air Crews were made available to American airmen on a limited basis. In MIS-X Evasion & Escape Report No. 1, compiled 21 October 1943, it details the account of Capt. Arthur Post of the 8th Photographic Squadron based at Port Moresby. On 21 June 1943, Post was forced to bail-out after being jumped by several Japanese Zeros not far from Rabaul. He spent the next four weeks with natives until being rescued and transported back to Australia 28 August 1943. He notes in his narrative that he carried the booklet in his jungle kit. He further elaborates that the booklet was inadequate both in its treatment of jungle foods and its glossary of Pidgin.

The second printing by Directorate of Intelligence, HQ Allied Air Forces, SWPA was with intention of a wider spread distribution to both RAAF and US aircrew in 1943. However, by November 1943 it was brought to attention that these booklets contain some information that should be considered "secret" as it was standard operating procedure. Since the booklet had been in circulation for some time and, no doubt, copies were in enemy hands it felt advisable to issue another revised type and recall those in circulation. This, in part, led to the draft of Survival Hints.

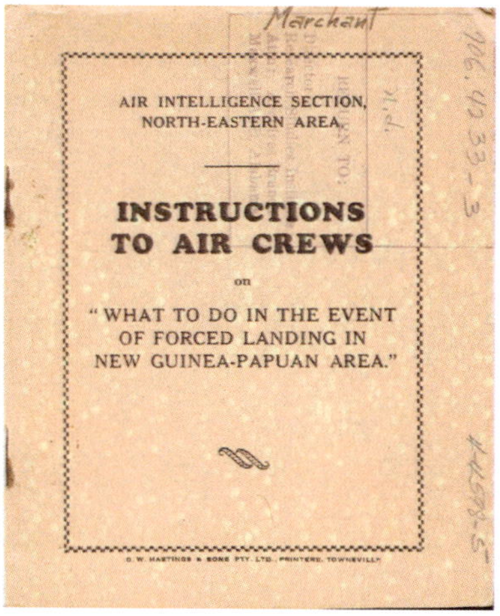

(Left and right) Instructions to Air Crews distributed to all requisitioning activities in the SWPA. (Left) First printing of 1942. (Right) Second printing of 1943. Each booklet measures 4-1/4"x 5-3/8".

Another notable publication by the Australian Army Education Service was a 1944 book called Living Off the Land: A Manual of Bushcraft. The Auckland Institute and Museum in New Zealand contributed the 72 page survival pamphlet Food is Where You Find It. This pocket guide described and illustrated edible and poisonous plants and animals in the western Pacific. It was first printed in August 1943 with subsequent impressions in October and November 1943 that added pages to each printing. To meet the needs of his command, Colonel W.G. Farrell of the USMC operating on Guadalcanal requested that this guide be distributed to both American and ANZAC forces. The 13th AF further requested that 10,000 copies be provided for that command. In 1945, the Air Department in Wellington issued a booklet like Instructions to Air Crews to the RNZAF called Survival Hints for Aircrew, Pacific Area.

(Right) Third Impression of Food is Where You Find It, 4-1/4"x 5-3/8".

(Left) In the wartime image, a rescued USAAF B-25 aircrew officer is exhibiting a copy of the booklet Food is Where You Find It, April 1944.

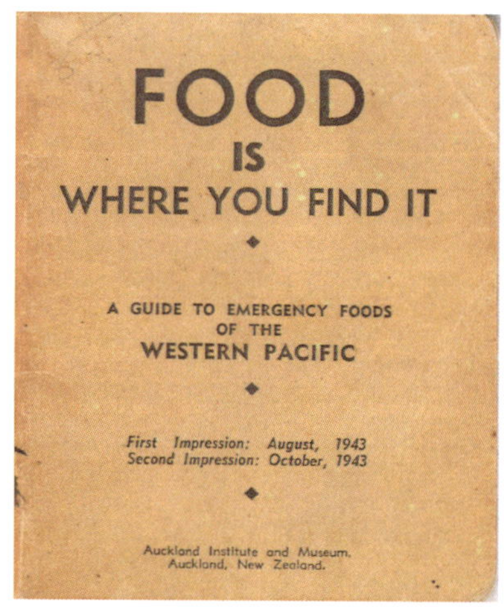

(Right) RNZAF equivalent to Instructions to Air Crews providing information upon an emergency landing, 4-1/4"x 5-1/2".

Other useful booklets were the War Department Pocket, Phrase and Language Guides, used extensively by airmen. The Pocket Guides put emphasis on native assistance for survival, rather than identifying the numerous animals and plants and constructing shelters through written instruction alone. It was thought that native assistance would help in evasion and escape. The two keys were communication and cultural appreciation. With respect, trust was built between troops and natives, which in turn helped foreigners navigate the terrain and avoid possible enemy contact.

These pocket and language guides were generated by President Roosevelt's staff in a project known as the M Project. Much of the work consisted of gathering intelligence around the world in secret. Individuals fluent in many languages listened to transmissions and translated material. Others traveled, investigating attitudes in foreign governments. Most applicable, however, were individuals associated with the M Project who had a working knowledge of native peoples and regions.

In 1942, just as the discussion commenced to circulate regional guides, an incident occurred in India that moved the topic to the forefront. The (unsubstantiated) story is that a US sergeant was driving his jeep through a crowded street and a cow was blocking his way. After an unsuccessful attempt at startling the cow out of the road with his jeep horn, the sergeant dismounted the vehicle and shot the cow. The locals, aghast at the sight of the sacred cow being shot, were incensed and surrounded the sergeant and the jeep. He was soon rescued by an MP patrol. Other instances were occurring in the Caribbean relating to US Military behavior in foreign territories.

This incident prompted the need for providing information on locals and their customs and beliefs to all visiting service members to avoid such incidents in the future. The template for these guides originated from work compiled by Anne Fuller. Fuller had worked at the Peabody Museum of Natural History, then volunteered to work in Washington. Her initial duties were to compile weekly summaries on the Near East. Craving more work, she was tasked with assembling a small handbook on Lebanon, an area in which she once lived. In this work she included notes on location, climate, history, and proper etiquette for service members arriving to the area. Lastly, she included a list of useful phrases in English-Arabic.

(Left) American soldier rubs noses in traditional greeting with native chief in New Guinea. Having a working knowledge in proper greetings, communication, and interaction significantly increased a survivor's chances of rescue.

(Below) In an effort to maintain good relations with indigenous peoples, pocket and language guides were produced for approximately 30 areas. The Pocket Guides were illustrated with sketches and distributed to the millions of service members going overseas. These were initially intended for instruction in behavior and knowledge of local etiquette.

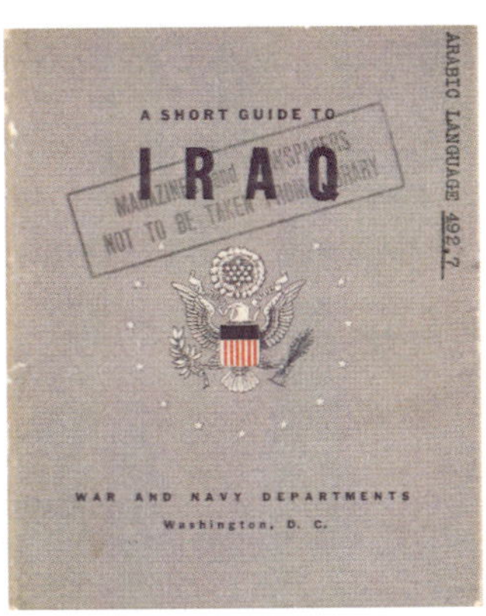

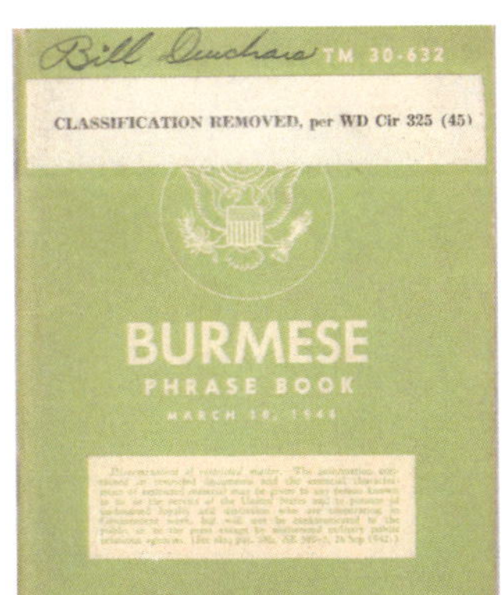

ADDENDA TO PREVIOUS VOLUMES

Back-pad kit referenced under NAF Dwg. No. 69233 was intended for pararaft-borne individuals, which at the time were only VF-Class aircraft or single engine fighters. Developed in parallel was a jungle emergency kit for aircrew referenced under the same drawing number and nomenclature. This was not the first instance where the BuAer referenced two model types under the same specification or drawing. The original specification for the pararaft M-524, as an example, included both a seat and back-type case, but the back type was dropped upon joint A/N standardization.

The additional jungle kit was not restricted in size. It was a quick attachable container, easily stored near a crew position. The original drawing of 30 July 1942 shows that it is similar to the USAAC pilot's emergency kit Dwg. No. 34G3416 in that it was to include personal armament. Listed was a .45 automatic pistol with holster, a pistol belt with magazine pocket, three magazines, and a five-inch hunting knife (Case & Co. #36 or equal). The armament complement was removed in a revision in January 1943, which falls in line with the authorization for the issuance of this type as personal equipment for all flying personnel. An official Boy Scout ax, previously included, was also deleted in January 1943. The compass originally cited was a night-marching dry suspension lensatic type or a Marine Corps marching compass; the January revision changed it to the match safe compass M-554.

This jungle kit contained more variety of components than the back type, such as a hammock with rope and a large section of mosquito netting (154 x 78 inches). The mosquito netting, by drawing, included a series of tie tapes around the perimeter like the concept of the USAAF quilted ponchos included in the B-4 and B-5 parachute kits. It also included an M-1910 canteen, ration unit M-539, official Boy Scout flashlight, two spare batteries (Eveready #950 Size D or equal), spare bulb, a fishing kit (Pflueger Hande-Pak #4005 with 50 assorted hooks and 50 yards of threaded line), saltshaker (1-1/4" diameter), six cubes solidified fuel in grease-resistant box (1-1/8"x 2-1/4"x 3-1/2"), first aid unit S2-1057, and instruction sheet. The fishing kit, spare bulb, and saltshaker were to be contained in a box 1-7/8 x 3-3/8 x 5 inches.

First aid unit S2-1057 consisted of one box of tannic acid gel in two ¾ oz. tubes, three vials of tincture of mild iodine, one four-inch compress bandage, a small battle dressing, 36-inch padded web tourniquet, and a glass or plastic bottle of 1-5/16 inch diameter containing 24 quinine tablets and two morphine syrettes. It was supplemented by a USAAC match safe 42A6931 containing 25 tablets of sulfathiazole, as cited in revised drawing.

(Below) Container manufactured with aircraft gray waterproof fabric, with carrying strap and two quick-attachable parachute snap assemblies, measuring 14" x 13-1/2" x 3-1/2". Container was to be legibly marked JUNGLE KIT; Carefully Read Instructions Within On Landing.

(Below) Slide fastener provided for access. Components secured by nonelastic webbing. Unit packed per drawing specifications. Due to the fragile nature of the M-539 ration, it is not secured in the webbing. Loop between canteen and ration and loop between match safe and fishing kit assembly were unassigned. Tannic acid has been replaced by boric acid; the bottle with quinine and morphine is absent.

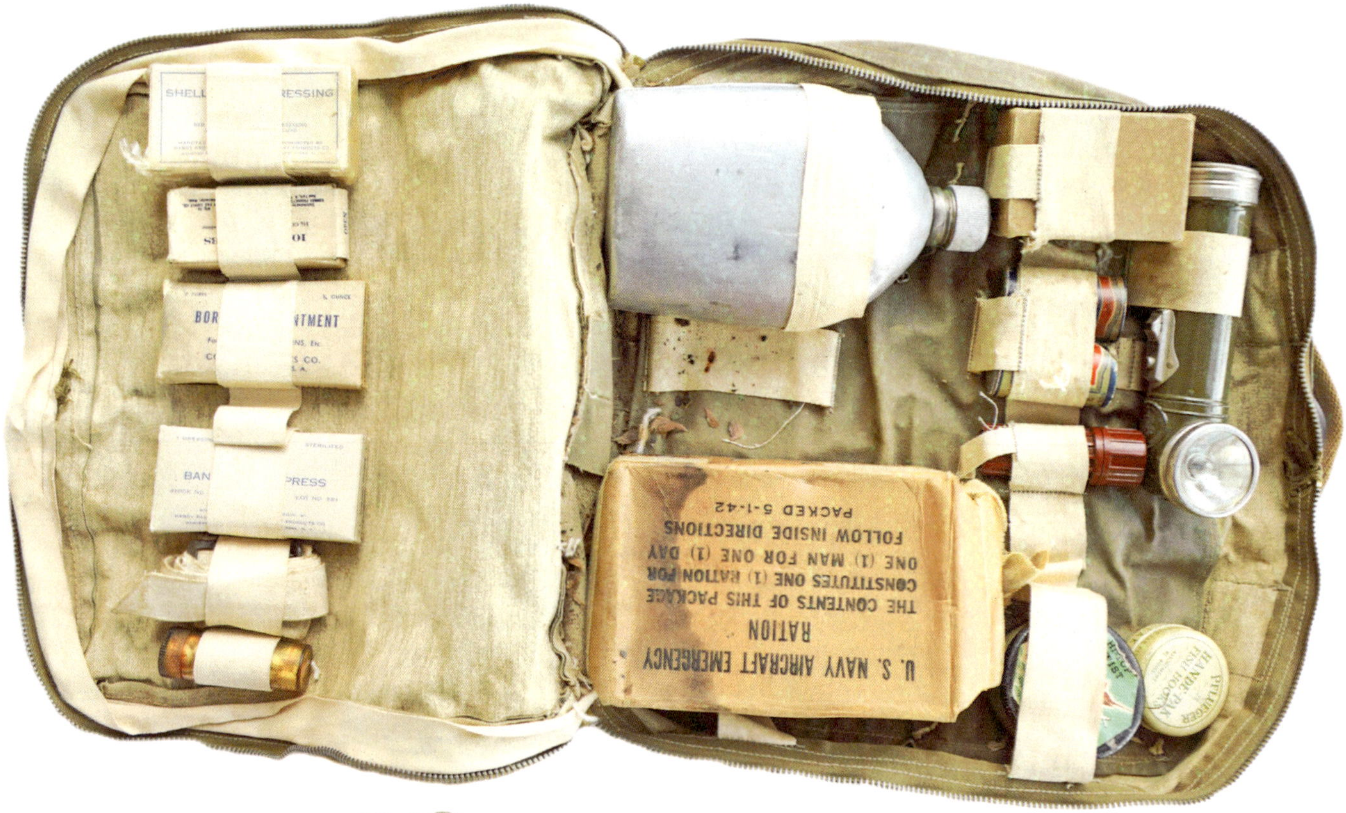

(Above and right) Headnet, hammock, and hammock rope. Not illustrated is the unique, large mosquito netting. Headnet is like USAAF type 43G3868 except crown is netting in lieu of fabric.

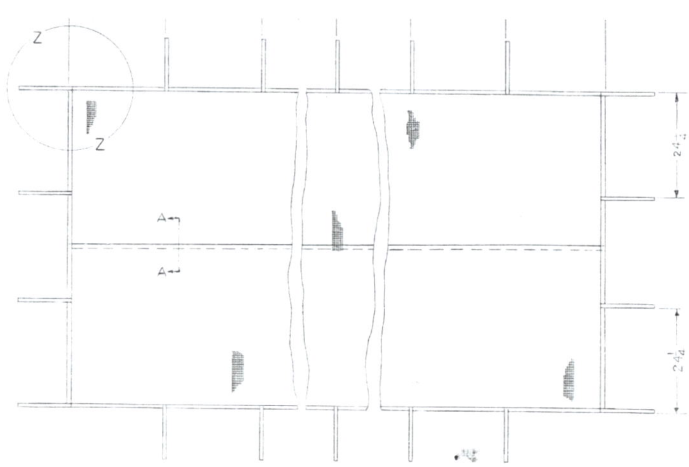

(Left) Drawing for large mosquito netting. Two pieces to be sewn in center and provided with tie tapes along perimeter.

Highlighted on page 152 of Volume 1 is a locally modified pararaft container that converted the AN series type case to the PK-1. This conversion was executed at Assembly and Repair facilities or parachute lofts, land- or carrier-based. Because it was to be a local task, the accessories were an M-592 and items from stock AN-R-2 pararafts. The intent was to comply with Technical Order 9-45 of 13 January 1945 for the Model-A pararaft unit.

ComAirPac Technical Bulletin No. 15TB-45 stated, line item #1: "Encl (1) describes a simple method for converting present pararaft cases to accommodate basic components of the back-pad kit." This conversion was recommended because of its simplicity and nonemployment of additional zippers and other materials that were not readily available.

Enclosure 1 included Dwg. No.1-1176-1, which detailed the modification process. All emergency gear, pockets for emergency gear, conduit for leg straps, and webbing oar retainers were to be removed from the pararaft container, leaving only the life raft lanyard intact. The seam cut and removal of dura-dot fasteners in cover of raft container. The new case was created by patching the leg strap slot and using scrap material from emergency gear pockets. The assembly was to use 18-inch tie tapes to secure components.

Illustrated on page 152 is a rendition of 15TB-45 conducted by Composite Squadron Ninety-Seven (VC-97). The Squadron members made some minor revisions to suit their preferences. Deviations can always be expected when resolution of recommendations is left to the creativity of those conducting the modifications. VC-97 noted that the modification of QAS parachute seat pack in accordance with 15TB-45 exaggerates the elevated position of the parachute in the seat of the plane to such an extent that in the Wildcat it prevented full back stick travel with seat above third position. A redistribution of the gear in the pararaft container lowers the pack so that full elevator control was allowed at all seat positions.

A drawing provided by VC-97 lists the following components and their placement: six-unit first aid kit (three packages), one can of water, one sea marker, one desalting kit with sponge, one desalting kit and mirror with "red-dot" (M-580A), one ration, sunburn cream, one sea marker and match safe compass, one modified fishing kit, two rust preventives, two cans of ration, one sea marker with line, paddles, gloves, repair kit, and bullet plugs stowed behind fishing kit.

The white fishing line and feathered lure were removed from the modified fishing kit; a burning lens, pack of safety pins, adhesive tape, and salt tablets were added. VC-97 noted that every pilot was equipped with a pyrotechnic kit, which was worn on a belt. For this reason and to conserve space in the container for additional emergency gear, pyrotechnics were not included. A belt type pyrotechnic kit can be seen on page 71 in Volume 2.

The Model-A eliminated the poncho to conserve space but it was retained in the suggested modification of 15TB-45. It was also retained in the deviated version by VC-97. Drawing 1-1176-1 prescribed the poncho to be packed and folded with the raft; it is shown on page 152 in Volume 1 with an arrow pointing to it. Drawing 1-1176-1 describes a modification of the poncho from the M-592 to reduce bulk closer to the dimensions of that revised type included with the poncho for the PK-1, page 146 Volume 1. The drawing simply states to "lay out the poncho in position shown and cut straight across bottom two inches below fifth button. Turn back one inch and sew in place."

(Below) Physical example complies with directions outlined in 15TB-45.

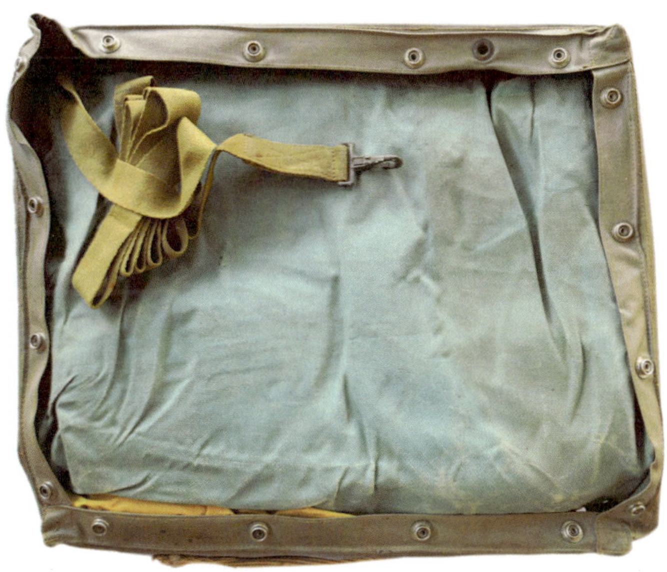

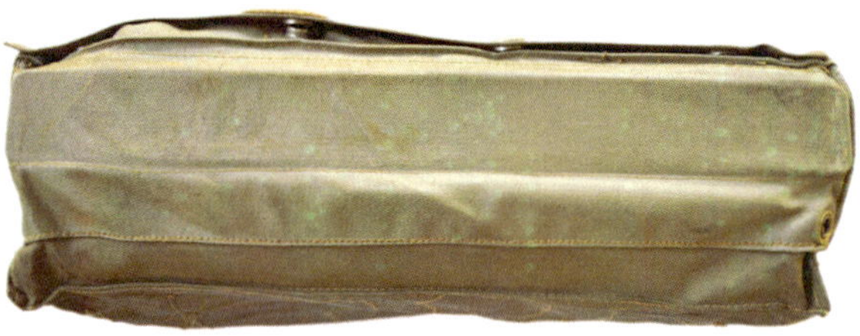

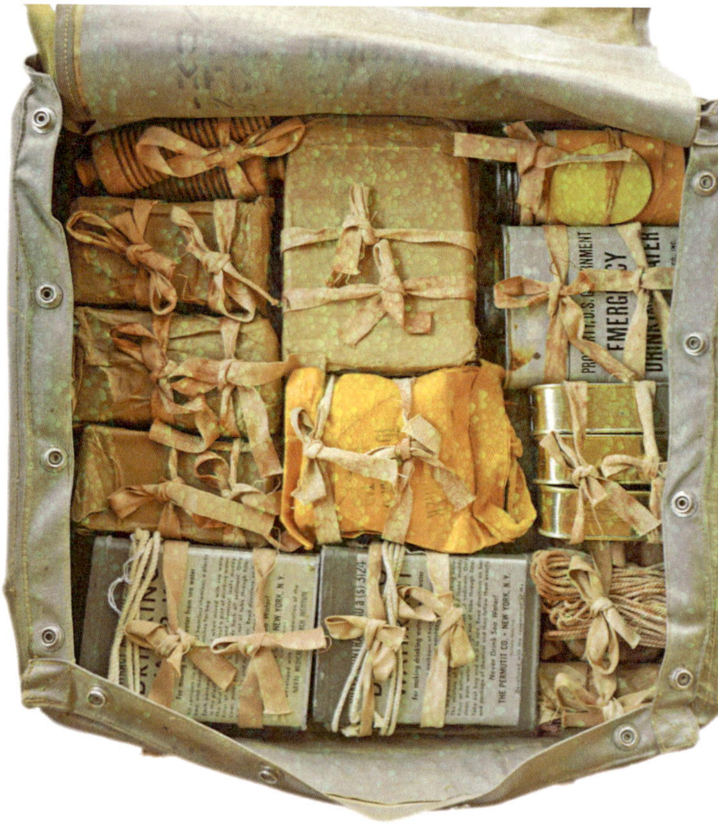

(Above) Example deviates to suit preference in component configuration. It complies in construction method. The standard case was almost completely deconstructed and rebuilt. The perimeter of the case remains but the cover and bottom were modified. One image shows the cover, leg strap conduit sealed with flap from inner pocket. Interestingly, it is stenciled MODEL-A, which suggests they used T.O. 9-45 following directions for accessories. With top cover removed, the raft is folded with the poncho inside. With the raft removed, 15TB-45 directs that an inner flap be installed, segregating the accessories from the raft for extra security to prevent loss. In this example an added flap was installed under it. Here, it is the original bottom to the raft cut down and resewn. To accommodate the accessories and raft, it was necessary to add extra depth to the case, which can be seen on the side profile; it amounts to about 1-1/4". For this application, waterproof airplane fabric was used. This material would have been readily available at any repair facility. The layout of the accessories in a single layer helped decrease bulk. In T.O. 9-45 and 15TB-45, many of the accessories were to be stacked.

(Left and below) Comparative view of size differences for the Emergency Equipment Container; MK-I, MK-II, MK-IV and MK-VII. With revision to M-3R, the Supply Pocket (far left) was revised, retaining attachment by lift-the-dot, but with an added grommet for a lanyard to prevent its loss once detached.

(Below) B-5 quilted poncho, 60"x 70". Construction is more complex than the B-4. Uses illustrated in Volume 3.

(Left) Comparative view of dimensions between the B-4 quilted poncho (yellow side up) and B-5 quilted poncho.

(Below) Quilted hood and wool ski socks.

(Below and far below) Quilted hood attaches to three snaps center of quit when used as poncho or arctic coat. When used for arctic sleeping bag and other method of coat, arrows index its attachment for circular fit.

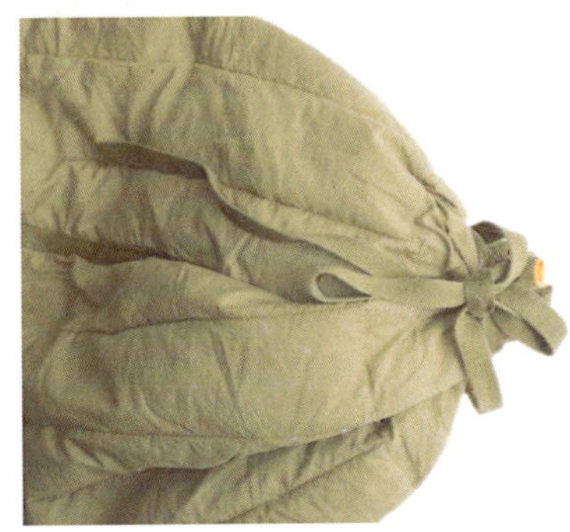

(Right and below) Adopted in February 1942 was the Accessory Container for USAAF multi-person life rafts, illustrated on page 70, Volume 1. (Right) Accessory Container bundle. Below right exhibits the container packed and secured to the floor of an A-3 life raft. Below is a typical carrying case for USAAF life rafts.

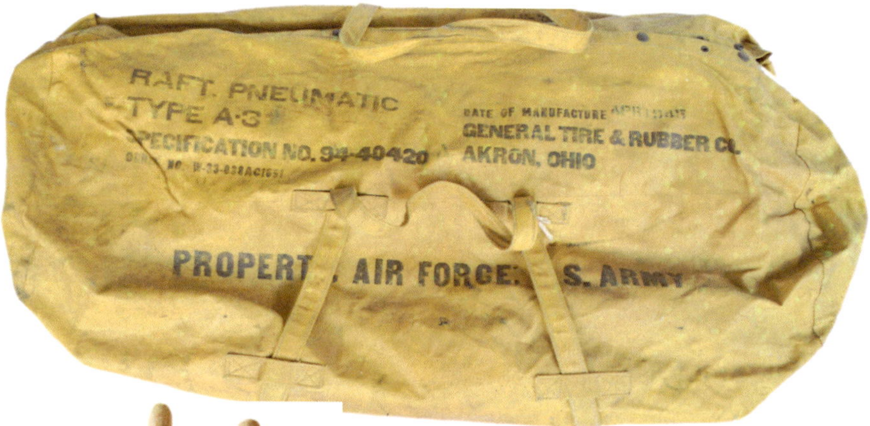

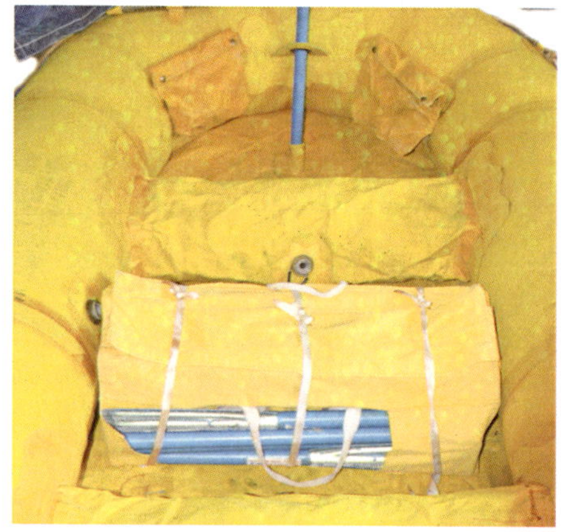

(Left) Natural wood oars provided with pre- and inter-war USAAF life rafts.

(Below) Parachute emergency kits were often locally revised, either with the combination of components or physical revisions to the container itself. A modified felt insert has replaced the standard type in this jungle kit 38J5271. All slots remained the same except the cut for an M-1917 bolo in lieu of the 9" machete. It includes two Air Corps emergency ration tins.

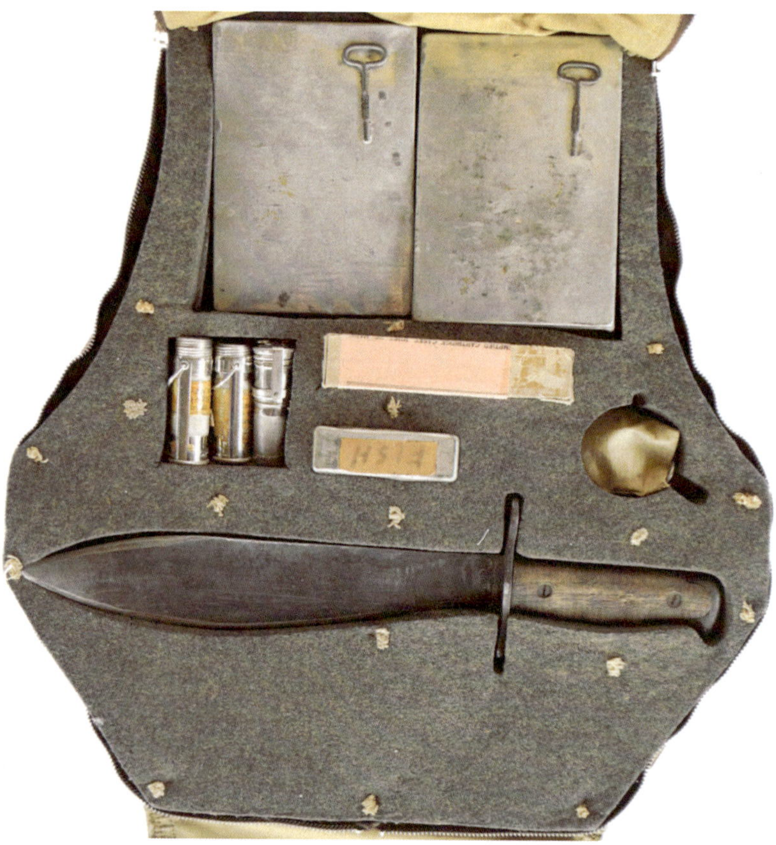

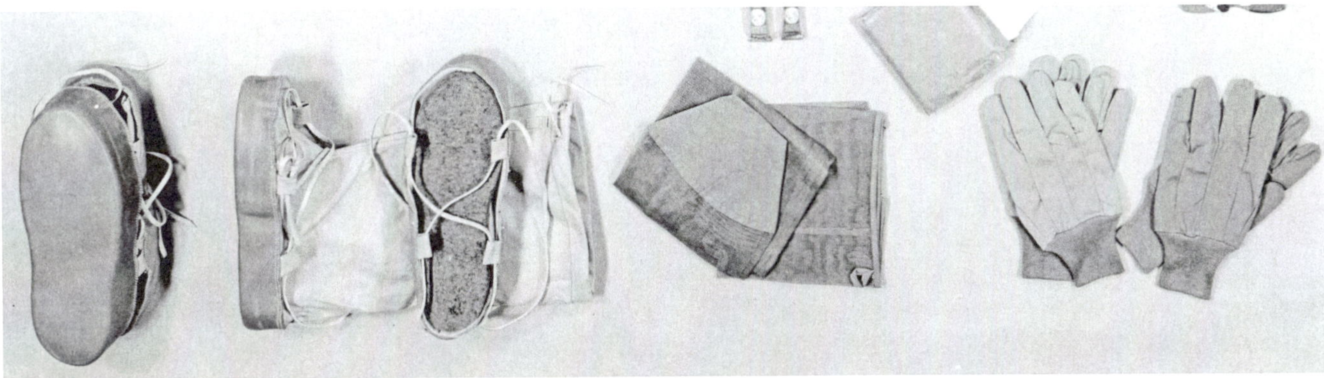

(Above) E-1 sustenance kit protective clothing included mukluk, gloves and mosquito head net. Unique to E-1 were mukluk, not cited in any stock catalogs or other sustenance kits. Specification states; Mukluk, medium-size complete with 1/2" felt innersole, (The Rasmussen Co. or equal). Gloves cited as; Canvas gloves treated as to be water repellent, (Sears & Roebuck catalog no. 33-3882 or equal). Originally cited as "Fine mesh mosquito proof head net", later revised to 43G3868 head net.

(Left) Simply cited as "Liner, AAF Dwg. 44G5060" was an insert for the E-14 sustenance kit. Formal nomenclature is Liner Assembly, Emergency Sustenance Drum. Inserted in drum and all accessories packed in it. Removed from drum, unit can be transported by shoulder straps provided.

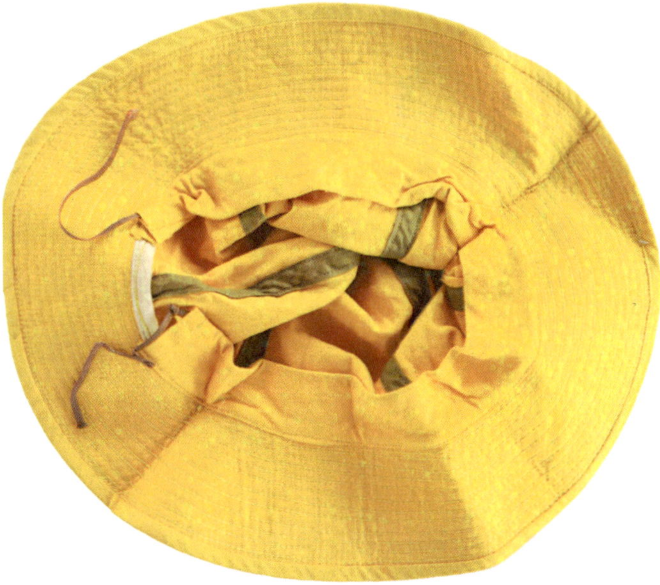

(Above and right) Predecessor to Sun Hat 44G5062 was a type manufactured by the R.A. Lloyd Co. Specifications for the E-9 and E-10 cite; Hat, Sun, six section crown, adjustable head size (6-3/4" to 7-3/4"), 2-3/4" stitched brim, cotton fabric dyed bright yellow, Model 6312, R.A. Lloyd Co. or equal. Also manufactured in brown, which was also provided in the E-9. Two hats were included in either unit, two yellow hats are seen in the assembly of the E-10, right.

(Below) Improved aeronautic first aid kit of 1939. First generation had stock class number 99517, which was for the container only. Stock number 99516 was used for complete unit. By January 1942 the stock class number had been revised to 97765 (complete) and 97770 (container only). Example on right is of a later production with revised stock number applied to container, 97770.

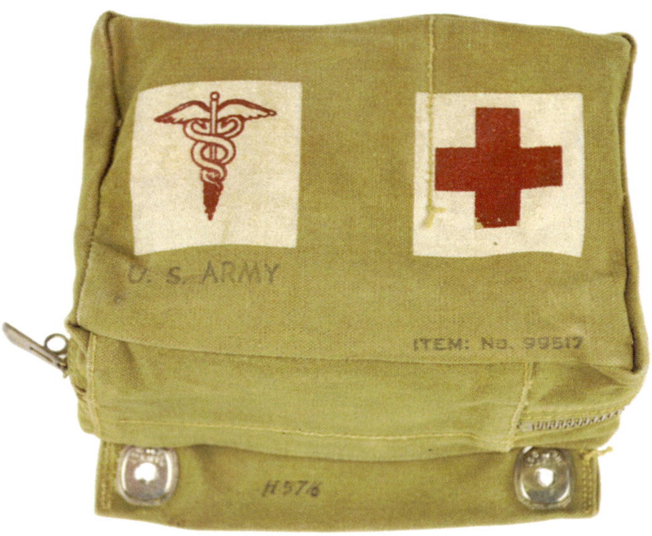

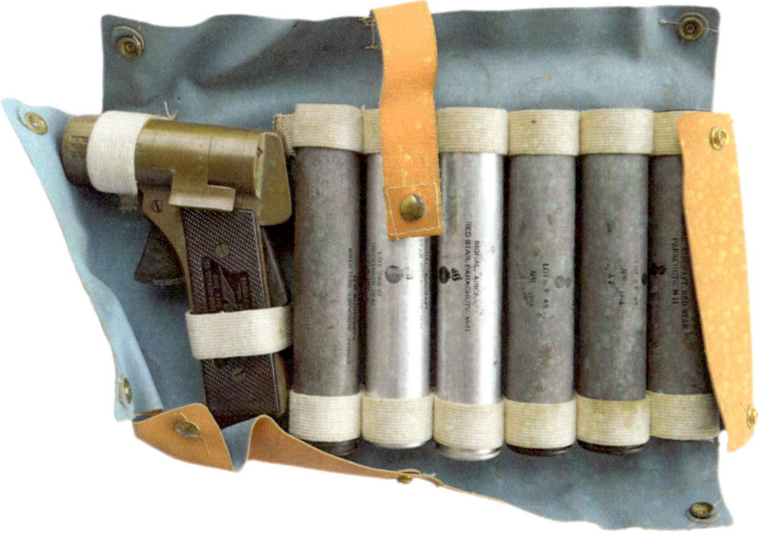

(Above) Volume 2, page 57 illustrates two USAAF distress signal kits 37G2993 for use in life rafts to accommodate the M8 and M9 pyrotechnic projectors. The original type included the M2 pistol, demonstrated here with outer container. Also shown is an unused tape for sealing the complete assembly.

(Right) Specification AN-S-10 indicates that the can was to be lithographed US Govt, Fluorescent Dye Sea Marker, Instructions For Use: Remove Cap and Empty Contents On Surface of Water, Specification AN-S-10, followed by batch number, name of manufacturer, government order number. The order number was either lithographed or rubber-stamped on container. However, the company responsible for the lithography could not suitably print the cans and returned them to the companies responsible for the packing of the dye, which resulted in an expedited version of the can simply marked US Government Sea Marking. AN-S-10 meeting all requirements in Specification AN-S-10. Order number indicates the Aviation Supply Office, BuAer, to include drawing from AN-S-10.

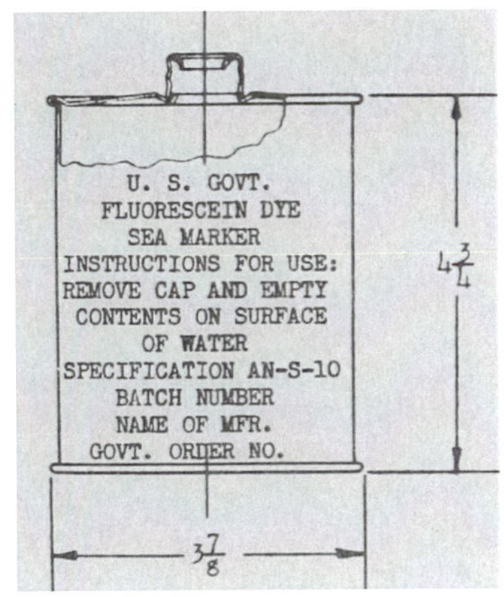

(Below) Cases for the round chromium mirror; square chamois type, round chamois and white cotton. Also illustrated: two locally printed instruction sheets folded and included with mirrors.

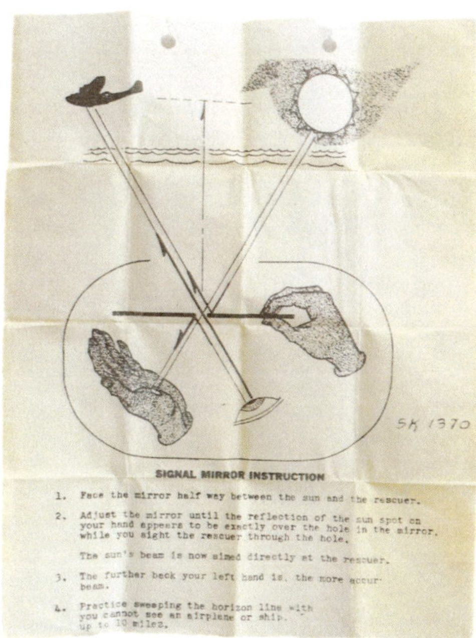

SK 1370

SIGNAL MIRROR INSTRUCTION

1. Face the mirror half way between the sun and the rescuer.
2. Adjust the mirror until the reflection of the sun spot on your hand appears to be exactly over the hole in the mirror, while you sight the rescuer through the hole.

   The sun's beam is now aimed directly at the rescuer.
3. The further back your left hand is, the more accur- beam.
4. Practice sweeping the horizon line with you cannot see an airplane or ship. up to 10 miles.

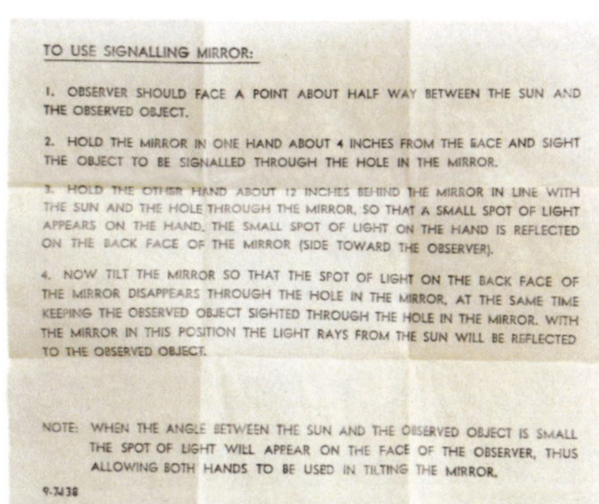

TO USE SIGNALLING MIRROR:

1. OBSERVER SHOULD FACE A POINT ABOUT HALF WAY BETWEEN THE SUN AND THE OBSERVED OBJECT.
2. HOLD THE MIRROR IN ONE HAND ABOUT 4 INCHES FROM THE FACE AND SIGHT THE OBJECT TO BE SIGNALLED THROUGH THE HOLE IN THE MIRROR.
3. HOLD THE OTHER HAND ABOUT 12 INCHES BEHIND THE MIRROR IN LINE WITH THE SUN AND THE HOLE THROUGH THE MIRROR, SO THAT A SMALL SPOT OF LIGHT APPEARS ON THE HAND. THE SMALL SPOT OF LIGHT ON THE HAND IS REFLECTED ON THE BACK FACE OF THE MIRROR (SIDE TOWARD THE OBSERVER).
4. NOW TILT THE MIRROR SO THAT THE SPOT OF LIGHT ON THE BACK FACE OF THE MIRROR DISAPPEARS THROUGH THE HOLE IN THE MIRROR, AT THE SAME TIME KEEPING THE OBSERVED OBJECT SIGHTED THROUGH THE HOLE IN THE MIRROR. WITH THE MIRROR IN THIS POSITION THE LIGHT RAYS FROM THE SUN WILL BE REFLECTED TO THE OBSERVED OBJECT.

NOTE: WHEN THE ANGLE BETWEEN THE SUN AND THE OBSERVED OBJECT IS SMALL THE SPOT OF LIGHT WILL APPEAR ON THE FACE OF THE OBSERVER, THUS ALLOWING BOTH HANDS TO BE USED IN TILTING THE MIRROR.

9-7438

(Above) Unusual BC-778 transmitter decal having the moniker Gibson Girl with the addition of the logo for Bendix Aviation. One such example is exhibited in the display room of the Liaison Committee for Emergency Rescue Equipment, purpose of this variation unknown. Additional accessory not illustrated in Volume-2 is container for headlamp M-308-A, exhibited here. Not an accessory for the SCR-578 but intended as a service device for the BC-778 is the Antenna A-98 (Phantom). It is a small box 2-½"x 1-½"x 2" designed to absorb the signal from the transmitter when testing function as to avoid reception of the SOS at external receiving stations. Being specially designed for the BC-778 it is additionally painted the orange/yellow color, it was not used for any other Signal Corps transmitter/receiver.

(Below) In Volume-2, the chapter for Pyrotechnic Flares it discusses the M1 Projector and the Driggs-Faber projector. No photo was provided for the M1, illustrated was that of a modified model M1A1 built during the war for training purposes. Very subtle changes were applied but here is the original M1 Projector as adopted 23 August 1928. In the process for finding a suitable pyrotechnic projector a model referred to as the Driggs-Faber was tested by the AAC, with modifications it was then adopted as the M2. As can be observed, the M2 shares the general characteristics of the Driggs-Faber model. (National Archives)

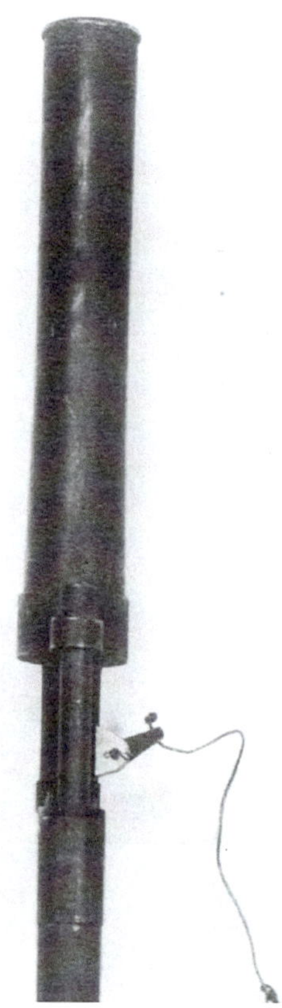

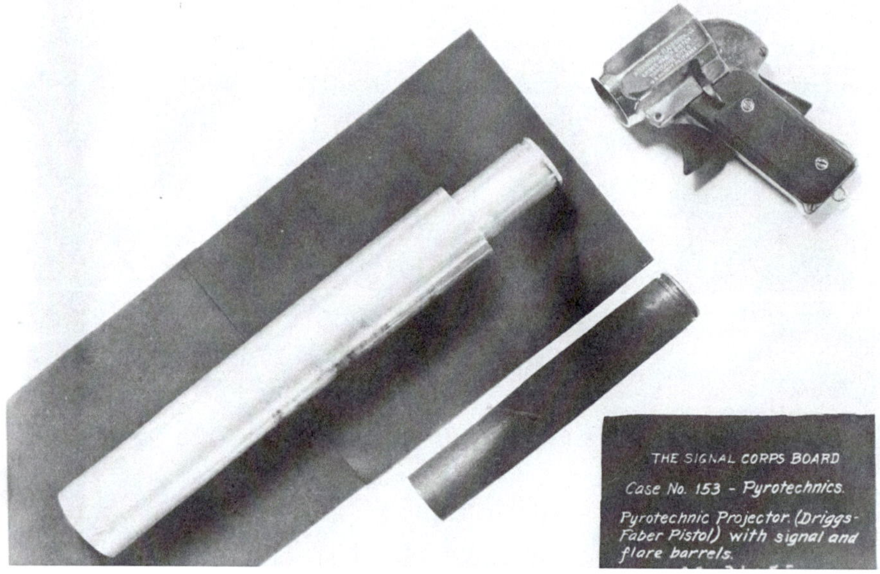

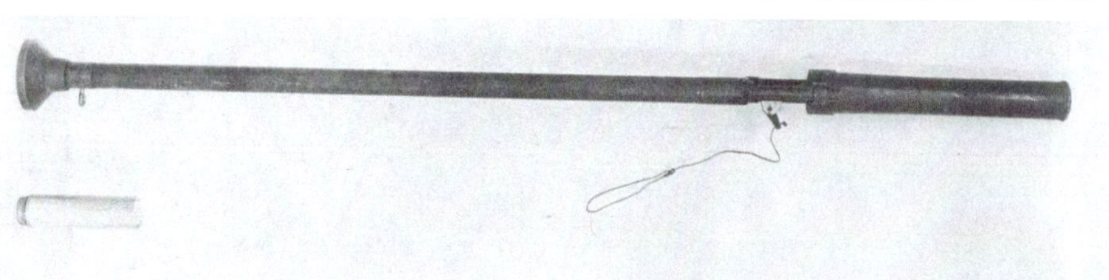

(Right) The US Navy obtained 20,000 M2 pistols as a temporary supplement until the AN-M8 became available, for aircraft use to accommodate the new interim 37mm signals, assigned stock number 2-P-3495. Standard for Bureau of Aeronautics up into 1941 was the 10-gauge Very signal pistol. Distinct markings are stamped on some of those pistols delivered to the US Navy: inspector initials GHB, serial number prefix NoN and anchor acceptance stamp.

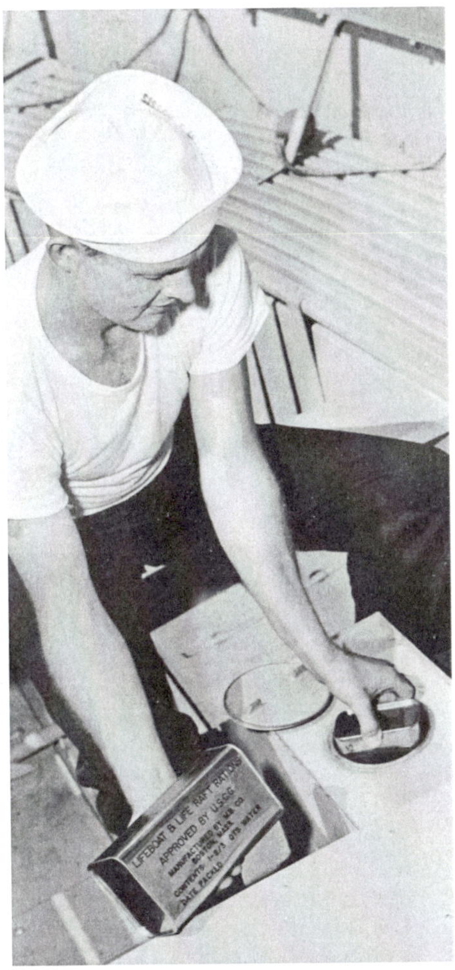

(Right) When observing USAAF sustenance kits exhibited in Volume-3, they included a variety of commercially available material. One such pyrotechnic device was that of the Shurlite Hiway Beacon, it comprised of a tin master container with three fuel burning hydro-carbon oil pods that when lit could be seen from a height of five miles. Matches for lighting are packed with the kit. This beacon was a component part to the E-8, E-10, E-12 and E-15 (x2). This same unit can be found in BuAer Class-31 materiel stocked with the ASO under R31-F-10 titled as Flares, Emergency Kit. It was intended as emergency lighting equipment for use by advance bases."

(Below) Vintage image illustrating Parachute Ration as packed in sealed tin container, layout of individual ration components is available on p.258 in Volume-3.

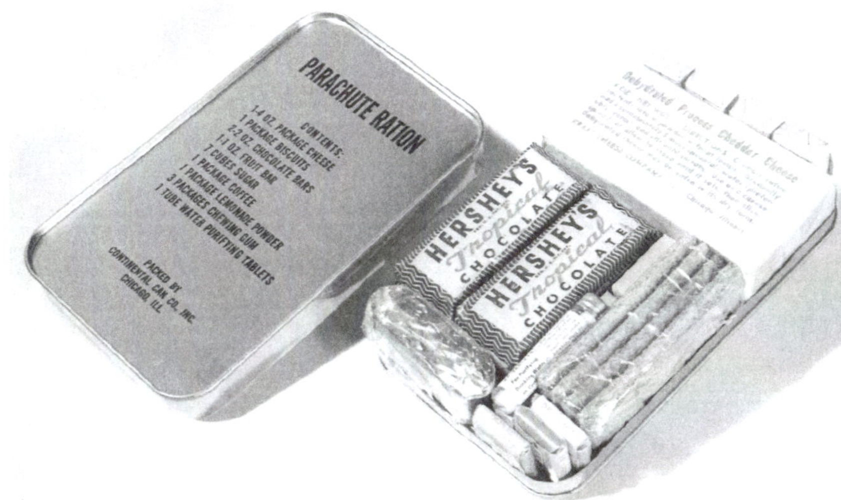

(Above) Highlighted in Volume-3 is a variety of U.S. Coast Guard approved rations, this vintage image lends perspective on how these tins are stored aboard a lifeboat. Food and water components are kept in a special hold with waterproof lid.

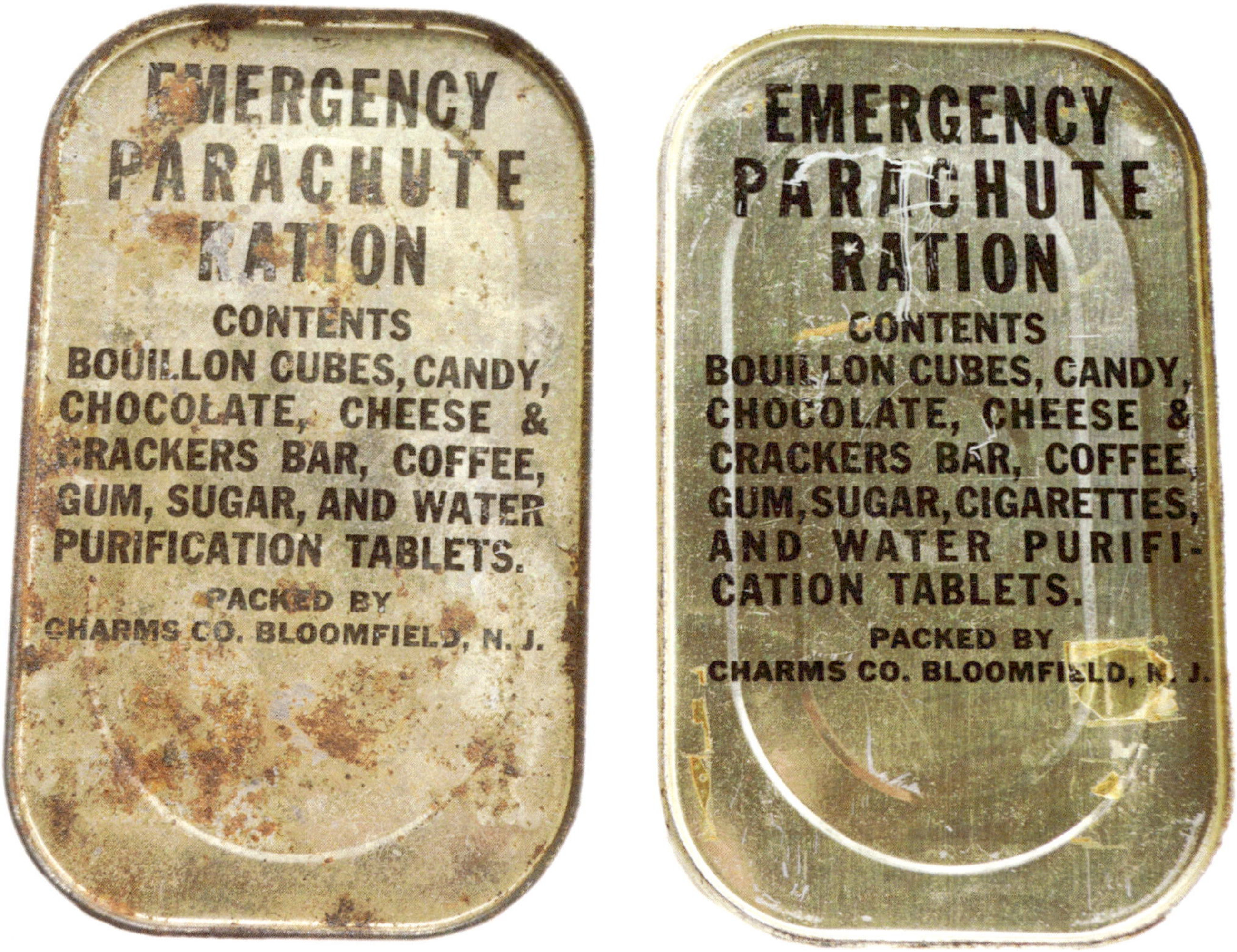

Initial expedited production of the Emergency Parachute Ration did not include cigarettes, they were then added upon standardization in July 1944. Initial production C-1 vests were provided with preliminary types until the new production became available. The pre-standardized emergency ration is pictured in Volume 2, p. 260. A physical example is exhibited here, it does not include listing of cigarettes.

Acknowledgements

This work could not have reached its potential without contributions and support from Robin Bartel and Paul Faulkner. Robin Bartel has been a friend and a peer for more than 25 years, and has contributed many artifacts for reference as well as solid support for the project, consistently urging me on. Paul Faulkner is a long-time friend who shares the same interests, and his knowledge of Naval aviation is unequalled. I would also like to thank Maryjane Stout, Kriss Usherwood, Charlie Flick, Frank Trzaska, Bob Baldwin, Mathieu Bianchi, Remi Bouche, Jeff Klein, Jerry Kolakowski, Ben Major, Steve Milnthorpe, George Petersen, Gregory Pons, Geoff Pringle and Mick Prodger for their contributions and graciously shared information. Significant assistance was provided by Brett Stolle of the National Museum of the United States Air Force, Bob Thomas of the National Naval Aviation Museum and Maranda Gilmore of the Air Force Historical Research Agency.

Author's Note

I have maintained an interest in WWII for as far back as I can recall, playing army in the woods in my home state of Oregon. It was on my 13th birthday, back in 1987, that I established myself as a collector. My gift was an M1 Carbine. Rummaging through my trunk of field gear, I separated the WWII items from everything else, and those, together with the rifle, formed the basis of my collection.

My interest only grew, accumulating more and more items until today, 30 years later, all aspects of United States WWII memorabilia remain a passion of mine. Some 20 years ago I obtained an empty US Navy M-592 emergency back pad kit – my first piece of WWII survival equipment – and so began my transition solely into that area. This initiated a quest to find out and obtain what went in the kit. I quickly found myself frustrated at the lack of information available, gathering more questions than answers. As the investigation moved forward I became more and more interested in the overall subject, accumulating artifacts but still operating with a very limited reference library. Influenced by peers in my local collecting community, I eagerly joined in conversations about their adventures across the country, networking and traveling to various archival holdings. It soon became apparent that I would need make those pilgrimages in order to find the answers to questions that I, and many others, shared. After a relatively short time I decided that if I was going to make that effort, I should put my findings in a reference book for anyone who may be interested.

My initial research efforts were at local university libraries then inter-library loans, until the environment allowed me to visit the National Archives. I have since been back to the NARA multiple times, accumulating documents and images. Not every answer to every question was found, though it was certainly through no lack of effort. Those answers simply may not exist.

It wasn't mandatory for government agencies or facilities to retain all documents, once they were determined obsolete or irrelevant, often ending in their destruction. What survived post-war downsizing and finished up in repositories was at the grace and courtesy of individuals at these government installations, for which I am eternally grateful! However, many records are incomplete, and it can be viewed as a stroke of luck that they still even exist. Information from various sources was sometimes contradictory but through continued accumulation of data, a compilation of redundancy pointed towards the facts. I have attempted to detail each genealogy as clearly as possible.

The overall experience has been very educational for me, and hopefully those who read these studies will find them helpful.

Dustin Clingenpeel
May, 2019

Index

www.ingramcontent.com/pod-product-compliance
Lightning Source LLC
LaVergne TN
LVRC101546220926
848486LV00041B/388